Frontiers in Drug Design & Discovery

(Volume 9)

Edited by

Atta-ur-Rahman, *FRS*
Honorary Life Fellow, Kings College, University of Cambridge, Cambridge, UK

&

M. Iqbal Choudhary
H.E.J. Research Institute of Chemistry, International Center for Chemical and Biological Sciences, University of Karachi, Karachi, Pakistan

Bentham Science Publishers Ltd.
Executive Suite Y - 2
PO Box 7917, Saif Zone
Sharjah, U.A.E.
Email: subscriptions@benthamscience.org

BENTHAM SCIENCE

CONTENTS

PREFACE

Modern drug discovery and development is truly an interdisciplinary science, that has benefitted on the one hand from the growing body of knowledge about the molecular basis of diseases and, on the other hand, from the fascinating new technologies which enable scientists to carry out the discovery phase with efficiency and cost effectiveness. Being an important translational science which addresses key human needs, the key emphasis has been on reducing the time from laboratory bench to clinical use. Global literature in the field of drug discovery and development is spread over journals and monographs of many disciplines, and it is often difficult for practitioners to keep pace with the most recent advances in the disciplines. The key goal of the book series *Frontiers in Drug Design and Development* is to provide updated information about important developments in this fast-moving field that will assist the practitioners to have better understanding of the recent advances.

The 9th volume of *Frontiers in Drug Design and Discovery* is a collection of seven scholarly written reviews, contributed by leading experts in the fields. Each review addresses an interesting aspect of drug discovery, drug delivery, and drug target identification in the broad areas of cancer, infections, and inflammatory diseases.

Kontos *et al.* have contributed a comprehensive review on the current understanding of various prognostic biomarkers of prostate cancer (CaP), including prostate-specific antigen (PSA). CaP is the most common cancer in men, and the second leading cause of death due to cancer in men globally. Early diagnosis of CaP is critically important to achieve therapeutic success. CaP is currently diagnosed through elevated levels of PSA and digital rectal examination followed by prostate biopsy. Prognostic biomarkers can play an important role in early diagnosis of the disease and in the selection of patients for specific treatment. PSA has served as a widely used biomarker of CaP and it has played an important role in lowering the mortality rate due to prostate cancer. However, PSA levels are now recognized as less than an ideal indicator of the disease state and over-diagnosis of CaP has been reported due to inherent limitations. The authors have provided an excellent review of the pros and cons of PSA, as well as the identification of new prognostic biomarkers for CaP, which can play an important role in the selection of treatment options.

Drug resistance in cancers is a major challenge in cancer therapy. Shrikant Kukreti's review outlines the current understanding of the factors and underlying mechanisms that enable cancer cells to resist anti-cancer drugs. Intrinsic and acquired resistance involve several mechanisms such as inactivation of drugs, drug efflux, alteration of drug targets, DNA damage repair, cell death inhibition, epithelial-mesenchymal transition, and the tumor microenvironment. The review describes the genetic and epigenetic factors that can induce drug resistance and contribute to the development of cancer progenitor cells, and resistance to conventional cancer therapies. The author provides an excellent overview of the understanding of the molecular basis and biomarkers of cancer resistance as well as diagnostic tools. The review highlights strategies to overcome drug resistance, and novel therapeutic approaches, thus providing a basis for developing effective treatments against drug-resistant cancers.

Kaushik and Sahi have focused their review on recent anticancer drug discovery and development, based on modulating the signaling pathways through G-protein-coupled receptors (GPCRs). The CPGRs form the largest group of membrane receptors in eukaryotes and act as signaling molecules to inform cells about the presence or absence of life-sustaining light or nutrients, as well as signals from other cells in their environment. Post genomic understanding of these receptors has greatly influenced modern drug discovery, particularly in relation to anti-cancer drugs. In this review, the authors have provided a comprehensive overview of the current understanding of GPCRs, their role in disease processes, and approaches towards targeting these proteins for therapeutic purposes.

The review contributed by Savic *et al.* focuses on the mechanism of actions of ribavirin (tribavirin), a broad spectrum antiviral medication. Ribavirin is among the oldest drugs used for the treatment of RSV infections, hepatitis C, and viral hemorrhagic fever. It is known to work equally well against both RNA and DNA viruses. Recently reported anti-cancer, anti-inflammatory, immunomodulatory, and immunosuppressive activities of this common drug have attracted major interest in its mechanisms of action, and in its use against various orphan and neglected diseases. Since the last four decades a lot of work has been done to elucidate the mechanism of action of ribavirin, but many questions remain unanswered. Savic *et al.* provide a systematic review of the literature on various sub-cellular, cellular, and enzymatic targets of ribavirn.

In the next chapter Das and Sarma focus on the applications of carbon nanotubes in pharmacy and medicine. Carbon nanotubes have a large surface areas and the capacity to absorb or conjugate various classes of drugs effectively, and can thus serve as excellent drug carriers. They can also conjugate with genes, vaccines and antibodies for other applications in medicine. The authors review the most recent developments in this field and provide a critical analysis of the prospects of the use of carbon nanotubes as drug delivery vehicles. Other aspects such as phamaco- and toxico-kinetic parameters, and the functionalization and compatibility of carbon nanotubes to various classes of biopharmaceuticals are also presented.

Tautomycin (TM), a natural product of bacterial origin, was identified as a highly specific inhibitor of protein phosphatase type 1 (PP1). PP1, along with protein phosphatase 2A (PP2A), is the major serine/threonine protein phosphatase. They are involved in the regulation of numerous cellular and biochemical functions. The roles of PP2A have been extensively studied using okadaic acid, which is a specific inhibitor of PP2A. Development of targeted inhibitors of PP1 has opened new avenues in drug discovery. Makoto Ubukata has reviewed the most recent literature on this topic, with special emphasis on neurological disorders in Minkowski space.

The last review in this volume by Mokry and Mokra is focused on the importance of phosphodiesterases (PDEs) as targets for drug discovery against various respiratory diseases. Phosphodiesterases regulate the intracellular levels of cyclic nucleotide second messengers. The authors have discussed various key specific and non-specific inhibitors of PDEs, such as siguazodan, rolipram, sildenafil, aminophylline, roflumilast, *etc.* and their combinations, in the context of clinical use against various respiratory diseases, as well as their benefits and risks.

We are most grateful to all the authors for the excellent contributions. We encourage readers to use the excellent materials presented here in developing their own research strategies, and as a vehicle for conversation with their peers. We are confident that this volume of the ebook series will receive wide appreciation from students, young researchers and established scientists. We would like to recognize the efforts of Mr. Omer Shafi (Assistant Manager Publications), Mr. Shehzad Naqvi (Senior Manager Publications) and the team leader Mr. Mahmood Alam (Director Publications) for putting together an excellent treatise of well written articles efficiently.

Atta-ur-Rahman, *FRS*
Kings College
University of Cambridge, Cambridge
UK

&

M. Iqbal Choudhary
H.E.J. Research Institute of Chemistry
International Center for Chemical and Biological Sciences
University of Karachi, Karachi
Pakistan

List of Contributors

Aman Chandra Kaushik	School of Biotechnology, Gautam Buddha University, Greater Noida, India The Shraga Segal Dept. of Microbiology, Immunology and Genetics, Faculty of Health Sciences, Ben-Gurion University of the Negev, Beer-Sheva, Israel
Andreas Scorilas	Department of Biochemistry and Molecular Biology, National and Kapodistrian University of Athens, Athens, Greece
Anupam Sarma	Department of Pharmaceutical Sciences, Dibrugarh University, Dibrugarh 786004, India
Christos K. Kontos	Department of Biochemistry and Molecular Biology, National and Kapodistrian University of Athens, Athens, Greece
Daniela Mokra	Biomedical Center Martin, Jessenius School of Medicine, Comenius University, Bratislava, Slovakia Department of Physiology, Jessenius School of Medicine, Comenius University, Bratislava, Slovakia
Danijela Savic	Department of Neurobiology, Institute for Biological Research "Sinisa Stankovic" University of Belgrade, Belgrade, Serbia
Irena Lavrnja	Department of Neurobiology, Institute for Biological Research "Sinisa Stankovic" University of Belgrade, Belgrade, Serbia
Ivana Bjelobaba	Department of Neurobiology, Institute for Biological Research "Sinisa Stankovic" University of Belgrade, Belgrade, Serbia
Juraj Mokry	Department of Pharmacology, Jessenius School of Medicine, Comenius University, Bratislava, Slovakia Biomedical Center Martin, Jessenius School of Medicine, Comenius University, Bratislava, Slovakia
Margaritis Avgeris	Department of Biochemistry and Molecular Biology, National and Kapodistrian University of Athens, Athens, Greece
Malay K. Das	Department of Pharmaceutical Sciences, Dibrugarh University, Dibrugarh 786004, India
Makoto Ubukata	Research Faculty of Agriculture, Hokkaido University, Sapporo, Japan
Mahima Kaushik	Cluster Innovation Centre, University of Delhi, Delhi, India Nucleic Acids Research Laboratory, Department of Chemistry, University of Delhi, Delhi, India
Mohan Kumar	Nucleic Acids Research Laboratory, Department of Chemistry, University of Delhi, Delhi, India
Shakti Sahi	School of Biotechnology, Gautam Buddha University, Greater Noida, India
Sanja Pekovic	Department of Neurobiology, Institute for Biological Research "Sinisa Stankovic" University of Belgrade, Belgrade, Serbia
Swati Chaudhary	Nucleic Acids Research Laboratory, Department of Chemistry, University of Delhi, Delhi, India
Sonia	Nucleic Acids Research Laboratory, Department of Chemistry, University of Delhi, Delhi, India

Shrikant Kukreti Nucleic Acids Research Laboratory, Department of Chemistry, University of Delhi, Delhi, India

Swati Mahendru Nucleic Acids Research Laboratory, Department of Chemistry, University of Delhi, Delhi, India

Saami Ahmed Nucleic Acids Research Laboratory, Department of Chemistry, University of Delhi, Delhi, India

Frontiers in Drug Design & Discovery, 2018, *Vol. 9*, 1-34

CHAPTER 1

Biomarkers with Prognostic Potential in Prostate Cancer

Christos K. Kontos[*], **Margaritis Avgeris** and **Andreas Scorilas**

Department of Biochemistry and Molecular Biology, National and Kapodistrian University of Athens, Athens, Greece

Abstract: Prostate cancer (CaP) is the most frequently diagnosed cancer among men and the second leading cause of cancer death for men in both economically developed and developing countries. CaP is usually diagnosed at an early stage with prostate biopsy, following a screening test showing elevated serum levels of prostate-specific antigen (PSA) and/or a positive digital rectal examination. Early CaP diagnosis is the main cause for the significant decrease of metastatic cases, observed during the last twenty years. On the other hand, the wide use of the PSA screening test led also to CaP overdiagnosis and overtreatment. Because of the large variation of CaP characteristics and clinicopathological features, accurate prognosis of CaP patients is a very important issue that determines treatment options. In this chapter, we review the most promising prognostic CaP biomarkers that have been suggested so far. In the emerging era of personalized medicine, these CaP biomarkers could be exploited in the clinics to increase CaP patients' survival and quality of life.

Keywords: Androgen receptor, Kallikrein-related peptidase, KLK, miRNA, Prognosis, Prostate carcinoma, Prostate-specific antigen, PSA, PCA3, Tumor marker.

INTRODUCTION

Prostate cancer (CaP) is the most commonly diagnosed cancer among men in Western countries and the second most frequent worldwide, next to lung cancer. The incidence of CaP in 2012 was estimated to be 1.1 million, while 307 thousand patients died from this malignancy [1]. Nevertheless, despite the fact that CaP can be a serious disease, it is rather not fatal for most patients. In fact, 3 million men diagnosed with CaP in the United States are CaP survivors, today. High survival rates of CaP patients can be attributed to early diagnosis, mainly achieved with

[*] **Corresponding author, Christos K. Kontos:** Department of Biochemistry and Molecular Biology, National and Kapodistrian University of Athens, Panepistimiopolis, Athens, Greece; Tel: (+30) 210-727-4616; Fax: (+30) 210-727-4158; E-mail: chkontos@biol.uoa.gr

Atta-ur-Rahman & M. Iqbal Choudhary (Eds.)

the use of prostate-specific antigen (PSA) as a diagnostic marker in clinical practice, and to slow progression of the disease in many cases [2].

In addition to its diagnostic utility, the PSA screening is used to monitor surgically treated CaP patients for recurrence. Therefore, elevated PSA in serum of patients having been submitted to radical prostatectomy is considered as biochemical relapse, which typically precedes clinical signs and symptoms of CaP recurrence. Nonetheless, a single increase in serum PSA levels in a patient with CaP history is not necessarily indicative of CaP relapse, as other pathological conditions may account for increased PSA, as well. Furthermore, no other molecular biomarkers are used in clinical practice for the prognosis and monitoring of CaP patients, or prediction of response or resistance to therapy [3].

One of the most critical factors for defining prognosis and for decision-making regarding CaP treatment is staging, which is based on the Gleason grading system, the serum PSA concentration, and any other tests revealing the malignant phenotype of CaP cells and how far CaP has spread at the time of initial diagnosis. The most common staging system for CaP is the TNM system of the American Joint Committee on Cancer (AJCC). Recently, an alternative staging system, called stage grouping, has been established; stage grouping combines the determined T, N, and M categories with the Gleason score and serum PSA level. As there are many treatment options for CaP patients, accurate prognosis is also a very important issue. Unfortunately, the currently used prognostic markers have resulted in CaP patient overtreatment so far, since age at the time of diagnosis and comorbidities are not taken into consideration for prognosis and treatment decision-making.

Researchers have focused their efforts on the discovery of novel protein and genetic biomarkers, so as to refine prognostication in CaP and predict systemic treatment of CaP patients who are likely to have a benefit. This chapter overviews current knowledge on promising prognostic and predictive CaP biomarkers (Table **1**), which can play a pivotal role in personalized medicine.

Table 1. List of most promising prognostic and predictive biomarkers in prostate cancer.

Protein-coding genes
AMACR
AR
CHGA
ENG
ETV1

(Table 1) contd.....

Protein-coding genes
ETV4
ETV5
FLI1
IGFBP1
IGFBP2
IGFBP3
IGFBP6
IGF-I
IGF-IR
IL-6
KLK11
KLK14
KLK15
KLK2
KLK3
KLK4
KLK5
PTEN
SFRP1
SFRP2
TGF-β1
TGF-β2
TGF-βRII
TMPRSS2:ERG (fusion)
Non-coding genes
Long non-coding RNAs (lncRNAs)
CBR3-AS1 (also known as: *PlncRNA1*)
CDKN2B-AS1 (also known as: *ANRIL*)
CTBP1-AS
GAS5
HOTAIR
MALAT1
MEG3
PCAT1

(Table 1) contd.....

Protein-coding genes
PCGEM1
PRNCR1
SCHLAP1
MicroRNAs (miRNAs)
miR-31
miR-96
miR-125b
miR-145
miR-205
miR-221
miR-222
miR-378

PROTEIN-CODING GENES AS MOLECULAR BIOMARKERS IN CAP

Kallikrein-related Peptidases (KLKs)

The tissue kallikrein (KLK1) and kallikrein-related peptidases (KLKs) are single-chain, secreted trypsin- or chymotrypsin-like serine peptidases. They are mainly synthesized and secreted by epithelial cells that form the glandular epithelia of many tissues, such as prostate, breast, ovary, colon, pancreas, skin, and brain. KLKs are secreted into body fluids, including sweat, milk, saliva, seminal plasma, and cerebrospinal fluid, or remain in the pericellular space [4]. KLKs are involved in many physiological processes, including remodeling of the extracellular-matrix, prohormone processing, neural plasticity, skin desquamation, regulation of blood pressure, and electrolyte balance. KLKs are signal transduction mediators, as they can activate cell-surface receptors by proteolytic cleavage and also participate in proteolytic cascades [5]. Although several studies have identified in the past *in vivo* substrates of KLKs, no prostate-specific targets have been characterized so far.

PSA (KLK3) is the most well-known member of the KLK family. Determination of serum concentration of total PSA (tPSA) is the most useful prognostic biomarker in CaP, as aforementioned. High preoperative tPSA values are related to advanced stages of the disease and hence predict poor patient prognosis [6]. Besides PSA, several other members of the KLK gene family (*KLK2*, *KLK4*, *KLK5*, *KLK11*, *KLK14*, and *KLK15*) have widely been studied as potential prognostic or treatment response-predictive biomarkers in CaP [7, 8]. Serum concentration of total KLK2 provides PSA-independent prognostic information. In particular, KLK2 levels can predict risk of biochemical relapse in men with serum PSA levels of ≤10 ng/mL. This is very important, because total PSA values 4.0 – 10.0 ng/mL constitute a diagnostic "gray zone" since they may indicate CaP, benign prostate hyperplasia (BPH), or prostate inflammation [9]. In this case, PSA not complexed with proteins, namely free PSA (fPSA), can also assist diagnosis, as men with elevated tPSA and low fPSA – which means high fPSA/tPSA – are more likely to suffer from CaP. Artificial neural networks combining tPSA, fPSA/tPSA, KLK2, KLK2/fPSA and KLK2/(fPSA/tPSA) as covariates possess increased diagnostic and prognostic value [10].

Moreover, increased intratumoral *KLK4* mRNA expression is associated with advanced stages of CaP and correlates with high tPSA in preoperative serum [11]. The investigation of the prognostic potential of *KLK5* and *KLK11* mRNA levels in needle biopsies from CaP patients revealed that both of them constitute promising, independent molecular biomarkers of prognosis in CaP. Another member of the KLK family with particular importance in CaP is *KLK15*, the mRNA expression of which is likely to predict treatment response of CaP cells [12 - 14]. Specific *KLK15* splice variants are aberrantly expressed in more aggressive prostate tumors, thus suggesting an unfavorable prognostic role of this KLK family member in CaP [15].

Androgen Receptor (AR)

The androgen receptor (AR) is a nuclear receptor superfamily member that becomes activated upon binding testosterone or its metabolite, dihydrotestosterone [16]. After its activation, AR can interact with and activate the non-receptor tyrosine kinase SRC in the cytoplasm [17] or can translocate to the nucleus and thereby act as a transcription factor [18]. Moreover, there is a membrane-associated AR isoform that mediates proliferative and survival signaling in CaP cells [19]. AR regulates the transcription of several genes and hence the levels of the respective proteins that are implicated in the proliferation and differentiation of CaP cells, by binding androgens in the epithelium and the prostate stroma [20, 21]. A profound decrease in the AR levels in the stroma underlying the malignant prostatic epithelium results in CaP progression and has significant prognostic value, especially in castration-resistant CaP patients [22, 23]. Interestingly, the mechanisms of loss of stromal AR expression in newly diagnosed prostate malignancy and hormone-resistant CaP are highly similar, thus suggesting that prostate tumors become hormone-independence early during the process of malignant transformation [22]. Dramatic modulations of AR expression in the cancerous epithelium and normal surrounding stroma are inversely associated with high Gleason score, distant metastasis, advanced stage of the disease, response to castration therapy, and biochemical relapse [22, 24 - 27]. In addition, AR mediates CaP relapse after androgen deprivation therapy [28].

Post-translational modifications of AR constitute another significant prognosticator in this malignancy. In more detail, AR phosphorylation at Ser 213 predicts poor overall survival of patients with hormone-refractory CaP [29] and phosphorylation at Ser 515 predicts biochemical relapse [30]. On the other hand, AR phosphorylation at Ser 308 and Ser 791 is indicative of favorable prognosis regarding overall survival of patients with castration-resistant CaP [31].

Secreted Frizzled-Related Proteins 1 (SFRP1) and 2 (SFRP2)

The biological significance and the potential value of Secreted frizzled-related protein 1 (SFRP1), also known as SARP2, as a prognostic or predictive biomarker in CaP has been investigated a lot in the recent past. SFRP1 expression is downregulated in many human malignancies, including CaP. Alterations of the expression balance between SFRP1 and β-catenin are associated with carcinogenesis and tumor progression [32].

SFRP1 counteracts WNT/β-catenin signaling by binding to the WNT proteins through its cysteine-rich domain, which is homologous to the one of frizzled receptors. Loss of SFRP1 may lead to activation of the WNT/β-catenin signaling pathway [33]. The members of the WNT protein family orchestrate and control many key cellular functions, including cell proliferation, migration, differentiation, synaptic activity, and embryogenesis. Furthermore, SFRP1 mediates stromal-to-epithelial paracrine signaling in CaP and, therefore, accounts for the capability of prostatic tumor stroma cells to provide a pro-proliferative paracrine signal to adjacent epithelial cells [34]. Experimental overexpression of SFRP1 in prostatic epithelial cells did not result in canonical WNT/beta-catenin signaling or activation of calcium/calmodulin-dependent protein kinase II gamma (CAMK2G), but led to sustained activation of JNK. Moreover, blocking JNK activity inhibited the SFRP1-induced proliferation of prostatic epithelial cells. Putting all this together, these data suggest that SFRP1 exerts its action through the non-canonical WNT/JNK pathway in prostate [32]. Besides compromising WNT/β-catenin signaling, treatment of a human prostatic epithelial cell line with SFRP1 increased cell proliferation and attenuated apoptosis, *in vitro* [34]. On the other hand, ablation of *SFRP1* expression is common in many distinct malignancies, including CaP [35]. In fact, promoter CpG hypermethylation as well as chromosomal deletions account for dramatic downregulation or even loss of expression of the *SFRP1* gene [35, 36]. Nonetheless, hypermethylation of *SFRP1* is rare in CaP tissues [37], implying that chromosomal deletion is likely to constitute the main cause for the transcriptional silencing of this gene in CaP.

A very recent study focusing on the prognostic and predictive potential of SFRP1 expression has clearly shown that SFRP1 and β-catenin levels are associated with the Gleason score, survival rate, and response of CaP patients to endocrine therapy. Additionally, the expression of the *SFRP1* gene, both at the mRNA and protein level, was significantly elevated in the androgen-dependent cell line LNCaP than in the androgen-independent cell lines DU 145 and PC-3. It is hence evident that SFRP1 expression, which is inversely associated with β-catenin expression, is a promising biomarker of favorable prognosis in CaP [38]. Accordingly, restoration of normal SFRP1 levels in CaP cell lines treated with genistein – achieved *via* DNA methylation or histone modifications – decreased their proliferation rates, invasion, and migration potential [39]. Similarly, DNA demethylation combined with EZH2 inhibition triggered restoration of the expression of the silenced tumor-suppressor *SFRP1* gene. Consequently, this had an additional inhibitory effect on CaP cell growth *in vitro*, hence rendering *SFRP1* a promising target for cancer cell-specific epigenetic therapy [40].

In addition to SFRP1, modulations of the expression of fourteen other WNT/β-catenin signaling-related genes have been observed in CaP. In particular, the expression of most of these genes was upregulated in CaP samples, in contrast with *SFRP1* and its paralog, *SFRP2*. This downregulation of *SFRP2* gene expression was mainly due to hypermethylation, which constitutes a common event in prostate carcinogenesis. Interestingly, *SFRP2* methylation in combination with other epigenetic markers represents a useful biomarker in CaP [41].

Phosphatase and Tensin Homolog (PTEN)

The phosphatase and tensin homolog (PTEN) is a tumor suppressor protein regulating cell growth and proliferation. By dephosphorylation of lipid-signaling intermediates, PTEN deactivates PI3K signaling, which plays a central role in CaP cell survival and progression to the androgen-refractory state [42]. Loss of PTEN and subsequent activation of AKT1 promotes prostate carcinogenesis and bone metastasis of CaP cells through stimulation of the CXCL12/CXCR4 signaling axis [43]. Thus, attenuation of PTEN expression definitely favors metastasis and is a potential biomarker for early detection of CaP metastasis [44, 45]. Furthermore, the fact that aberrant PI3K signaling is associated with metastasis and inferior overall survival of CaP patients suggests that inhibition of the PI3K signaling pathway using AKT1 inhibitors could be beneficial for CaP patients, preventing the development of bone metastases [43, 46].

Homozygous deletion of the *PTEN* gene or loss of heterozygosity followed by a second inactivation event are common in CaP [47 - 49]. *PTEN* haploinsufficiency is also likely to contribute to the transition from preneoplastic prostatic intra-epithelial neoplasia to cancer [49]. Interphase FISH analysis of *PTEN* in histological sections of malignant prostate tumors demonstrated that loss of *PTEN* gene copies is associated with advanced stages of CaP and that it also predicts biochemical relapse [50, 51]. *PTEN* gene deletion results in decreased PTEN protein expression both in localized and metastatic CaP [52]. Altered copy numbers of *PTEN* plus other genes, including allelic gain of *MYC*, possess prognostic significance for recurrence [53]. A recent study aiming at the discrimination of CaP patients with local risk from those with systemic risk, based on the application of array-based comparative genomic hybridization (aCGH) on prostate tissues and positive lymph nodes after radical prostatectomy, revealed that a combination of loss of chromosome 10q (*PTEN*) and gain on 8q (*MYC*) is associated with locally aggressive and metastatic disease [54]. Risk stratification of CaP patients, based on specific gene copy number changes detected in pre-treatment biopsies, may improve the use of systemic therapies to target subclinical metastases or locally recurrent disease and ameliorate patients' outcome.

PTEN protein loss, assessed by IHC, is associated with adverse clinicopathological features of prostate tumors, including high Gleason score and advanced pathologic stage. In fact, ablation of PTEN protein expression is not always due to genomic loss [55]. Absence of PTEN protein expression at the time of prostate biopsy is a predictor of metastasis, higher CaP-specific mortality, castration-resistant CaP, and response to androgen deprivation therapy after radical prostatectomy [56]. Moreover, in conservatively managed CaP patients with localized disease, negative PTEN expression status is a prognosticator, independent of the Gleason score, PSA, Ki-67, and tumor extent [57]. Moreover, in patients with clinically localized CaP treated by prostatectomy, absence of PTEN expression is an indicator of poor prognosis, independently predicting an increased risk of recurrence and adding prognostic value to other clinicopathological factors [58]. Thus, it becomes evident that immunohistochemical assessment of PTEN protein expression in prostate tissue samples could be extremely useful in clinical routine and deserves more evaluation as an early marker of aggressive CaP.

Insulin-like Growth Factor (IGF) System

The insulin-like growth factor (IGF) system consists of three circulating ligands (IGF-I, IGF-II, and insulin), three cell-surface receptors (IGF-IR, IGF-IIR, and insulin receptor) mediating the biological effects of IGFs, six high-affinity IGF-binding proteins (IGFBP1 – IGFBP6), a large group of IGFBP proteases and the most recently discovered group of low-affinity IGFBP-related proteins (IGFBP-rPs) [59, 60]. IGF-IR, a transmembrane tyrosine kinase receptor with higher affinity for IGF-I than for IGF-II and insulin, mainly mediates the biological functions of IGFs. IGF-IR is a tetramer, consisting of identical extracellular α-subunits conferring ligand binding specificity and two identical transmembrane β-subunits possessing the tyrosine kinase activity [61, 62]. IGF-IIR, also known as IGF-II/mannose-6 phosphate (M6P) receptor, is a monomer that has no tyrosine kinase activity [63]. Binding of IGF-I and IGF-II to IGF-IR results in elevation of the tyrosine kinase activity of the β-subunits though conformational changes, and finally activates the MAPK and PI3K/AKT1/mTOR intracellular signaling pathways [64, 65]. Less than 5% of the circulating IGFs are free, while most IGFs are complexed with IGFBPs, mainly with IGFBP3 (>90%) [66, 67]. IGFBPs regulate IGFs/IGF-Rs axis by increasing the half-life time of IGFs, by modulating bioavailability of IGFs, and by modulating their binding to IGF-Rs.

The IGF system is crucial in fetal and linear growth and development of skeleton and multiple organs, as well as in cellular homeostasis, regulating cell proliferation and apoptosis. Moreover, the IGF system is implicated in several human malignancies, including prostate, breast, colon, pancreatic, and liver cancer as well as melanoma and glioblastoma [59, 68 - 70]. Focusing on CaP, IGFs exert their mitogenic and apoptotic action on both normal and malignant prostate cells, modulating CaP development and progression [63, 71]. Epidemiological studies from late 1990s have shown that high circulating IGF-I levels are associated with stronger risk of developing CaP [72 - 74], mainly in younger men [75, 76]. In general, high IGF-I serum levels in younger men predict CaP development; yet, IGF-I levels are not so informative for male population screening and early CaP diagnosis. Similarly, overexpression of the *IGF-IR* gene is common in CaP in contrast with normal prostate epithelium, as well as in metastatic and androgen-independent CaP, compared to primary tumors [77, 78].

Focusing on IGFBPs, IGFBP2 and IGFBP3 have a growth inhibitory effect on normal prostate epithelial cells and seem to have a direct predictive value for CaP prognosis. Elevated IGFBP2 and IGFBP3 levels have been reported in CaP patients [79, 80]. However, in patients with clinically localized CaP, IGFBP2 levels are inversely correlated with aggressive disease features and disease progression, including high tumor volume, high Gleason score, extracapsular extension and invasion of seminal vessels, as well as with poor prognostic outcome [81]. Additionally, lower circulating IGFBP3 levels correlate with bone metastasis of CaP [79, 82]. Finally, IGFBP6 has been proposed as a serum diagnostic marker for CaP, while increased IGFBP1 levels in patients with metastatic CaP are associated with shorter time to androgen-independent disease and poor overall survival [83].

Transforming Growth Factor Beta (TGF-β) Isoforms

Transforming growth factor beta (TGF-β) is a master regulator of the prostate physiology. In the normal prostate, this secreted pleiotropic protein participates in the control of cell growth, proliferation, differentiation, and apoptosis. TGF-β is also a potent immunosuppressor, secreted by cells of the immune system [84]. Once activated, this ligand binds to the type II receptor (TGF-βRII), which in turn recruits type I receptor (TGF-βRI). Next, the activated TGF-βRI phosphorylates its downstream targets, the members of the SMAD family of signal transducers, and hence mediates SMAD signaling. TGF-β can also induce other non-SMAD signaling pathways such as the MAPK pathway. On the other hand, perturbation of TGF-β signaling is involved in autoimmunity, inflammation, and cancer [85].

Three different TGF-β isoforms are currently known: TGF-β1, TGF-β2, and TGF-β3. TGF-β1 is overexpressed in CaP cells and facilitates their growth and metastasis, both by enhancing angiogenesis and by counteracting immune response directed against cancer cells. Furthermore, CaP-derived TGF-β1 induces the expression of DNMTs in CaP, which in turn methylate the *TGFBR1* and *TGFBR2* genes [86]. Therefore, by gradually losing their TGF-β receptors, CaP cells acquire resistance to the antiproliferative and proapoptotic effects of TGF-β1. The loss of TGF-βRI and TGF-βRII may also be the reason for the ability of some malignant prostate neoplasms to escape castration-induced apoptosis. In accordance, high TGF-β1 protein levels along with loss of TGF-βRII expression are related to angiogenesis and metastasis, and also predict poor prognosis for CaP patients [87]. According to Perry *et al.* [88], plasma TGF-β1 levels are similar between CaP patients and normal controls, and are not associated with PSA, clinical and pathologic stages, or Gleason grade. Unlike plasma TGF-β1 levels, urinary TGF-β1 are higher in patients with CaP, as shown in the same study. Nevertheless, for those subjected to radical prostatectomy, elevated preoperative plasma TGF-β1 concentration is associated with positive regional lymph nodes, presumed occult metastases at the time of primary treatment, and disease progression [89]. Moreover, TGF-β1 overexpression by CaP cells is associated with pathological features of the tumor and biochemical progression [87]. TGF-β1 is also overexpressed in metastatic primary CaP, compared to non-metastatic CaP [90].

Concerning TGF-β2, this TGF-β isoform is likely to constitutively activate NF-κB in PC-3 cells. As a result, TGF-β2 hinder CaP cell apoptosis and favors sustained CaP cell survival. As *TGFB2* silencing *via* specific siRNAs was shown to result in PC-3 cell death, TGF-β2 could be exploited as a potential therapeutic target to inhibit growth of tumor cells or to sensitize them to cytotoxic drugs [91]. Moreover, plasma TGF-β2 concentration is higher in CaP patients and has thus been suggested as a useful biomarker in this malignancy [88].

Interleukin-6 (IL-6)

Interleukin-6 (IL-6) is a cytokine with pleiotropic action: it participates in the regulation of the immune system, hematopoiesis, acute phase responses, cell proliferation, and cell differentiation [92]. By binding its receptor (IL6R), it activates JAK/STAT and MAPK signaling pathways. Moreover, IL-6 regulates VEGFA expression and enhances CaP cell growth [92 - 96]. IL-6 is overexpressed in prostatic tissues at early stage of CaP and is also implicated in autocrine and paracrine loops that trigger neuroendocrine differentiation in prostate [95 - 97].

Interestingly, preoperative levels of IL-6 and its soluble receptor sIL6R [official symbol: IL6ST (interleukin 6 signal transducer)] in serum are associated with the extent of CaP, high tumor volume, high Gleason score, and positive nodal status [89, 98]. Serum IL-6 level is also associated with the clinical stage of CaP and holds significant prognostic value [99]. In patients with clinically localized CaP treated by radical prostatectomy, the preoperative plasma concentrations of IL-6 and sIL6R are independent predictors of biochemical relapse after radical prostatectomy, most probably because of the existence of occult metastases at the time of surgery [89, 100, 101]. In support of this conclusion, plasma IL-6 and IL-6sR levels are dramatically elevated in CaP patients with bone metastases [100]. More importantly, plasma IL-6 concentration is a significant prognosticator in CaP patients with hormone-refractory metastatic disease [102]. These findings suggest that IL-6 concentration in serum or plasma could be incorporated into multiparametric prognostic models. Furthermore, quantification of IL-6 levels could be used for monitoring of response to docetaxel, as a substantial decline in IL-6 levels early after treatment onset indicate response of CaP patients to this anticancer agent [103]. Besides that, Wu *et al.* demonstrated that IL-6 constitutes a therapeutic target in mice, as high IL-6 levels confer radiation resistance, whereas IL-6 inhibition sensitizes mouse CaP cells to radiation therapy [104]. Future research efforts will undoubtedly shed more light in the potential clinical utility of IL-6 inhibition in human CaP.

Chromogranin A (CHGA)

Chromogranin A (CHGA), also known as parathyroid secretory protein 1, is a member of the chromogranin/secretogranin family of neuroendocrine secretory proteins [105]. CHGA is localized in the secretory vesicles of neurons and endocrine cells. It is a precursor of many functional biologically active peptides, such as vasostatins 1 and 2, pancreastatin, parastatin, catestatin, and chromofungin. Vasostatin 1, vasostatin 2, pancreastatin, and parastatin modulate negatively the neuroendocrine system with either an autocrine or a paracrine mode of action. With regard to catestatin and chromofungin, both these peptides have antimicrobial properties [106].

CHGA protein overexpression has been observed in neuroendocrine malignancies and CaP with neuroendocrine differentiation [107, 108]. A recent study has shown that neuroendocrine differentiation, indirectly assessed by CHGA immunostaining, predicts existence of distant metastases and worse overall survival in newly diagnosed CaP patients treated with definitive radiotherapy [109]. Thus, CHGA has been suggested as a promising biomarker of CaP prognosis. Strong CHGA immunostaining in prostate tissue is related to high Gleason score [110]. Besides that, CHGA protein expression in needle biopsies of CaP at the time of diagnosis is an independent prognostic factor of patients' survival [111, 112] and correlates with circulating levels of CHGA, especially in CaP patients with metastatic disease [113]. Elevated plasma CHGA levels are very common in patients with hormone-refractory CaP and predict unfavorable prognosis [114], mainly in patients with castration-resistant metastatic CaP treated with abiraterone [115] or enzalutamide [116]. In patients with non-metastatic CaP, high circulating levels of CHGA predict biochemical progression [117] and, in combination with PSA, inferior patient survival after endocrine therapy [118]. Apparently, CHGA is a very promising blood-based biomarker in CaP and merits more clinical evaluation.

Erythroblast Transformation-specific (ETS) Transcription Factors

The erythroblast transformation-specific (ETS) transcription factors regulate embryonic development, cell proliferation, differentiation, angiogenesis, inflammation, and apoptosis. In PPC-1 tumor cells of prostatic tissue origin, the ETS transcription factors inhibit anchorage-independent cell growth and survival as well as cell invasiveness [119]. Interestingly, recurrent gene fusions occurring between the 5'-untranslated region (5'-UTR) of the androgen-regulated transmembrane serine protease 2 (*TMPRSS2*) gene and the ETS family genes *ERG*, *ETV1*, *ETV4*, *ETV5*, and *FLI1* have been identified in the majority of CaP tissue specimens [120 - 126]. The ERG transcription factor plays a central role in CaP progression. Its action comprises the activation of the transcription of the *MYC* oncogene and the repression of prostate epithelial differentiation genes, including PSA and prostein (solute carrier family 45 member 3, SLC45A3). Besides abolishing prostate epithelial differentiation [127, 128], increased ERG activity stimulates prostaglandin-mediated signaling, thereby enhancing tumor progression [129].

In CaP patients subjected to prostatectomy, the detection of fusion transcripts originating from the formation of a *TMPRSS2:ERG* chimeric gene strongly predicts patient relapse, independently of histological grade, tumor stage, and serum PSA levels [130, 131]. On the contrary, the findings of another study support the notion that the *TMPRSS2:ERG* rearrangement in primary tumors predicts longer progression-free and overall survival in CaP patients treated by prostatectomy [132 - 135]. However, it is not associated with Gleason score, pathological tumor extent, diagnostic PSA levels, cell proliferation activity, and genomic amplification of AR, in hormone-refractory malignant neoplasms [133]. This controversy regarding the prognostic role of the *TMPRSS2:ERG* fusion in CaP could be attributed to the fact that *ERG* gene expression is AR-dependent and, therefore, prostate tumors with ERG overexpression due to this genomic rearrangement progress in an androgen-rich environment but show better response to androgen suppression [136]; still, there are also some studies questioning the prognostic value of this chimeric gene [137, 138].

Other 5′-fusion partners of ETS transcription factors that are occasionally present in prostate tumors include *SLC45A3* [123, 125, 139, 140], *DDX5* [141], *OR51E2* [123], *ERVK-17* [121, 123, 124], *UBTF* [123], *NDRG1* [140, 142], and the strongly expressed housekeeping gene *HNRPA2B1* [121, 124]. Nevertheless, the potential significance of chimeric genes – resulting from gene arrangements between ETS family genes and the aforementioned genes – as molecular biomarkers of prognosis in CaP has not been extensively studied so far.

α-methylacyl-CoA Racemase (AMACR)

α-methylacyl-CoA racemase (AMACR) is an enzyme regulating the entry of branched-chain lipids into the peroxisomal and mitochondrial β-oxidation pathways [143]. Remarkable upregulation of AMACR expression has been observed in CaP, in comparison with BPH and normal prostatic tissue, as clearly demonstrated by a meta-analysis of DNA microarray data [144]. In addition to its important diagnostic value in CaP characterized by high sensitivity and specificity [145, 146], AMACR expression is an important molecular biomarker of favorable prognosis in CaP. Low AMACR expression in localized tumors of the prostate predicts biochemical relapse poor overall survival of CaP patients [147]. Still, additional validation is needed to establish the prognostic utility of AMACR in CaP.

Endoglin (ENG)

Endoglin (ENG) is a major homodimeric transmembrane glycoprotein of the vascular endothelium. It is also a component of the TGF-β receptor complex and therefore binds TGF-β1 and TGF-β3 with high affinity. ENG has been suggested as a biomarker of ongoing angiogenesis in CaP, as it is mainly expressed in proliferating endothelial cells of newly formed tumor vessels [148]. In non-metastatic CaP patients treated with radical prostatectomy and neoadjuvant hormonal therapy, high microvessel density in the prostate tumor, assessed by intense ENG immunostaining, predicts independently biochemical relapse [149]. ENG immunopositivity is also related to high Gleason score, metastasis in the regional and distant lymph nodes, and advanced stage of the disease [148]. These findings imply potential prognostic value for ENG regarding overall survival of CaP patients [148, 150].

NON-CODING GENES AS MOLECULAR BIOMARKERS IN CAP

Long Non-coding RNAs (lncRNAs)

Long non-coding RNA (lncRNA) molecules are transcriptional units without protein-coding potential. lncRNAs have similar length and splicing signals with mRNAs, as well as a more tissue-specific pattern than mRNAs [151]. These RNA molecules regulate gene expression both at the transcriptional and post-transcriptional level. This function is accomplished *via* epigenetic mechanisms and mainly through the recruitment of chromatin-remodeling or histone-modifying protein complexes, transcription factors, and RNA polymerases at specific genomic sites, alternative splicing and other post-transcriptional modifications, nuclear import, and translational control [152 - 154]. Moreover, lncRNAs can act as precursors of small non-coding RNAs (sncRNAs), such as microRNAs (miRNAs), and as molecular sponges for miRNAs [155].

Deregulated lncRNA expression has already been reported in several human malignancies, including CaP [156, 157]. Several lncRNAs are implicated in carcinogenesis and progression of CaP, and are hence associated with patients' prognosis and treatment outcome [158 - 161]. Prostate cancer associated 3 (*PCA3*) is a lncRNA, produced almost exclusively in the prostate and overexpressed in CaP tissues and CaP-derived metastases, as compared to BPH tissues. PCA3 participates in the control of CaP cell survival, partly through modulating AR signaling, and exerts its main action in the nuclei and microsomal cell fractions [162]. Despite the fact that castration-resistant prostate tumors show loss of *PCA3* expression, the question whether disease progression *per se* affects *PCA3* expression remains unanswered. *PCA3* expression stratifies patients based on prostatectomy tumor volume and Gleason score. This, it is useful as a biomarker for selecting those patients who have a low volume/low grade tumor and who may benefit from active surveillance and nerve-sparing surgery [163]. However, the potential association of *PCA3* score with CaP clinicopathological factors is unclear, since several studies have failed so far to demonstrate such associations or any prognostic significance for this long non-coding RNA in CaP

[164]. In conclusion, despite the already proven clinical significance of the *PCA3* assay for CaP diagnosis and the decrease of unnecessary biopsies [165], the prognostic potential of *PCA3* expression merits further investigation.

PCGEM1 and prostate cancer associated non-coding RNA 1 (*PRNCR1*) are two prostate-specific and androgen-regulated lncRNAs, showing remarkable upregulation in prostate tumors. *PCGEM1* promotes cell growth and inhibits apoptosis, while *PRNCR1* is implicated in cell viability through AR recruitment to the promoter of androgen-responsive genes [166 - 168]. Several others lncRNAs can regulate AR activity. More precisely, CTBP1 antisense RNA (*CTBP1-AS*) is an androgen-responsive lncRNA showing overexpression in prostate tumors, promoting prostate cell growth by regulating AR expression and antagonizing *CTBP1* expression, a known AR corepressor [169]. Similarly, high expression of *CBR3* antisense RNA 1 (*CBR3-AS1*; also known as *PlncRNA1*) has been observed in CaP, compared to BPH or normal prostate tissues, and its depletion results in decreased CaP cell proliferation and AR activity [170]. Furthermore, HOX transcript antisense RNA (*HOTAIR*) levels are upregulated in castration-resistant CaP and metastatic prostate tumors, facilitating the androgen-independent AR activity through the stabilization of AR [171]. prostate cancer associated transcript 1 (*PCAT1*) represents another prostate-specific lncRNA, highly upregulated in localized and metastatic CaP, promoting cell growth and disease progression by repressing *BRCA2* expression and enhancing *MYC* expression [172 - 174]. *CDKN2B* antisense RNA 1 (*CDKN2B-AS1*; also known as *ANRIL*), another lncRNA that is overexpressed in preneoplastic prostate tissues and CaP, facilitates tumor development and progression through the transcriptional silencing of the *INK4b-ARF-INK4a* locus [175]. Recently, a tumor suppressor role has been attributed to growth arrest specific 5 (*GAS5*) and maternally expressed 3 (*MEG3*) lncRNAs, which. *GAS5* and *MEG3* are downregulated in CaP and promote apoptotic death by suppressing AR signaling and enhancing p53 activity, respectively [176, 177].

Focusing on the clinical significance of lncRNAs in CaP, urine levels of metastasis associated lung adenocarcinoma transcript 1 (*MALAT1*) can assist CaP diagnosis in PSA "gray zone" (4-10 ng/mL), with better diagnostic accuracy than the one of fPSA) [178]. Similarly, *MALAT1* and a *MALAT1*-derived miniRNA (MD-miniRNA) are differentially expressed in plasma and serum of CaP patients compared to normal controls; thus, their exploitation could significantly improve the diagnostic specificity [179]. Additionally, SWI/SNF complex antagonist associated with prostate cancer 1 (*SCHLAP1*) expression represents a promising prognostic marker in CaP. Elevated *SCHLAP1* levels are associated with significantly high risk for biochemical and clinical recurrence, disease progression to advanced stages and metastasis, aggressive disease, and disease-specific mortality [180, 181]. Finally, nuclear paraspeckle assembly transcript 1 (*NEAT1*) overexpression has been associated with biochemical recurrence, metastatic spread, and resistance to anti-androgen therapies [182].

MicroRNAs (miRNAs)

MicroRNAs (miRNAs) are small non-coding RNAs of about 19-24 nucleotides that regulate protein-coding gene expression by inhibiting translation and/or triggering degradation of targeted transcripts. miRNAs bind target sites residing mostly in the 3′-untranslated region (UTR) of mRNAs through imperfect base pairing. miRNAs are involved in several pathways that are important for prostate physiology, while alterations in their expression levels contribute to prostate carcinogenesis [183]. As miRNAs serve as phenotypic signatures of different cancers, several of them appear as candidate diagnostic or prognostic biomarkers.

The first high-throughput study examining the diagnostic and prognostic implications of microRNA profiling in CaP was reported in 2010. Among the fifteen miRNAs (miR-16, miR-31, miR-96, miR-125b, miR-145, miR-149, miR-181b, miR-182, miR-182*, miR-183, miR-184, miR-205, miR-221, miR-222, and miR-375) that were differentially expressed between cancerous and normal prostate tissue, miR-31, miR-96 and miR-205 were the only ones to be associated with Gleason score, while miR-125b, miR-205 and miR-222 were associated with tumor staging [184]. One of the most promising prognostic biomarkers for CaP is the prostate-specific miR-221, an oncogenic miRNA that inhibits IRF2 and SOCS3. miR-221 is implicated in proliferation, apoptosis, and invasion of CaP cells [185]. Low miR-221 levels are correlated with clinicopathological factors, including the Gleason score and the clinical recurrence [186]. Patients without locally confined prostate tumors showing low intratumoral miR-221 expression have an elevated risk for post-operative recurrence. Progressive loss of miR-221 expression has been characterized as a hallmark of metastasis, which explains the high prognostic value of miR-221, highlighted by independent studies [185, 186].

Another miRNA that appears as promising prognostic biomarker in CaP is miR-145. This tumor-suppressor miRNA has been shown to *in vitro* regulate androgen-dependent prostate cell growth by suppressing AR. Low miR-145 levels in CaP are correlated with high Gleason score, presence of metastases, advanced clinical stage, and androgen deprivation therapy response. The same study concluded that miR-145 constitutes a reliable predictor of biochemical relapse and inferior DFS of CaP patients [187]. Similarly, loss of miR-378 expression was noticed in aggressive prostate tumors and was a prognosticator of short-term relapse of CaP patients, independent of treatment [188].

In addition to the abovementioned miRNAs, there are several others that have been suggested as candidate molecular biomarkers of prognosis in CaP. miR-224 is such an example. This CaP-related miRNA regulates the expression of calcium/calmodulin-dependent protein kinase 2 (CAMKK2) and apelin (APLN), a peptide that functions as an endogenous ligand for the G protein-coupled receptor APJ and thus activates distinct tissue-specific signaling pathways [189, 190]. Defects in the function of the miR-224/APLN axis are considered to enhance prostate carcinogenesis and to favor aggressive progression of this malignancy. In accordance with this finding, low miR-224 levels were significantly correlated with advanced clinical stage and metastasis in CaP [190]. Nevertheless, the prognostic significance of miR-224 expression in CaP was not superior to the prognostic value of already established CaP markers being used for prognostic purposes.

MOLECULAR BIOMARKERS DERIVED FROM EXOSOMES AND OTHER EXTRACELLULAR VESICLES

Extracellular vesicles (EVs) are small (40 – 2,000 nm) mediators of intercellular communication, being involved in the transmission of biological signals between cells, so as to regulate a diverse range of biological processes. EVs may be broadly classified into exosomes, microvesicles (MVs) and apoptotic bodies according to their size, origin, morphology, and mode of release [191, 192]. EVs are secreted by most cell types and under various normal and pathological conditions into the extracellular environment and body fluids. They comprise proteins, nucleic acids, lipids, and various metabolites. The biological functions of EVs vary from immune response, cell communication, and extracellular matrix remodeling in normal physiology to cell survival, cell migration, tumor progression, and angiogenesis in human malignancies [193, 194]. The content of EVs is protected and stable, due to encapsulation and association with binding proteins/lipoproteins. Consequently, identification of non-invasive biomarkers derived from EVs seems very promising [195].

Recently, EVs (mainly exosomes) have received considerable attention as rich sources of novel biomarkers in urologic malignancies, including CaP [192, 196, 197]. In this regard, increased RNA levels of *PCA3* and *TMPRSS2:EGR* fusion gene have been reported in urine exosomes of CaP patients. Moreover, all hormone-treated patients were negative for *PCA3* and *TMPRSS2:EGR* transcripts in contrast with the untreated cohort, supporting the notion that they could be exploited for post-treatment monitoring of CaP patients [198, 199]. Additionally, the specific detection of prostate/CaP-derived exosomes both in plasma and urine samples highlighted their elevated concentration in CaP patients compared to the normal control cohort [200, 201]. Focusing on proteins found in exosomes, significantly higher levels of PTEN and BIRC5 (also known as Survivin) have been observed in CaP patients compared to normal controls and BPH [202, 203], while elevated exosomal levels of integrin subunits beta 1 (ITGB1) and alpha 3 (ITGA3) were found in metastatic CaP patients [204].

Exosome-derived miRNAs have attracted also considerable attention as novel biomarker candidates in CaP [205, 206]. The concentration of several miRNAs in plasma and serum exosomes of CaP patients is significantly altered, including miR-141 and miR-375, which have been proposed as novel diagnostic and prognostic biomarkers in CaP [207, 208]. Elevated exosomal levels of miR-141 and miR-375 levels are associated with high Gleason score, positive nodal status, and metastatic CaP. Additionally, the study of plasma exosome miR-375 and miR-1290 in castration-resistant CaP revealed their prognostic capabilities regarding patients' overall survival; thus, they have been proposed as candidate biomarkers for treatment decision in castration-resistant CaP [209]. Similarly, serum exosome miRNA profiles for discrimination of CaP from BPH and for advanced-stage CaP detection have also been reported [210, 211]. Finally, significantly lower exosomal miR-34a levels have been found in CaP patients compared to healthy individuals; low exosomal miR-34a concentration has also been associated with resistance to treatment with docetaxel [212].

CONCLUSION

The currently used staging system and prognostic markers in CaP have severe limitations and certainly lead to overtreatment of this malignancy, especially in men of 66 years and older with a limited life expectancy. Apparently, there is a need for novel biomarkers that would improve the accuracy of patient prognosis and decision-making regarding patient treatment. Of course, the discovery and clinical application of new, prognostic and predictive molecular biomarkers could fill this gap between prediction of outcome and true patient outcome. Promising molecular tumor biomarkers could be also included in multivariate panels of biomarkers. In the new era of personalized medicine, molecular biomarkers with high clinical value will definitely contribute to a better stratification of CaP patients.

KEY ISSUES

- Besides PSA, several other members of the kallikrein-related peptidase (KLK) family, including *KLK2*, *KLK4*, *KLK5*, *KLK11*, and *KLK15*, have been intensively studied for their ability to assist CaP prognosis and/or therapy response monitoring.
- Remarkable alterations in AR expression in the malignant epithelium and normal surrounding stroma are inversely associated with Gleason score, distant metastasis, tumor stage, response to castration therapy, and biochemical relapse.

- Although the diagnostic significance of the PCA3 test for the detection of CaP and the possible decrease of unnecessary biopsies has been well established so far, the potential of *PCA3* expression as a prognostic biomarker in CaP remains doubtful.
- SFRP1 and β-catenin expression levels are associated with the Gleason score, survival rate, and response of CaP patients to endocrine therapy; *SFRP2* methylation in combination with other epigenetic markers is a promising biomarker in CaP.
- Loss of PTEN protein expression at the time of prostate biopsy predicts time to development of metastasis, cancer-specific survival, castration-resistant CaP, and response to androgen deprivation therapy after radical prostatectomy.
- For CaP patients treated with radical prostatectomy, elevated preoperative plasma TGF-β1 levels are associated with positive regional lymph nodes, presumed occult metastases at the time of primary treatment, and disease progression; TGF-β1 overexpression by prostate tumor cells is associated with pathological features of the tumor and biochemical progression.
- In CaP patients undergoing radical prostatectomy due to clinically localized CaP, the preoperative levels of IL-6 and sIL6R in plasma independently predict biochemical progression following surgery, most likely because of occult metastases at the time of surgery.
- CHGA expression in CaP needle biopsies at the time of diagnosis is an independent prognostic factor of survival and also correlates with circulating CHGA levels, particularly in CaP patients with metastatic disease.
- The prognostic potential of chimeric genes containing ETS family members such as the *TMPRSS2:ERG* fusion is controversial and merits further investigation.
- Several miRNAs, including miR-145, miR-221, and miR-224, have been proposed as putative prognostic biomarkers in CaP.

CONSENT FOR PUBLICATION

Not applicable.

CONFLICT OF INTEREST

The author (editor) declares no conflict of interest, financial or otherwise.

ACKNOWLEDGEMENTS

Declared none.

REFERENCE

[1] Torre LA, Bray F, Siegel RL, Ferlay J, Lortet-Tieulent J, Jemal A. Global cancer statistics, 2012. CA

Cancer J Clin 2015; 65(2): 87-108.
[http://dx.doi.org/10.3322/caac.21262] [PMID: 25651787]

[2] Antonarakis ES, Feng Z, Trock BJ, *et al.* The natural history of metastatic progression in men with prostate-specific antigen recurrence after radical prostatectomy: long-term follow-up. BJU Int 2012; 109(1): 32-9.
[http://dx.doi.org/10.1111/j.1464-410X.2011.10422.x] [PMID: 21777360]

[3] Lin DW. Beyond PSA: utility of novel tumor markers in the setting of elevated PSA. Urol Oncol 2009; 27(3): 315-21.
[http://dx.doi.org/10.1016/j.urolonc.2009.01.026] [PMID: 19414121]

[4] Shaw JL, Diamandis EP. Distribution of 15 human kallikreins in tissues and biological fluids. Clin Chem 2007; 53(8): 1423-32.
[http://dx.doi.org/10.1373/clinchem.2007.088104] [PMID: 17573418]

[5] Borgoño CA, Diamandis EP. The emerging roles of human tissue kallikreins in cancer. Nat Rev Cancer 2004; 4(11): 876-90.
[http://dx.doi.org/10.1038/nrc1474] [PMID: 15516960]

[6] Sardana G, Dowell B, Diamandis EP. Emerging biomarkers for the diagnosis and prognosis of prostate cancer. Clin Chem 2008; 54(12): 1951-60.
[http://dx.doi.org/10.1373/clinchem.2008.110668] [PMID: 18927246]

[7] Emami N, Diamandis EP. Utility of kallikrein-related peptidases (KLKs) as cancer biomarkers. Clin Chem 2008; 54(10): 1600-7.
[http://dx.doi.org/10.1373/clinchem.2008.105189] [PMID: 18687738]

[8] Kontos CK, Scorilas A. Kallikrein-related peptidases (KLKs): a gene family of novel cancer biomarkers. Clin Chem Lab Med 2012; 50(11): 1877-91.
[http://dx.doi.org/10.1515/cclm-2012-0247] [PMID: 23093268]

[9] Steuber T, Vickers AJ, Serio AM, *et al.* Comparison of free and total forms of serum human kallikrein 2 and prostate-specific antigen for prediction of locally advanced and recurrent prostate cancer. Clin Chem 2007; 53(2): 233-40.
[http://dx.doi.org/10.1373/clinchem.2006.074963] [PMID: 17185368]

[10] Stephan C, Cammann H, Jung K. Artificial neural networks: has the time come for their use in prostate cancer patients? Nat Clin Pract Urol 2005; 2(6): 262-3.
[http://dx.doi.org/10.1038/ncpuro0207] [PMID: 16474814]

[11] Avgeris M, Stravodimos K, Scorilas A. Kallikrein-related peptidase 4 gene (KLK4) in prostate tumors: quantitative expression analysis and evaluation of its clinical significance. Prostate 2011; 71(16): 1780-9.
[http://dx.doi.org/10.1002/pros.21395] [PMID: 21520157]

[12] Scorilas A, Gregorakis AK. mRNA expression analysis of human kallikrein 11 (KLK11) may be useful in the discrimination of benign prostatic hyperplasia from prostate cancer after needle prostate biopsy. Biol Chem 2006; 387(6): 789-93.
[http://dx.doi.org/10.1515/BC.2006.099] [PMID: 16800741]

[13] Korbakis D, Gregorakis AK, Scorilas A. Quantitative analysis of human kallikrein 5 (KLK5) expression in prostate needle biopsies: an independent cancer biomarker. Clin Chem 2009; 55(5): 904-13.
[http://dx.doi.org/10.1373/clinchem.2008.103788] [PMID: 19299547]

[14] Stavropoulou P, Gregorakis AK, Plebani M, Scorilas A. Expression analysis and prognostic significance of human kallikrein 11 in prostate cancer. Clin Chim Acta 2005; 357(2): 190-5.
[http://dx.doi.org/10.1016/j.cccn.2005.03.026] [PMID: 15893744]

[15] Mavridis K, Avgeris M, Koutalellis G, Stravodimos K, Scorilas A. Expression analysis and study of the KLK15 mRNA splice variants in prostate cancer and benign prostatic hyperplasia. Cancer Sci

2010; 101(3): 693-9.
[http://dx.doi.org/10.1111/j.1349-7006.2009.01450.x] [PMID: 20067463]

[16] Roy AK, Lavrovsky Y, Song CS, *et al.* Regulation of androgen action. Vitam Horm 1999; 55: 309-52. [http://dx.doi.org/10.1016/S0083-6729(08)60938-3] [PMID: 9949684]

[17] Heinlein CA, Chang C. The roles of androgen receptors and androgen-binding proteins in nongenomic androgen actions. Mol Endocrinol 2002; 16(10): 2181-7.
[http://dx.doi.org/10.1210/me.2002-0070] [PMID: 12351684]

[18] Lee DK, Chang C. Molecular communication between androgen receptor and general transcription machinery. J Steroid Biochem Mol Biol 2003; 84(1): 41-9.
[http://dx.doi.org/10.1016/S0960-0760(03)00005-0] [PMID: 12648523]

[19] Yang X, Guo Z, Sun F, *et al.* Novel membrane-associated androgen receptor splice variant potentiates proliferative and survival responses in prostate cancer cells. J Biol Chem 2011; 286(41): 36152-60. [http://dx.doi.org/10.1074/jbc.M111.265124] [PMID: 21878636]

[20] Cunha GR, Hayward SW, Wang YZ, Ricke WA. Role of the stromal microenvironment in carcinogenesis of the prostate. Int J Cancer 2003; 107(1): 1-10.
[http://dx.doi.org/10.1002/ijc.11335] [PMID: 12925950]

[21] Culig Z, Klocker H, Bartsch G, Steiner H, Hobisch A. Androgen receptors in prostate cancer. J Urol 2003; 170(4 Pt 1): 1363-9.
[http://dx.doi.org/10.1097/01.ju.0000075099.20662.7f] [PMID: 14501770]

[22] Olapade-Olaopa EO, MacKay EH, Taub NA, Sandhu DP, Terry TR, Habib FK. Malignant transformation of human prostatic epithelium is associated with the loss of androgen receptor immunoreactivity in the surrounding stroma. Clin Cancer Res 1999; 5(3): 569-76.
[PMID: 10100708]

[23] Komiya A, Yasuda K, Watanabe A, Fujiuchi Y, Tsuzuki T, Fuse H. The prognostic significance of loss of the androgen receptor and neuroendocrine differentiation in prostate biopsy specimens among castration-resistant prostate cancer patients. Mol Clin Oncol 2013; 1(2): 257-62.
[http://dx.doi.org/10.3892/mco.2013.69] [PMID: 24649157]

[24] Wikström P, Marusic J, Stattin P, Bergh A. Low stroma androgen receptor level in normal and tumor prostate tissue is related to poor outcome in prostate cancer patients. Prostate 2009; 69(8): 799-809. [http://dx.doi.org/10.1002/pros.20927] [PMID: 19189305]

[25] Li Y, Li CX, Ye H, *et al.* Decrease in stromal androgen receptor associates with androgen-independent disease and promotes prostate cancer cell proliferation and invasion. J Cell Mol Med 2008; 12(6B): 2790-8.
[http://dx.doi.org/10.1111/j.1582-4934.2008.00279.x] [PMID: 18266956]

[26] Ricciardelli C, Choong CS, Buchanan G, *et al.* Androgen receptor levels in prostate cancer epithelial and peritumoral stromal cells identify non-organ confined disease. Prostate 2005; 63(1): 19-28. [http://dx.doi.org/10.1002/pros.20154] [PMID: 15378523]

[27] Henshall SM, Quinn DI, Lee CS, *et al.* Altered expression of androgen receptor in the malignant epithelium and adjacent stroma is associated with early relapse in prostate cancer. Cancer Res 2001; 61(2): 423-7.
[PMID: 11212224]

[28] Gregory CW, He B, Johnson RT, *et al.* A mechanism for androgen receptor-mediated prostate cancer recurrence after androgen deprivation therapy. Cancer Res 2001; 61(11): 4315-9.
[PMID: 11389051]

[29] McCall P, Gemmell LK, Mukherjee R, Bartlett JM, Edwards J. Phosphorylation of the androgen receptor is associated with reduced survival in hormone-refractory prostate cancer patients. Br J Cancer 2008; 98(6): 1094-101.
[http://dx.doi.org/10.1038/sj.bjc.6604152] [PMID: 18349820]

[30] Willder JM, Heng SJ, McCall P, *et al.* Androgen receptor phosphorylation at serine 515 by Cdk1 predicts biochemical relapse in prostate cancer patients. Br J Cancer 2013; 108(1): 139-48. [PMID: 23321516]

[31] McCall P, Adams CE, Willder JM, *et al.* Androgen receptor phosphorylation at serine 308 and serine 791 predicts enhanced survival in castrate resistant prostate cancer patients. Int J Mol Sci 2013; 14(8): 16656-71. [http://dx.doi.org/10.3390/ijms140816656] [PMID: 23945560]

[32] Joesting MS, Cheever TR, Volzing KG, *et al.* Secreted frizzled related protein 1 is a paracrine modulator of epithelial branching morphogenesis, proliferation, and secretory gene expression in the prostate. Dev Biol 2008; 317(1): 161-73. [http://dx.doi.org/10.1016/j.ydbio.2008.02.021] [PMID: 18371946]

[33] Kawano Y, Diez S, Uysal-Onganer P, Darrington RS, Waxman J, Kypta RM. Secreted Frizzled-related protein-1 is a negative regulator of androgen receptor activity in prostate cancer. Br J Cancer 2009; 100(7): 1165-74. [http://dx.doi.org/10.1038/sj.bjc.6604976] [PMID: 19277043]

[34] Joesting MS, Perrin S, Elenbaas B, *et al.* Identification of SFRP1 as a candidate mediator of stromal-to-epithelial signaling in prostate cancer. Cancer Res 2005; 65(22): 10423-30. [http://dx.doi.org/10.1158/0008-5472.CAN-05-0824] [PMID: 16288033]

[35] Lodygin D, Epanchintsev A, Menssen A, Diebold J, Hermeking H. Functional epigenomics identifies genes frequently silenced in prostate cancer. Cancer Res 2005; 65(10): 4218-27. [http://dx.doi.org/10.1158/0008-5472.CAN-04-4407] [PMID: 15899813]

[36] Stoehr R, Wissmann C, Suzuki H, *et al.* Deletions of chromosome 8p and loss of sFRP1 expression are progression markers of papillary bladder cancer. Lab Invest 2004; 84(4): 465-78. [http://dx.doi.org/10.1038/labinvest.3700068] [PMID: 14968126]

[37] Florl AR, Steinhoff C, Müller M, *et al.* Coordinate hypermethylation at specific genes in prostate carcinoma precedes LINE-1 hypomethylation. Br J Cancer 2004; 91(5): 985-94. [http://dx.doi.org/10.1038/sj.bjc.6602030] [PMID: 15292941]

[38] Zheng L, Sun D, Fan W, Zhang Z, Li Q, Jiang T. Diagnostic value of SFRP1 as a favorable predictive and prognostic biomarker in patients with prostate cancer. PLoS One 2015; 10(2): e0118276. [http://dx.doi.org/10.1371/journal.pone.0118276] [PMID: 25719802]

[39] Hirata H, Hinoda Y, Shahryari V, *et al.* Genistein downregulates onco-miR-1260b and upregulates sFRP1 and Smad4 *via* demethylation and histone modification in prostate cancer cells. Br J Cancer 2014; 110(6): 1645-54. [http://dx.doi.org/10.1038/bjc.2014.48] [PMID: 24504368]

[40] Takeshima H, Wakabayashi M, Hattori N, Yamashita S, Ushijima T. Identification of coexistence of DNA methylation and H3K27me3 specifically in cancer cells as a promising target for epigenetic therapy. Carcinogenesis 2015; 36(2): 192-201. [http://dx.doi.org/10.1093/carcin/bgu238] [PMID: 25477340]

[41] Perry AS, O'Hurley G, Raheem OA, *et al.* Gene expression and epigenetic discovery screen reveal methylation of SFRP2 in prostate cancer. Int J Cancer 2013; 132(8): 1771-80. [http://dx.doi.org/10.1002/ijc.27798] [PMID: 22915211]

[42] Murillo H, Huang H, Schmidt LJ, Smith DI, Tindall DJ. Role of PI3K signaling in survival and progression of LNCaP prostate cancer cells to the androgen refractory state. Endocrinology 2001; 142(11): 4795-805. [http://dx.doi.org/10.1210/endo.142.11.8467] [PMID: 11606446]

[43] Conley-LaComb MK, Saliganan A, Kandagatla P, Chen YQ, Cher ML, Chinni SR. PTEN loss mediated Akt activation promotes prostate tumor growth and metastasis *via* CXCL12/CXCR4 signaling. Mol Cancer 2013; 12(1): 85.

[http://dx.doi.org/10.1186/1476-4598-12-85] [PMID: 23902739]

[44] Schmitz M, Grignard G, Margue C, *et al.* Complete loss of PTEN expression as a possible early prognostic marker for prostate cancer metastasis. Int J Cancer 2007; 120(6): 1284-92. [http://dx.doi.org/10.1002/ijc.22359] [PMID: 17163422]

[45] Koksal IT, Dirice E, Yasar D, *et al.* The assessment of PTEN tumor suppressor gene in combination with Gleason scoring and serum PSA to evaluate progression of prostate carcinoma. Urol Oncol 2004; 22(4): 307-12. [http://dx.doi.org/10.1016/j.urolonc.2004.01.009] [PMID: 15283888]

[46] Saal LH, Johansson P, Holm K, *et al.* Poor prognosis in carcinoma is associated with a gene expression signature of aberrant PTEN tumor suppressor pathway activity. Proc Natl Acad Sci USA 2007; 104(18): 7564-9. [http://dx.doi.org/10.1073/pnas.0702507104] [PMID: 17452630]

[47] Cairns P, Okami K, Halachmi S, *et al.* Frequent inactivation of PTEN/MMAC1 in primary prostate cancer. Cancer Res 1997; 57(22): 4997-5000. [PMID: 9371490]

[48] Li J, Yen C, Liaw D, *et al.* PTEN, a putative protein tyrosine phosphatase gene mutated in human brain, breast, and prostate cancer. Science 1997; 275(5308): 1943-7. [http://dx.doi.org/10.1126/science.275.5308.1943] [PMID: 9072974]

[49] Yoshimoto M, Cutz JC, Nuin PA, *et al.* Interphase FISH analysis of PTEN in histologic sections shows genomic deletions in 68% of primary prostate cancer and 23% of high-grade prostatic intra-epithelial neoplasias. Cancer Genet Cytogenet 2006; 169(2): 128-37. [http://dx.doi.org/10.1016/j.cancergencyto.2006.04.003] [PMID: 16938570]

[50] Yoshimoto M, Cunha IW, Coudry RA, *et al.* FISH analysis of 107 prostate cancers shows that PTEN genomic deletion is associated with poor clinical outcome. Br J Cancer 2007; 97(5): 678-85. [http://dx.doi.org/10.1038/sj.bjc.6603924] [PMID: 17700571]

[51] Heselmeyer-Haddad KM, Berroa Garcia LY, Bradley A, *et al.* Single-cell genetic analysis reveals insights into clonal development of prostate cancers and indicates loss of PTEN as a marker of poor prognosis. Am J Pathol 2014; 184(10): 2671-86. [http://dx.doi.org/10.1016/j.ajpath.2014.06.030] [PMID: 25131421]

[52] Han B, Mehra R, Lonigro RJ, *et al.* Fluorescence in situ hybridization study shows association of PTEN deletion with ERG rearrangement during prostate cancer progression. Mod Pathol 2009; 22(8): 1083-93. [http://dx.doi.org/10.1038/modpathol.2009.69] [PMID: 19407851]

[53] Zafarana G, Ishkanian AS, Malloff CA, *et al.* Copy number alterations of c-MYC and PTEN are prognostic factors for relapse after prostate cancer radiotherapy. Cancer 2012; 118(16): 4053-62. [http://dx.doi.org/10.1002/cncr.26729] [PMID: 22281794]

[54] Lapointe J, Li C, Giacomini CP, *et al.* Genomic profiling reveals alternative genetic pathways of prostate tumorigenesis. Cancer Res 2007; 67(18): 8504-10. [http://dx.doi.org/10.1158/0008-5472.CAN-07-0673] [PMID: 17875689]

[55] Lotan TL, Gurel B, Sutcliffe S, *et al.* PTEN protein loss by immunostaining: analytic validation and prognostic indicator for a high risk surgical cohort of prostate cancer patients. Clin Cancer Res 2011; 17(20): 6563-73. [http://dx.doi.org/10.1158/1078-0432.CCR-11-1244] [PMID: 21878536]

[56] Mithal P, Allott E, Gerber L, *et al.* PTEN loss in biopsy tissue predicts poor clinical outcomes in prostate cancer. Int J Urol 2014; 21(12): 1209-14. [http://dx.doi.org/10.1111/iju.12571] [PMID: 25099119]

[57] Cuzick J, Yang ZH, Fisher G, *et al.* Transatlantic Prostate Group. Prognostic value of PTEN loss in men with conservatively managed localised prostate cancer. Br J Cancer 2013; 108(12): 2582-9.

[http://dx.doi.org/10.1038/bjc.2013.248] [PMID: 23695019]

[58] Chaux A, Peskoe SB, Gonzalez-Roibon N, *et al.* Loss of PTEN expression is associated with increased risk of recurrence after prostatectomy for clinically localized prostate cancer. Mod Pathol 2012; 25(11): 1543-9.
[http://dx.doi.org/10.1038/modpathol.2012.104] [PMID: 22684219]

[59] Pollak MN, Schernhammer ES, Hankinson SE. Insulin-like growth factors and neoplasia. Nat Rev Cancer 2004; 4(7): 505-18.
[http://dx.doi.org/10.1038/nrc1387] [PMID: 15229476]

[60] Foulstone E, Prince S, Zaccheo O, *et al.* Insulin-like growth factor ligands, receptors, and binding proteins in cancer. J Pathol 2005; 205(2): 145-53.
[http://dx.doi.org/10.1002/path.1712] [PMID: 15641016]

[61] Ullrich A, Gray A, Tam AW, *et al.* Insulin-like growth factor I receptor primary structure: comparison with insulin receptor suggests structural determinants that define functional specificity. EMBO J 1986; 5(10): 2503-12.
[PMID: 2877871]

[62] LeRoith D, Werner H, Beitner-Johnson D, Roberts CT Jr. Molecular and cellular aspects of the insulin-like growth factor I receptor. Endocr Rev 1995; 16(2): 143-63.
[http://dx.doi.org/10.1210/edrv-16-2-143] [PMID: 7540132]

[63] Gennigens C, Menetrier-Caux C, Droz JP. Insulin-Like Growth Factor (IGF) family and prostate cancer. Crit Rev Oncol Hematol 2006; 58(2): 124-45.
[http://dx.doi.org/10.1016/j.critrevonc.2005.10.003] [PMID: 16387509]

[64] Riedemann J, Macaulay VM. IGF1R signalling and its inhibition. Endocr Relat Cancer 2006; 13 (Suppl. 1): S33-43.
[http://dx.doi.org/10.1677/erc.1.01280] [PMID: 17259557]

[65] Reynolds AR, Kyprianou N. Growth factor signalling in prostatic growth: significance in tumour development and therapeutic targeting. Br J Pharmacol 2006; 147 (Suppl. 2): S144-52.
[http://dx.doi.org/10.1038/sj.bjp.0706635] [PMID: 16465179]

[66] Rajaram S, Baylink DJ, Mohan S. Insulin-like growth factor-binding proteins in serum and other biological fluids: regulation and functions. Endocr Rev 1997; 18(6): 801-31.
[PMID: 9408744]

[67] Mohan S, Baylink DJ. IGF-binding proteins are multifunctional and act *via* IGF-dependent and -independent mechanisms. J Endocrinol 2002; 175(1): 19-31.
[http://dx.doi.org/10.1677/joe.0.1750019] [PMID: 12379487]

[68] LeRoith D, Roberts CT Jr. The insulin-like growth factor system and cancer. Cancer Lett 2003; 195(2): 127-37.
[http://dx.doi.org/10.1016/S0304-3835(03)00159-9] [PMID: 12767520]

[69] Sachdev D, Yee D. Disrupting insulin-like growth factor signaling as a potential cancer therapy. Mol Cancer Ther 2007; 6(1): 1-12.
[http://dx.doi.org/10.1158/1535-7163.MCT-06-0080] [PMID: 17237261]

[70] Fürstenberger G, Senn HJ. Insulin-like growth factors and cancer. Lancet Oncol 2002; 3(5): 298-302.
[http://dx.doi.org/10.1016/S1470-2045(02)00731-3] [PMID: 12067807]

[71] Meinbach DS, Lokeshwar BL. Insulin-like growth factors and their binding proteins in prostate cancer: cause or consequence? Urol Oncol 2006; 24(4): 294-306.
[http://dx.doi.org/10.1016/j.urolonc.2005.12.004] [PMID: 16818181]

[72] Chan JM, Stampfer MJ, Giovannucci E, *et al.* Plasma insulin-like growth factor-I and prostate cancer risk: a prospective study. Science 1998; 279(5350): 563-6.
[http://dx.doi.org/10.1126/science.279.5350.563] [PMID: 9438850]

[73] Cohen P. Serum insulin-like growth factor-I levels and prostate cancer risk--interpreting the evidence. J Natl Cancer Inst 1998; 90(12): 876-9.
[http://dx.doi.org/10.1093/jnci/90.12.876] [PMID: 9637133]

[74] Wolk A, Mantzoros CS, Andersson SO, *et al.* Insulin-like growth factor 1 and prostate cancer risk: a population-based, case-control study. J Natl Cancer Inst 1998; 90(12): 911-5.
[http://dx.doi.org/10.1093/jnci/90.12.911] [PMID: 9637140]

[75] Harman SM, Metter EJ, Blackman MR, Landis PK, Carter HB. Baltimore Longitudinal Study on Aging. Serum levels of insulin-like growth factor I (IGF-I), IGF-II, IGF-binding protein-3, and prostate-specific antigen as predictors of clinical prostate cancer. J Clin Endocrinol Metab 2000; 85(11): 4258-65.
[http://dx.doi.org/10.1210/jcem.85.11.6990] [PMID: 11095464]

[76] Stattin P, Bylund A, Rinaldi S, *et al.* Plasma insulin-like growth factor-I, insulin-like growth factor-binding proteins, and prostate cancer risk: a prospective study. J Natl Cancer Inst 2000; 92(23): 1910-7.
[http://dx.doi.org/10.1093/jnci/92.23.1910] [PMID: 11106682]

[77] Krueckl SL, Sikes RA, Edlund NM, *et al.* Increased insulin-like growth factor I receptor expression and signaling are components of androgen-independent progression in a lineage-derived prostate cancer progression model. Cancer Res 2004; 64(23): 8620-9.
[http://dx.doi.org/10.1158/0008-5472.CAN-04-2446] [PMID: 15574769]

[78] Hellawell GO, Turner GD, Davies DR, Poulsom R, Brewster SF, Macaulay VM. Expression of the type 1 insulin-like growth factor receptor is up-regulated in primary prostate cancer and commonly persists in metastatic disease. Cancer Res 2002; 62(10): 2942-50.
[PMID: 12019176]

[79] Shariat SF, Lamb DJ, Kattan MW, *et al.* Association of preoperative plasma levels of insulin-like growth factor I and insulin-like growth factor binding proteins-2 and -3 with prostate cancer invasion, progression, and metastasis. J Clin Oncol 2002; 20(3): 833-41.
[http://dx.doi.org/10.1200/JCO.2002.20.3.833] [PMID: 11821468]

[80] Rowlands MA, Tilling K, Holly JM, *et al.* Insulin-like growth factors (IGFs) and IGF-binding proteins in active monitoring of localized prostate cancer: a population-based observational study. Cancer Causes Control 2013; 24(1): 39-45.
[http://dx.doi.org/10.1007/s10552-012-0087-7] [PMID: 23085814]

[81] Inman BA, Harel F, Audet JF, *et al.* Insulin-like growth factor binding protein 2: an androgen-dependent predictor of prostate cancer survival. Eur Urol 2005; 47(5): 695-702.
[http://dx.doi.org/10.1016/j.eururo.2004.12.015] [PMID: 15826765]

[82] Shariat SF, Karam JA, Margulis V, Karakiewicz PI. New blood-based biomarkers for the diagnosis, staging and prognosis of prostate cancer. BJU Int 2008; 101(6): 675-83.
[http://dx.doi.org/10.1111/j.1464-410X.2007.07283.x] [PMID: 17941930]

[83] Heidegger I, Massoner P, Sampson N, Klocker H. The insulin-like growth factor (IGF) axis as an anticancer target in prostate cancer. Cancer Lett 2015; 367(2): 113-21.
[http://dx.doi.org/10.1016/j.canlet.2015.07.026] [PMID: 26231734]

[84] Massagué J. TGFbeta in Cancer. Cell 2008; 134(2): 215-30.
[http://dx.doi.org/10.1016/j.cell.2008.07.001] [PMID: 18662538]

[85] Moustakas A, Pardali K, Gaal A, Heldin CH. Mechanisms of TGF-beta signaling in regulation of cell growth and differentiation. Immunol Lett 2002; 82(1-2): 85-91.
[http://dx.doi.org/10.1016/S0165-2478(02)00023-8] [PMID: 12008039]

[86] Zhang Q, Chen L, Helfand BT, *et al.* TGF-β regulates DNA methyltransferase expression in prostate cancer, correlates with aggressive capabilities, and predicts disease recurrence. PLoS One 2011; 6(9): e25168.

[http://dx.doi.org/10.1371/journal.pone.0025168] [PMID: 21980391]

[87] Shariat SF, Menesses-Diaz A, Kim IY, Muramoto M, Wheeler TM, Slawin KM. Tissue expression of transforming growth factor-beta1 and its receptors: correlation with pathologic features and biochemical progression in patients undergoing radical prostatectomy. Urology 2004; 63(6): 1191-7. [http://dx.doi.org/10.1016/j.urology.2003.12.015] [PMID: 15183988]

[88] Perry KT, Anthony CT, Case T, Steiner MS. Transforming growth factor beta as a clinical biomarker for prostate cancer. Urology 1997; 49(1): 151-5. [http://dx.doi.org/10.1016/S0090-4295(96)00426-8] [PMID: 9000208]

[89] Shariat SF, Kattan MW, Traxel E, *et al.* Association of pre- and postoperative plasma levels of transforming growth factor beta(1) and interleukin 6 and its soluble receptor with prostate cancer progression. Clin Cancer Res 2004; 10(6): 1992-9. [http://dx.doi.org/10.1158/1078-0432.CCR-0768-03] [PMID: 15041717]

[90] Richardsen E, Uglehus RD, Due J, Busch C, Busund LT. COX-2 is overexpressed in primary prostate cancer with metastatic potential and may predict survival. A comparison study between COX-2, TGF-beta, IL-10 and Ki67. Cancer Epidemiol 2010; 34(3): 316-22. [http://dx.doi.org/10.1016/j.canep.2010.03.019] [PMID: 20409773]

[91] Lu T, Burdelya LG, Swiatkowski SM, *et al.* Secreted transforming growth factor beta2 activates NF-kappaB, blocks apoptosis, and is essential for the survival of some tumor cells. Proc Natl Acad Sci USA 2004; 101(18): 7112-7. [http://dx.doi.org/10.1073/pnas.0402048101] [PMID: 15118089]

[92] Kishimoto T, Akira S, Taga T. Interleukin-6 and its receptor: a paradigm for cytokines. Science 1992; 258(5082): 593-7. [http://dx.doi.org/10.1126/science.1411569] [PMID: 1411569]

[93] Siegsmund MJ, Yamazaki H, Pastan I. Interleukin 6 receptor mRNA in prostate carcinomas and benign prostate hyperplasia. J Urol 1994; 151(5): 1396-9. [http://dx.doi.org/10.1016/S0022-5347(17)35267-9] [PMID: 7512667]

[94] Okamoto M, Lee C, Oyasu R. Interleukin-6 as a paracrine and autocrine growth factor in human prostatic carcinoma cells in vitro. Cancer Res 1997; 57(1): 141-6. [PMID: 8988055]

[95] Spiotto MT, Chung TD. STAT3 mediates IL-6-induced neuroendocrine differentiation in prostate cancer cells. Prostate 2000; 42(3): 186-95. [http://dx.doi.org/10.1002/(SICI)1097-0045(20000215)42:3<186::AID-PROS4>3.0.CO;2-E] [PMID: 10639189]

[96] Culig Z, Steiner H, Bartsch G, Hobisch A. Interleukin-6 regulation of prostate cancer cell growth. J Cell Biochem 2005; 95(3): 497-505. [http://dx.doi.org/10.1002/jcb.20477] [PMID: 15838876]

[97] Hobisch A, Rogatsch H, Hittmair A, *et al.* Immunohistochemical localization of interleukin-6 and its receptor in benign, premalignant and malignant prostate tissue. J Pathol 2000; 191(3): 239-44. [http://dx.doi.org/10.1002/1096-9896(2000)9999:9999<::AID-PATH633>3.0.CO;2-X] [PMID: 10878544]

[98] Michalaki V, Syrigos K, Charles P, Waxman J. Serum levels of IL-6 and TNF-alpha correlate with clinicopathological features and patient survival in patients with prostate cancer. Br J Cancer 2004; 90(12): 2312-6. [http://dx.doi.org/10.1038/sj.bjc.6601814] [PMID: 15150588]

[99] Nakashima J, Tachibana M, Horiguchi Y, *et al.* Serum interleukin 6 as a prognostic factor in patients with prostate cancer. Clin Cancer Res 2000; 6(7): 2702-6. [PMID: 10914713]

[100] Shariat SF, Andrews B, Kattan MW, Kim J, Wheeler TM, Slawin KM. Plasma levels of interleukin-6

and its soluble receptor are associated with prostate cancer progression and metastasis. Urology 2001; 58(6): 1008-15.
[http://dx.doi.org/10.1016/S0090-4295(01)01405-4] [PMID: 11744478]

[101] Kattan MW, Shariat SF, Andrews B, *et al.* The addition of interleukin-6 soluble receptor and transforming growth factor beta1 improves a preoperative nomogram for predicting biochemical progression in patients with clinically localized prostate cancer. J Clin Oncol 2003; 21(19): 3573-9.
[http://dx.doi.org/10.1200/JCO.2003.12.037] [PMID: 12913106]

[102] George DJ, Halabi S, Shepard TF, *et al.* The prognostic significance of plasma interleukin-6 levels in patients with metastatic hormone-refractory prostate cancer: results from cancer and leukemia group B 9480. Clin Cancer Res 2005; 11(5): 1815-20.
[http://dx.doi.org/10.1158/1078-0432.CCR-04-1560] [PMID: 15756004]

[103] Ignatoski KM, Friedman J, Escara-Wilke J, *et al.* Change in markers of bone metabolism with chemotherapy for advanced prostate cancer: interleukin-6 response is a potential early indicator of response to therapy. J Interferon Cytokine Res 2009; 29(2): 105-12.
[http://dx.doi.org/10.1089/jir.2008.0024] [PMID: 19014338]

[104] Wu CT, Chen MF, Chen WC, Hsieh CC. The role of IL-6 in the radiation response of prostate cancer. Radiat Oncol 2013; 8: 159.
[http://dx.doi.org/10.1186/1748-717X-8-159] [PMID: 23806095]

[105] Helman LJ, Ahn TG, Levine MA, *et al.* Molecular cloning and primary structure of human chromogranin A (secretory protein I) cDNA. J Biol Chem 1988; 263(23): 11559-63.
[PMID: 3403545]

[106] D'amico MA, Ghinassi B, Izzicupo P, Manzoli L, Di Baldassarre A. Biological function and clinical relevance of chromogranin A and derived peptides. Endocr Connect 2014; 3(2): R45-54.
[http://dx.doi.org/10.1530/EC-14-0027] [PMID: 24671122]

[107] Syversen U, Ramstad H, Gamme K, Qvigstad G, Falkmer S, Waldum HL. Clinical significance of elevated serum chromogranin A levels. Scand J Gastroenterol 2004; 39(10): 969-73.
[http://dx.doi.org/10.1080/00365520410003362] [PMID: 15513336]

[108] Kadmon D, Thompson TC, Lynch GR, Scardino PT. Elevated plasma chromogranin-A concentrations in prostatic carcinoma. J Urol 1991; 146(2): 358-61.
[http://dx.doi.org/10.1016/S0022-5347(17)37793-5] [PMID: 1856931]

[109] Krauss DJ, Amin M, Stone B, *et al.* Chromogranin A staining as a prognostic variable in newly diagnosed Gleason score 7-10 prostate cancer treated with definitive radiotherapy. Prostate 2014; 74(5): 520-7.
[http://dx.doi.org/10.1002/pros.22771] [PMID: 24375481]

[110] Grimaldi F, Valotto C, Barbina G, *et al.* The possible role of chromogranin A as a prognostic factor in organ-confined prostate cancer. Int J Biol Markers 2006; 21(4): 229-34.
[http://dx.doi.org/10.5301/JBM.2008.5160] [PMID: 17177161]

[111] Cheville JC, Tindall D, Boelter C, *et al.* Metastatic prostate carcinoma to bone: clinical and pathologic features associated with cancer-specific survival. Cancer 2002; 95(5): 1028-36.
[http://dx.doi.org/10.1002/cncr.10788] [PMID: 12209687]

[112] Berruti A, Bollito E, Cracco CM, *et al.* The prognostic role of immunohistochemical chromogranin a expression in prostate cancer patients is significantly modified by androgen-deprivation therapy. Prostate 2010; 70(7): 718-26.
[PMID: 20087896]

[113] Bollito E, Berruti A, Bellina M, *et al.* Relationship between neuroendocrine features and prognostic parameters in human prostate adenocarcinoma. Ann Oncol 2001; 12 (Suppl. 2): S159-64.
[http://dx.doi.org/10.1093/annonc/12.suppl_2.S159] [PMID: 11762345]

[114] Berruti A, Mosca A, Tucci M, *et al.* Independent prognostic role of circulating chromogranin A in

prostate cancer patients with hormone-refractory disease. Endocr Relat Cancer 2005; 12(1): 109-17. [http://dx.doi.org/10.1677/erc.1.00876] [PMID: 15788643]

[115] Burgio SL, Conteduca V, Menna C, *et al.* Chromogranin A predicts outcome in prostate cancer patients treated with abiraterone. Endocr Relat Cancer 2014; 21(3): 487-93. [http://dx.doi.org/10.1530/ERC-14-0071] [PMID: 24741024]

[116] Conteduca V, Burgio SL, Menna C, *et al.* Chromogranin A is a potential prognostic marker in prostate cancer patients treated with enzalutamide. Prostate 2014; 74(16): 1691-6. [http://dx.doi.org/10.1002/pros.22890] [PMID: 25262981]

[117] Alessandro S, Vincenzo G, Maria AG, *et al.* Chromogranin A and biochemical progression-free survival in prostate adenocarcinomas submitted to radical prostatectomy. Endocr Relat Cancer 2007; 14(3): 625-32. [http://dx.doi.org/10.1677/ERC-07-0089] [PMID: 17914093]

[118] Isshiki S, Akakura K, Komiya A, Suzuki H, Kamiya N, Ito H. Chromogranin a concentration as a serum marker to predict prognosis after endocrine therapy for prostate cancer. J Urol 2002; 167(2 Pt 1): 512-5. [http://dx.doi.org/10.1097/00005392-200202000-00014] [PMID: 11792908]

[119] Foos G, Hauser CA. Altered Ets transcription factor activity in prostate tumor cells inhibits anchorage-independent growth, survival, and invasiveness. Oncogene 2000; 19(48): 5507-16. [http://dx.doi.org/10.1038/sj.onc.1203946] [PMID: 11114728]

[120] Tomlins SA, Rhodes DR, Perner S, *et al.* Recurrent fusion of TMPRSS2 and ETS transcription factor genes in prostate cancer. Science 2005; 310(5748): 644-8. [http://dx.doi.org/10.1126/science.1117679] [PMID: 16254181]

[121] Tomlins SA, Laxman B, Dhanasekaran SM, *et al.* Distinct classes of chromosomal rearrangements create oncogenic ETS gene fusions in prostate cancer. Nature 2007; 448(7153): 595-9. [http://dx.doi.org/10.1038/nature06024] [PMID: 17671502]

[122] Tomlins SA, Mehra R, Rhodes DR, *et al.* TMPRSS2:ETV4 gene fusions define a third molecular subtype of prostate cancer. Cancer Res 2006; 66(7): 3396-400. [http://dx.doi.org/10.1158/0008-5472.CAN-06-0168] [PMID: 16585160]

[123] Barros-Silva JD, Paulo P, Bakken AC, *et al.* Novel 5′ fusion partners of ETV1 and ETV4 in prostate cancer. Neoplasia 2013; 15(7): 720-6. [http://dx.doi.org/10.1593/neo.13232] [PMID: 23814484]

[124] Helgeson BE, Tomlins SA, Shah N, *et al.* Characterization of TMPRSS2:ETV5 and SLC45A3:ETV5 gene fusions in prostate cancer. Cancer Res 2008; 68(1): 73-80. [http://dx.doi.org/10.1158/0008-5472.CAN-07-5352] [PMID: 18172298]

[125] Paulo P, Barros-Silva JD, Ribeiro FR, *et al.* FLI1 is a novel ETS transcription factor involved in gene fusions in prostate cancer. Genes Chromosomes Cancer 2012; 51(3): 240-9. [http://dx.doi.org/10.1002/gcc.20948] [PMID: 22081504]

[126] Mehra R, Tomlins SA, Shen R, *et al.* Comprehensive assessment of TMPRSS2 and ETS family gene aberrations in clinically localized prostate cancer. Mod Pathol 2007; 20(5): 538-44. [http://dx.doi.org/10.1038/modpathol.3800769] [PMID: 17334343]

[127] Sun C, Dobi A, Mohamed A, *et al.* TMPRSS2-ERG fusion, a common genomic alteration in prostate cancer activates C-MYC and abrogates prostate epithelial differentiation. Oncogene 2008; 27(40): 5348-53. [http://dx.doi.org/10.1038/onc.2008.183] [PMID: 18542058]

[128] Perner S, Rupp NJ, Braun M, *et al.* Loss of SLC45A3 protein (prostein) expression in prostate cancer is associated with SLC45A3-ERG gene rearrangement and an unfavorable clinical course. Int J Cancer 2013; 132(4): 807-12. [http://dx.doi.org/10.1002/ijc.27733] [PMID: 22821757]

[129] Mohamed AA, Tan SH, Sun C, *et al.* ERG oncogene modulates prostaglandin signaling in prostate cancer cells. Cancer Biol Ther 2011; 11(4): 410-7. [http://dx.doi.org/10.4161/cbt.11.4.14180] [PMID: 21178489]

[130] Nam RK, Sugar L, Wang Z, *et al.* Expression of TMPRSS2:ERG gene fusion in prostate cancer cells is an important prognostic factor for cancer progression. Cancer Biol Ther 2007; 6(1): 40-5. [http://dx.doi.org/10.4161/cbt.6.1.3489] [PMID: 17172822]

[131] Nam RK, Sugar L, Yang W, *et al.* Expression of the TMPRSS2:ERG fusion gene predicts cancer recurrence after surgery for localised prostate cancer. Br J Cancer 2007; 97(12): 1690-5. [http://dx.doi.org/10.1038/sj.bjc.6604054] [PMID: 17971772]

[132] Kim SH, Kim SH, Joung JY, *et al.* Overexpression of ERG and Wild-Type PTEN Are Associated with Favorable Clinical Prognosis and Low Biochemical Recurrence in Prostate Cancer. PLoS One 2015; 10(4): e0122498. [http://dx.doi.org/10.1371/journal.pone.0122498] [PMID: 25897494]

[133] Saramäki OR, Harjula AE, Martikainen PM, Vessella RL, Tammela TL, Visakorpi T. TMPRSS2:ERG fusion identifies a subgroup of prostate cancers with a favorable prognosis. Clin Cancer Res 2008; 14(11): 3395-400. [http://dx.doi.org/10.1158/1078-0432.CCR-07-2051] [PMID: 18519769]

[134] Fleischmann A, Saramäki OR, Zlobec I, *et al.* Prevalence and prognostic significance of TMPRSS2-ERG gene fusion in lymph node positive prostate cancers. Prostate 2014; 74(16): 1647-54. [http://dx.doi.org/10.1002/pros.22882] [PMID: 25252136]

[135] Hermans KG, Boormans JL, Gasi D, *et al.* Overexpression of prostate-specific TMPRSS2(exon 0)-ERG fusion transcripts corresponds with favorable prognosis of prostate cancer. Clin Cancer Res 2009; 15(20): 6398-403. [http://dx.doi.org/10.1158/1078-0432.CCR-09-1176] [PMID: 19825963]

[136] Taris M, Irani J, Blanchet P, Multigner L, Cathelineau X, Fromont G. ERG expression in prostate cancer: the prognostic paradox. Prostate 2014; 74(15): 1481-7. [http://dx.doi.org/10.1002/pros.22863] [PMID: 25175352]

[137] Leinonen KA, Tolonen TT, Bracken H, *et al.* Association of SPINK1 expression and TMPRSS2:ERG fusion with prognosis in endocrine-treated prostate cancer. Clin Cancer Res 2010; 16(10): 2845-51. [http://dx.doi.org/10.1158/1078-0432.CCR-09-2505] [PMID: 20442300]

[138] Minner S, Enodien M, Sirma H, *et al.* ERG status is unrelated to PSA recurrence in radically operated prostate cancer in the absence of antihormonal therapy. Clin Cancer Res 2011; 17(18): 5878-88. [http://dx.doi.org/10.1158/1078-0432.CCR-11-1251] [PMID: 21791629]

[139] Rickman DS, Pflueger D, Moss B, *et al.* SLC45A3-ELK4 is a novel and frequent erythroblast transformation-specific fusion transcript in prostate cancer. Cancer Res 2009; 69(7): 2734-8. [http://dx.doi.org/10.1158/0008-5472.CAN-08-4926] [PMID: 19293179]

[140] Esgueva R, Perner S, J LaFargue C, *et al.* Prevalence of TMPRSS2-ERG and SLC45A3-ERG gene fusions in a large prostatectomy cohort. Mod Pathol 2010; 23(4): 539-46. [http://dx.doi.org/10.1038/modpathol.2009.193] [PMID: 20118910]

[141] Han B, Mehra R, Dhanasekaran SM, *et al.* A fluorescence in situ hybridization screen for E26 transformation-specific aberrations: identification of DDX5-ETV4 fusion protein in prostate cancer. Cancer Res 2008; 68(18): 7629-37. [http://dx.doi.org/10.1158/0008-5472.CAN-08-2014] [PMID: 18794152]

[142] Pflueger D, Rickman DS, Sboner A, *et al.* N-myc downstream regulated gene 1 (NDRG1) is fused to ERG in prostate cancer. Neoplasia 2009; 11(8): 804-11. [http://dx.doi.org/10.1593/neo.09572] [PMID: 19649210]

[143] Lloyd MD, Darley DJ, Wierzbicki AS, Threadgill MD. Alpha-methylacyl-CoA racemase--an 'obscure' metabolic enzyme takes centre stage. FEBS J 2008; 275(6): 1089-102.

[http://dx.doi.org/10.1111/j.1742-4658.2008.06290.x] [PMID: 18279392]

[144] Rhodes DR, Barrette TR, Rubin MA, Ghosh D, Chinnaiyan AM. Meta-analysis of microarrays: interstudy validation of gene expression profiles reveals pathway dysregulation in prostate cancer. Cancer Res 2002; 62(15): 4427-33.
[PMID: 12154050]

[145] Jiang Z, Wu CL, Woda BA, *et al.* Alpha-methylacyl-CoA racemase: a multi-institutional study of a new prostate cancer marker. Histopathology 2004; 45(3): 218-25.
[http://dx.doi.org/10.1111/j.1365-2559.2004.01930.x] [PMID: 15330799]

[146] Sreekumar A, Laxman B, Rhodes DR, *et al.* Humoral immune response to alpha-methylacyl-CoA racemase and prostate cancer. J Natl Cancer Inst 2004; 96(11): 834-43.
[http://dx.doi.org/10.1093/jnci/djh145] [PMID: 15173267]

[147] Rubin MA, Bismar TA, Andrén O, *et al.* Decreased alpha-methylacyl CoA racemase expression in localized prostate cancer is associated with an increased rate of biochemical recurrence and cancer-specific death. Cancer Epidemiol Biomarkers Prev 2005; 14(6): 1424-32.
[http://dx.doi.org/10.1158/1055-9965.EPI-04-0801] [PMID: 15941951]

[148] El-Gohary YM, Silverman JF, Olson PR, *et al.* Endoglin (CD105) and vascular endothelial growth factor as prognostic markers in prostatic adenocarcinoma. Am J Clin Pathol 2007; 127(4): 572-9.
[http://dx.doi.org/10.1309/X6NXYE57DLUE2NQ8] [PMID: 17369132]

[149] Miyata Y, Mitsunari K, Asai A, Takehara K, Mochizuki Y, Sakai H. Pathological significance and prognostic role of microvessel density, evaluated using CD31, CD34, and CD105 in prostate cancer patients after radical prostatectomy with neoadjuvant therapy. Prostate 2015; 75(1): 84-91.
[http://dx.doi.org/10.1002/pros.22894] [PMID: 25307287]

[150] Wikström P, Lissbrant IF, Stattin P, Egevad L, Bergh A. Endoglin (CD105) is expressed on immature blood vessels and is a marker for survival in prostate cancer. Prostate 2002; 51(4): 268-75.
[http://dx.doi.org/10.1002/pros.10083] [PMID: 11987155]

[151] Derrien T, Johnson R, Bussotti G, *et al.* The GENCODE v7 catalog of human long noncoding RNAs: analysis of their gene structure, evolution, and expression. Genome Res 2012; 22(9): 1775-89.
[http://dx.doi.org/10.1101/gr.132159.111] [PMID: 22955988]

[152] Wilusz JE, Sunwoo H, Spector DL. Long noncoding RNAs: functional surprises from the RNA world. Genes Dev 2009; 23(13): 1494-504.
[http://dx.doi.org/10.1101/gad.1800909] [PMID: 19571179]

[153] Gibb EA, Brown CJ, Lam WL. The functional role of long non-coding RNA in human carcinomas. Mol Cancer 2011; 10: 38.
[http://dx.doi.org/10.1186/1476-4598-10-38] [PMID: 21489289]

[154] Gupta RA, Shah N, Wang KC, *et al.* Long non-coding RNA HOTAIR reprograms chromatin state to promote cancer metastasis. Nature 2010; 464(7291): 1071-6.
[http://dx.doi.org/10.1038/nature08975] [PMID: 20393566]

[155] Guttman M, Rinn JL. Modular regulatory principles of large non-coding RNAs. Nature 2012; 482(7385): 339-46.
[http://dx.doi.org/10.1038/nature10887] [PMID: 22337053]

[156] Prensner JR, Chinnaiyan AM. The emergence of lncRNAs in cancer biology. Cancer Discov 2011; 1(5): 391-407.
[http://dx.doi.org/10.1158/2159-8290.CD-11-0209] [PMID: 22096659]

[157] Shi X, Sun M, Liu H, Yao Y, Song Y. Long non-coding RNAs: a new frontier in the study of human diseases. Cancer Lett 2013; 339(2): 159-66.
[http://dx.doi.org/10.1016/j.canlet.2013.06.013] [PMID: 23791884]

[158] Martens-Uzunova ES, Böttcher R, Croce CM, Jenster G, Visakorpi T, Calin GA. Long noncoding RNA in prostate, bladder, and kidney cancer. Eur Urol 2014; 65(6): 1140-51.

[http://dx.doi.org/10.1016/j.eururo.2013.12.003] [PMID: 24373479]

[159] Malik B, Feng FY. Long noncoding RNAs in prostate cancer: overview and clinical implications. Asian J Androl 2016; 18(4): 568-74.
[http://dx.doi.org/10.4103/1008-682X.177123] [PMID: 27072044]

[160] Cheng W, Zhang Z, Wang J. Long noncoding RNAs: new players in prostate cancer. Cancer Lett 2013; 339(1): 8-14.
[http://dx.doi.org/10.1016/j.canlet.2013.07.008] [PMID: 23856029]

[161] Chen J, Miao Z, Xue B, Shan Y, Weng G, Shen B. Long Non-coding RNAs in Urologic Malignancies: Functional Roles and Clinical Translation. J Cancer 2016; 7(13): 1842-55.
[http://dx.doi.org/10.7150/jca.15876] [PMID: 27698924]

[162] Ferreira LB, Palumbo A, de Mello KD, *et al.* PCA3 noncoding RNA is involved in the control of prostate-cancer cell survival and modulates androgen receptor signaling. BMC Cancer 2012; 12: 507.
[http://dx.doi.org/10.1186/1471-2407-12-507] [PMID: 23130941]

[163] Durand X, Xylinas E, Radulescu C, *et al.* The value of urinary prostate cancer gene 3 (PCA3) scores in predicting pathological features at radical prostatectomy. BJU Int 2012; 110(1): 43-9.
[http://dx.doi.org/10.1111/j.1464-410X.2011.10682.x] [PMID: 22221521]

[164] Augustin H, Mayrhofer K, Pummer K, Mannweiler S. Relationship between prostate cancer gene 3 (PCA3) and characteristics of tumor aggressiveness. Prostate 2013; 73(2): 203-10.
[http://dx.doi.org/10.1002/pros.22558] [PMID: 22782916]

[165] Haese A, de la Taille A, van Poppel H, *et al.* Clinical utility of the PCA3 urine assay in European men scheduled for repeat biopsy. Eur Urol 2008; 54(5): 1081-8.
[http://dx.doi.org/10.1016/j.eururo.2008.06.071] [PMID: 18602209]

[166] Srikantan V, Zou Z, Petrovics G, *et al.* PCGEM1, a prostate-specific gene, is overexpressed in prostate cancer. Proc Natl Acad Sci USA 2000; 97(22): 12216-21.
[http://dx.doi.org/10.1073/pnas.97.22.12216] [PMID: 11050243]

[167] Petrovics G, Zhang W, Makarem M, *et al.* Elevated expression of PCGEM1, a prostate-specific gene with cell growth-promoting function, is associated with high-risk prostate cancer patients. Oncogene 2004; 23(2): 605-11.
[http://dx.doi.org/10.1038/sj.onc.1207069] [PMID: 14724589]

[168] Yang L, Lin C, Jin C, *et al.* lncRNA-dependent mechanisms of androgen-receptor-regulated gene activation programs. Nature 2013; 500(7464): 598-602.
[http://dx.doi.org/10.1038/nature12451] [PMID: 23945587]

[169] Takayama K, Horie-Inoue K, Katayama S, *et al.* Androgen-responsive long noncoding RNA CTBP1-AS promotes prostate cancer. EMBO J 2013; 32(12): 1665-80.
[http://dx.doi.org/10.1038/emboj.2013.99] [PMID: 23644382]

[170] Cui Z, Ren S, Lu J, *et al.* The prostate cancer-up-regulated long noncoding RNA PlncRNA-1 modulates apoptosis and proliferation through reciprocal regulation of androgen receptor. Urol Oncol 2013; 31(7): 1117-23.
[http://dx.doi.org/10.1016/j.urolonc.2011.11.030] [PMID: 22264502]

[171] Zhang A, Zhao JC, Kim J, *et al.* LncRNA HOTAIR Enhances the Androgen-Receptor-Mediated Transcriptional Program and Drives Castration-Resistant Prostate Cancer. Cell Reports 2015; 13(1): 209-21.
[http://dx.doi.org/10.1016/j.celrep.2015.08.069] [PMID: 26411689]

[172] Prensner JR, Iyer MK, Balbin OA, *et al.* Transcriptome sequencing across a prostate cancer cohort identifies PCAT-1, an unannotated lincRNA implicated in disease progression. Nat Biotechnol 2011; 29(8): 742-9.
[http://dx.doi.org/10.1038/nbt.1914] [PMID: 21804560]

[173] Prensner JR, Chen W, Han S, *et al.* The long non-coding RNA PCAT-1 promotes prostate cancer cell

proliferation through cMyc. Neoplasia 2014; 16(11): 900-8.
[http://dx.doi.org/10.1016/j.neo.2014.09.001] [PMID: 25425964]

[174] Prensner JR, Chen W, Iyer MK, *et al.* PCAT-1, a long noncoding RNA, regulates BRCA2 and controls homologous recombination in cancer. Cancer Res 2014; 74(6): 1651-60.
[http://dx.doi.org/10.1158/0008-5472.CAN-13-3159] [PMID: 24473064]

[175] Yap KL, Li S, Muñoz-Cabello AM, *et al.* Molecular interplay of the noncoding RNA ANRIL and methylated histone H3 lysine 27 by polycomb CBX7 in transcriptional silencing of INK4a. Mol Cell 2010; 38(5): 662-74.
[http://dx.doi.org/10.1016/j.molcel.2010.03.021] [PMID: 20541999]

[176] Mourtada-Maarabouni M, Pickard MR, Hedge VL, Farzaneh F, Williams GT. GAS5, a non-protei--coding RNA, controls apoptosis and is downregulated in breast cancer. Oncogene 2009; 28(2): 195-208.
[http://dx.doi.org/10.1038/onc.2008.373] [PMID: 18836484]

[177] Zhou Y, Zhang X, Klibanski A. MEG3 noncoding RNA: a tumor suppressor. J Mol Endocrinol 2012; 48(3): R45-53.
[http://dx.doi.org/10.1530/JME-12-0008] [PMID: 22393162]

[178] Wang F, Ren S, Chen R, *et al.* Development and prospective multicenter evaluation of the long noncoding RNA MALAT-1 as a diagnostic urinary biomarker for prostate cancer. Oncotarget 2014; 5(22): 11091-102.
[http://dx.doi.org/10.18632/oncotarget.2691] [PMID: 25526029]

[179] Ren S, Wang F, Shen J, *et al.* Long non-coding RNA metastasis associated in lung adenocarcinoma transcript 1 derived miniRNA as a novel plasma-based biomarker for diagnosing prostate cancer. Eur J Cancer 2013; 49(13): 2949-59.
[http://dx.doi.org/10.1016/j.ejca.2013.04.026] [PMID: 23726266]

[180] Prensner JR, Iyer MK, Sahu A, *et al.* The long noncoding RNA SChLAP1 promotes aggressive prostate cancer and antagonizes the SWI/SNF complex. Nat Genet 2013; 45(11): 1392-8.
[http://dx.doi.org/10.1038/ng.2771] [PMID: 24076601]

[181] Mehra R, Udager AM, Ahearn TU, *et al.* Overexpression of the Long Non-coding RNA SChLAP1 Independently Predicts Lethal Prostate Cancer. Eur Urol 2016; 70(4): 549-52.
[http://dx.doi.org/10.1016/j.eururo.2015.12.003] [PMID: 26724257]

[182] Chakravarty D, Sboner A, Nair SS, *et al.* The oestrogen receptor alpha-regulated lncRNA NEAT1 is a critical modulator of prostate cancer. Nat Commun 2014; 5: 5383.
[http://dx.doi.org/10.1038/ncomms6383] [PMID: 25415230]

[183] Coppola V, De Maria R, Bonci D. MicroRNAs and prostate cancer. Endocr Relat Cancer 2010; 17(1): F1-F17.
[http://dx.doi.org/10.1677/ERC-09-0172] [PMID: 19779034]

[184] Schaefer A, Jung M, Mollenkopf HJ, *et al.* Diagnostic and prognostic implications of microRNA profiling in prostate carcinoma. Int J Cancer 2010; 126(5): 1166-76.
[PMID: 19676045]

[185] Kneitz B, Krebs M, Kalogirou C, *et al.* Survival in patients with high-risk prostate cancer is predicted by miR-221, which regulates proliferation, apoptosis, and invasion of prostate cancer cells by inhibiting IRF2 and SOCS3. Cancer Res 2014; 74(9): 2591-603.
[http://dx.doi.org/10.1158/0008-5472.CAN-13-1606] [PMID: 24607843]

[186] Spahn M, Kneitz S, Scholz CJ, *et al.* Expression of microRNA-221 is progressively reduced in aggressive prostate cancer and metastasis and predicts clinical recurrence. Int J Cancer 2010; 127(2): 394-403.
[PMID: 19585579]

[187] Avgeris M, Stravodimos K, Fragoulis EG, Scorilas A. The loss of the tumour-suppressor miR-145

results in the shorter disease-free survival of prostate cancer patients. Br J Cancer 2013; 108(12): 2573-81.
[http://dx.doi.org/10.1038/bjc.2013.250] [PMID: 23703249]

[188] Avgeris M, Stravodimos K, Scorilas A. Loss of miR-378 in prostate cancer, a common regulator of KLK2 and KLK4, correlates with aggressive disease phenotype and predicts the short-term relapse of the patients. Biol Chem 2014; 395(9): 1095-104.
[http://dx.doi.org/10.1515/hsz-2014-0150] [PMID: 25153390]

[189] Fu H, He HC, Han ZD, *et al.* MicroRNA-224 and its target CAMKK2 synergistically influence tumor progression and patient prognosis in prostate cancer. Tumour Biol 2015; 36(3): 1983-91.
[http://dx.doi.org/10.1007/s13277-014-2805-0] [PMID: 25394900]

[190] Wan Y, Zeng ZC, Xi M, *et al.* Dysregulated microRNA-224/apelin axis associated with aggressive progression and poor prognosis in patients with prostate cancer. Hum Pathol 2015; 46(2): 295-303.
[http://dx.doi.org/10.1016/j.humpath.2014.10.027] [PMID: 25532941]

[191] Raposo G, Stoorvogel W. Extracellular vesicles: exosomes, microvesicles, and friends. J Cell Biol 2013; 200(4): 373-83.
[http://dx.doi.org/10.1083/jcb.201211138] [PMID: 23420871]

[192] Nawaz M, Camussi G, Valadi H, *et al.* The emerging role of extracellular vesicles as biomarkers for urogenital cancers. Nat Rev Urol 2014; 11(12): 688-701.
[http://dx.doi.org/10.1038/nrurol.2014.301] [PMID: 25403245]

[193] Théry C, Zitvogel L, Amigorena S. Exosomes: composition, biogenesis and function. Nat Rev Immunol 2002; 2(8): 569-79.
[PMID: 12154376]

[194] Junker K, Heinzelmann J, Beckham C, Ochiya T, Jenster G. Extracellular Vesicles and Their Role in Urologic Malignancies. Eur Urol 2016; 70(2): 323-31.
[http://dx.doi.org/10.1016/j.eururo.2016.02.046] [PMID: 26924769]

[195] Cappello F, Logozzi M, Campanella C, *et al.* Exosome levels in human body fluids: A tumor marker by themselves? Eur J Pharm Sci 2017; 96: 93-8.
[http://dx.doi.org/10.1016/j.ejps.2016.09.010] [PMID: 27640113]

[196] Dijkstra S, Mulders PF, Schalken JA. Clinical use of novel urine and blood based prostate cancer biomarkers: a review. Clin Biochem 2014; 47(10-11): 889-96.
[http://dx.doi.org/10.1016/j.clinbiochem.2013.10.023] [PMID: 24177197]

[197] Duijvesz D, Luider T, Bangma CH, Jenster G. Exosomes as biomarker treasure chests for prostate cancer. Eur Urol 2011; 59(5): 823-31.
[http://dx.doi.org/10.1016/j.eururo.2010.12.031] [PMID: 21196075]

[198] Nilsson J, Skog J, Nordstrand A, *et al.* Prostate cancer-derived urine exosomes: a novel approach to biomarkers for prostate cancer. Br J Cancer 2009; 100(10): 1603-7.
[http://dx.doi.org/10.1038/sj.bjc.6605058] [PMID: 19401683]

[199] Dijkstra S, Birker IL, Smit FP, *et al.* Prostate cancer biomarker profiles in urinary sediments and exosomes. J Urol 2014; 191(4): 1132-8.
[http://dx.doi.org/10.1016/j.juro.2013.11.001] [PMID: 24211598]

[200] Tavoosidana G, Ronquist G, Darmanis S, *et al.* Multiple recognition assay reveals prostasomes as promising plasma biomarkers for prostate cancer. Proc Natl Acad Sci USA 2011; 108(21): 8809-14.
[http://dx.doi.org/10.1073/pnas.1019330108] [PMID: 21555566]

[201] Duijvesz D, Versluis CY, van der Fels CA, *et al.* Immuno-based detection of extracellular vesicles in urine as diagnostic marker for prostate cancer. Int J Cancer 2015; 137(12): 2869-78.
[http://dx.doi.org/10.1002/ijc.29664] [PMID: 26139298]

[202] Gabriel K, Ingram A, Austin R, *et al.* Regulation of the tumor suppressor PTEN through exosomes: a diagnostic potential for prostate cancer. PLoS One 2013; 8(7): e70047.

[http://dx.doi.org/10.1371/journal.pone.0070047] [PMID: 23936141]

[203] Khan S, Jutzy JM, Valenzuela MM, *et al.* Plasma-derived exosomal survivin, a plausible biomarker for early detection of prostate cancer. PLoS One 2012; 7(10): e46737. [http://dx.doi.org/10.1371/journal.pone.0046737] [PMID: 23091600]

[204] Bijnsdorp IV, Geldof AA, Lavaei M, Piersma SR, van Moorselaar RJ, Jimenez CR. Exosomal ITGA3 interferes with non-cancerous prostate cell functions and is increased in urine exosomes of metastatic prostate cancer patients. J Extracell Vesicles 2013; 2: 2. [http://dx.doi.org/10.3402/jev.v2i0.22097] [PMID: 24371517]

[205] Valentino A, Reclusa P, Sirera R, *et al.* Exosomal microRNAs in liquid biopsies: future biomarkers for prostate cancer. Clin Transl Oncol 2017; 19(6): 651-7. [http://dx.doi.org/10.1007/s12094-016-1599-5] [PMID: 28054319]

[206] Endzeliņš E, Melne V, Kalniņa Z, *et al.* Diagnostic, prognostic and predictive value of cell-free miRNAs in prostate cancer: a systematic review. Mol Cancer 2016; 15(1): 41. [http://dx.doi.org/10.1186/s12943-016-0523-5] [PMID: 27189160]

[207] Bryant RJ, Pawlowski T, Catto JW, *et al.* Changes in circulating microRNA levels associated with prostate cancer. Br J Cancer 2012; 106(4): 768-74. [http://dx.doi.org/10.1038/bjc.2011.595] [PMID: 22240788]

[208] Brase JC, Johannes M, Schlomm T, *et al.* Circulating miRNAs are correlated with tumor progression in prostate cancer. Int J Cancer 2011; 128(3): 608-16. [http://dx.doi.org/10.1002/ijc.25376] [PMID: 20473869]

[209] Huang X, Yuan T, Liang M, *et al.* Exosomal miR-1290 and miR-375 as prognostic markers in castration-resistant prostate cancer. Eur Urol 2015; 67(1): 33-41. [http://dx.doi.org/10.1016/j.eururo.2014.07.035] [PMID: 25129854]

[210] Mahn R, Heukamp LC, Rogenhofer S, von Ruecker A, Müller SC, Ellinger J. Circulating microRNAs (miRNA) in serum of patients with prostate cancer. Urology 2011; 11(5): 1265 e9-1265 e16. [http://dx.doi.org/10.1016/j.urology.2011.01.020] [PMID: 21539977]

[211] Lodes MJ, Caraballo M, Suciu D, Munro S, Kumar A, Anderson B. Detection of cancer with serum miRNAs on an oligonucleotide microarray. PLoS One 2009; 4(7): e6229. [http://dx.doi.org/10.1371/journal.pone.0006229] [PMID: 19597549]

[212] Corcoran C, Rani S, O'Driscoll L. miR-34a is an intracellular and exosomal predictive biomarker for response to docetaxel with clinical relevance to prostate cancer progression. Prostate 2014; 74(13): 1320-34. [http://dx.doi.org/10.1002/pros.22848] [PMID: 25053345]

CHAPTER 2

Overview of Chemoresistance in Cancerous Cells

Mahima Kaushik[1,2,*], **Swati Mahendru**[2], **Mohan Kumar**[2], **Swati Chaudhary**[2], **Saami Ahmed**[2], **Sonia** [2] and **Shrikant Kukreti**[2,*]

[1] *Cluster Innovation Centre, University of Delhi, Delhi, India*

[2] *Nucleic Acids Research Laboratory, Department of Chemistry, University of Delhi, Delhi, India*

Abstract: Due to the continuous burden of cancer, an extensive research on the factors responsible for an increase in cancer cases and its effective curative measures are utmost essential. Cancer has the potential to resist drug activity resulting in limited chemotherapeutic effectiveness, which leads to mortality in most of the cases. Drug resistance creates a major hurdle in cancer therapy research and can be classified as intrinsic and acquired resistance. Factors which are responsible for developing the resistance are generally present in the oncogenic cells even before chemotherapy, in case of intrinsic resistance, whereas acquired resistance develops during or after chemotherapeutic treatment. Chemoresistance in cancerous cells can be comprehensively ascribed to genetic factors, epigenetic changes as well as tumor microenvironment. This chapter aims to provide a paradigm to understand the factors and mechanisms underlying the cause of resistance to cancer chemotherapeutic drugs. Also, recent strategies to overcome chemoresistance along with novel interventions involving the understanding of molecular signatures of cancer, chemoresistance biomarkers and companion diagnostics have been reviewed. Knowledge thus gained would benefit us in developing effective strategies to help us overcome the burden of chemoresistance in cancerous cells.

Keywords: Apoptosis Inhibition, Cancer Therapies, Chemoresistance, Chemotherapy, Drug resistance, Drug companion diagnostics, DNA repair, Drug transport, Drug targets, Drug metabolism, DNA Methylation, Histone Modification, miRNA, Targeted therapy, Tumor Microenvironment, Mechanisms of chemoresistance, miRNA and cancer, miRNA as biomarker, Novel approaches to overcome chemoresistance.

INTRODUCTION TO CANCER AND CHEMOTHERAPY

Cancer is considered as one of the main causes of death worldwide in spite of so much advancement in cancer treatments [1]. Cell growth and proliferation are the

* **Corresponding author Mahima Kaushik and Shrikant Kukreti:** Cluster Innovation Centre, University of Delhi, Delhi, India; Nucleic Acids Research Laboratory, Department of Chemistry, University of Delhi, Delhi, India; Tel: +91-11-27666726; E-mail: kaushikmahima@yahoo.com

Atta-ur-Rahman & M. Iqbal Choudhary (Eds.)

pre-requisites for the development of an individual but certain factors lead to the abnormal growth of the cell, which is termed as cancer. Cancerous cells damage the normal check points in the cell cycle that control and limit the cell division. Though cancer is mostly genetically inherited disease, however, an individual's lifestyle is also a major cause for the onset of abnormal cell growth and proliferation.

Despite many efforts to fight against cancer, very little progress has been made in the complete treatment of malignant neoplasms, which are a continuous growth of abnormal cells that have a tendency to develop into tumors later. Main reason behind it is, metastasis *i.e.* very fast spreading of cancer in the other body parts. Tumor metastasis, related preclinical research, combination therapies and clinical trial designs have been very recently reviewed in detail [2]. For cancer treatment, a large number of drugs have been developed till now, but most of them are non-selective, have severe side effects, show drug-resistance and are costly. In spite of a lot of advancements in research, its transition into medicines has remained quite slow [3]. Therefore, an effective anticancer therapy is still a major challenge among the scientific community. Most common methods of cancer treatment include surgery, chemotherapy, radiation therapy, and targeted therapy.

Chemotherapeutic drugs such as alkylating agents, antimetabolites and natural products have been used for the treatment of various types of cancers [4]. Cancers have the potential to develop resistance against drug activity, which results into drugs being ineffective, leading to death in many cases. Drug resistance creates a major hurdle in cancer therapy research, and it can be divided into intrinsic and acquired resistance. In intrinsic resistance, mediating factors of resistance are present in the tumor cells even before the start of chemotherapy, possibly due to some genetic characteristics. On the contrary, acquired resistance develops during or after the treatment, which were initially sensitive to drugs. Acquired resistance can be caused by the development of new mutations during treatment, as well as through various other adaptive responses, such as an increased expression of the therapeutic target and activation of alternative compensatory signaling pathways [5]. Cancerous tumors comprise of the mixed population of cancer sensitive cells and cancer resistant cells. Chemotherapeutic drugs target and kill the drug sensitive cells only, while the drug-resistant cells remain as it is. Some scientists suggest that a specific drug encounters resistance from 1 in 10^6-10^7 cancerous cells present in a tumor. These cells are responsible for repopulating the tumor and poor prognosis for cancer patients [6, 7]. The response of normal cells to drugs is similar but cancer cells respond differently due to different gene makeup. Several factors such as genetic and epigenetic mechanisms as well as tumor microenvironment underlying the drug resistance by cancer cells have already been uncovered [8]. Various components of cells like microRNA, DNA, drug transporter protein, which are drug targets and are responsible for sensitivity of cancer cells towards cytotoxic drugs, sometimes also play an antagonistic role for the treatment. MicroRNAs play an important role in sensitivity and as well as resistance to chemotherapy. The fate of microRNAs targeted chemotherapeutic agents depends on the extent of expression of the different type of microRNAs in a cancer cell. The miRNAs coding genes involved in oncogenesis generally diminish the translation, resulting into reduction of mRNA stability [9, 10]. The miRNAs can also act as intermediates, thus affecting a specific response to treatments in various oncogenes or tumor suppressor genes. MicroRNAs can also be good choice to be used as biomarkers in predicting drug resistance in chemotherapy. Expression of certain target microRNA increases the sensitivity of cancer cells towards drug, whereas up-regulation or down-regulation of another specific microRNA can make cancer cells resistant to the same drug [11]. Tumor microenvironment (TME) is the cellular environment, in which tumor exists, which is mainly composed of three classes of stromal components: cells of hematopoietic origin, cells of mesenchymal origin, and non-cellular components [specifically Extracellular matrix (ECM)] [12]. TME facilitates innate as well as acquired resistance to cancer cells from chemotherapeutics and it differs greatly from the modifications at the genomic and/ or epigenomic levels of cancer cells. Soluble factor-mediated drug resistance (SFM-DR) and cell adhesion-mediated drug resistance (CAM-DR) are the two overlapping mechanisms of drug resistance associated with tumor microenvironment [13].

It is also known that each cytotoxic drug has different modes of interaction with cancer cells, therefore sometimes cancer cells use alternate biochemical pathways and bypass the action of the drug. These mechanisms are generally based on qualitative or quantitative alterations. Quantitative alterations involve changes in expression of drug targets, whereas qualitative alterations involve mutations in relevant genes or pathways [14]. Drug resistance in patients can be treated with targeted molecular therapies which have shown remarkable effects [15]. The mechanisms of action, cross-resistance, and overlapping drug toxicity should be taken into consideration, while designing the combination regimen to prevent chemoresistance. Combination therapy is one of the important methods for prevention of drug resistance. In this method, combination of two or more compatible drugs is given to the patient, which can counter the resistance offered by the cancer cells. Researchers have developed several combinations of cytotoxic drugs which are much more effective than their alone constituents. Another important approach to overcome cancer drug resistance is based on nanoparticle drug delivery systems [16].

Following sections will elaborate on various cancer therapies, factors and mechanisms underlying the cause of resistance to cancer chemotherapeutic drugs as well as novel approaches to overcome the chemoresistance in cancer cells.

COMMON CAUSES OF CANCER

Carcinogens, physical inactivity, obesity, unhealthy diets, viral infections, heredity, exposure to solar radiations, and ageing are some of the common causes of cancer. Tobacco and alcohol are also well known chemical carcinogens. Tobacco smoke alone contains more than fifty carcinogens, therefore it is associated with a number of cancers including lung, stomach, bladder and kidney [17]. Exposure to chemicals like benzene, nickel, cadmium, vinyl chloride, benzidine, asbestos and N-nitrosamines can also result in to cancer development. These factors have negative impact on immune systems and endocrine systems and hence contribute to cancer risk [18].

Hereditary cancers like breast and prostate cancer are generally caused by genetic mutations, which are carried on to the next generation. One example of such gene is breast cancer gene 1 (*BRCA*1), which is a tumor suppressor gene and mutation in *BRCA*1 gene is reported to cause hereditary breast cancer [19]. Somatic mutations due to ageing, exposure to Ultra-Violet radiations from sunlight, or any chemical can give rise to many diseases including cancer. Somatic mutations are the acquired genetic mutations of a cell that are transmitted to the daughter cells during the cell division. Tissue specific accumulation of these somatic mutations with age is an important risk factor for cancer [20]. Exposure to solar Ultra-Violet radiations is also an important cause for the development of non-melanoma and melanoma skin cancer in humans. Out of two major types of non-melanoma skin cancer, squamous cell carcinomas (SCC) are associated with occupational solar exposure, whereas basal cell carcinomas (BCC) are associated with non-occupational solar exposure [21].

Viruses are small contagious agents that replicate inside the living cells and seize the normal functioning of cell. There are six classes of genomically different viruses, which can cause tumors in humans [22]. Out of these six classes, four classes, *i.e.* Epstein-Barr virus, human papilloma virus, hepatitis B virus, and human herpes virus-8 are 'DNA viruses', whereas human T lymphotrophic virus type 1 and hepatitis C are 'RNA viruses'.

CANCER THERAPIES

Various strategies have been utilized for the management and treatment of cancer throughout the world, depending on their type, size, position *etc.* Some of them (Fig. **1**) have been elaborated in the following sections.

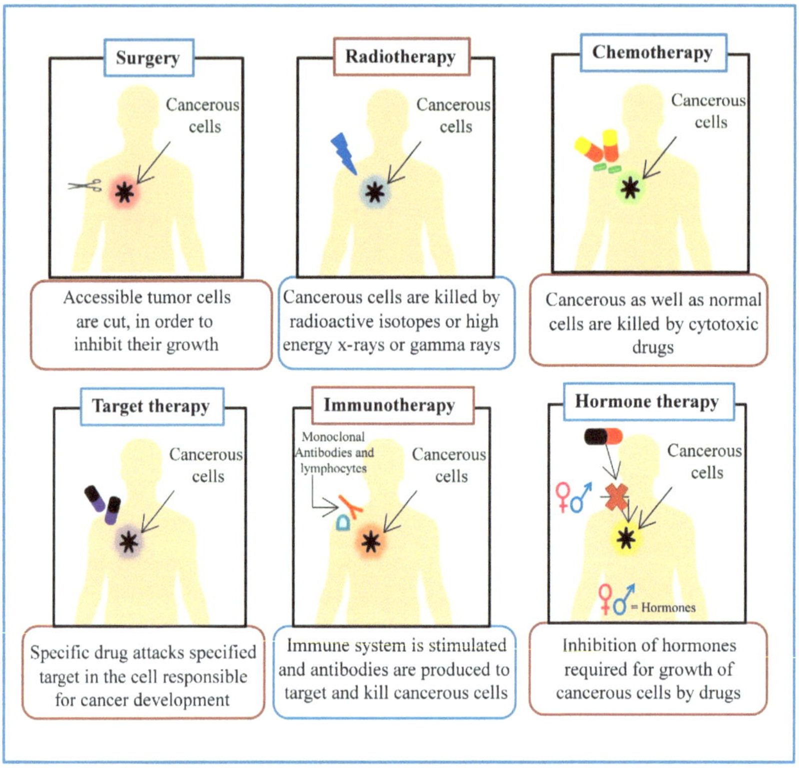

Fig. (1). Various types of cancer treatments.

(i). Surgery

From past several decades, surgery had been used most extensively for the treatment of early stage (stage I and II) colon and rectal cancer [23]. Surgical methods are employed not only to diagnose cancer, but also to identify its size and the extent to which it is spread. In invasive surgery, a surgeon excises the tumor along with some surrounding healthy tissue. Less invasive surgery minimizes the harm to normal tissues and hence is preferred. Depending upon the type of cancer, its stage and place of occurrence, various types of surgeries are used, such as preventive (prophylactic), diagnostic, staging, curative, and debulking surgery.

(ii). Radiation Therapy

Along with other cancer treatments, another one of the most frequently used technique for more than last 100 years, is radiotherapy. In radiation therapy, high-energy of X-rays, gamma rays, electron beams, or proton particles are employed to eradicate cancerous cells [24]. The radiation therapy works on the principle of damaging the DNA either directly or by generating the free radicals that further break down DNA beyond repair, which inhibits cell division and eventually leads to cancer cell's death [24] by activating apoptotic signals. This technique is more effective in curing cancerous cells, when they are easily accessible and are located away from sensitive organs of the body. Major side effect of radiotherapy is that it fails to distinguish between normal and cancerous cells [25].

Depending on various factors like type, size and location of cancer, patient's age and his medical history, the type of radiation is prescribed by the doctors. Radiation therapy along with other types of treatment like surgery or chemotherapy is usually given simultaneously with palliative intentions. Cancerous cells may respond differently to the radiation therapy, depending on their type. Leukemia, Lymphoma, and Germ cell tumors can mostly be removed by modest exposure to radiations. Epithelial cancers require higher doses of radiations, whereas Renal and Melanoma cancerous cells are often resistant to radiotherapy. Radiation therapy is a painless therapy and low dose treatments do not cause much significant side effects. However, higher doses may cause acute and cumulative longer side effects.

(iii). Targeted Therapy

Targeted therapies have significantly revolutionized the cancer treatment, as their mechanism of action and toxicities differ from the conventional and more preferred chemotherapy. These have now been used for many cancers like breast, pancreas and lung [26]. Target therapy is often called as 'molecular targeted therapy', as a specific drug is used in this, which interferes with specific proteins in oncogenesis. These drugs not only stop malignant cell growth, but they also attempt to induce their death and restrict the spreading of cancer. Target therapy is often confused with chemotherapy, but these are different from each other in several ways:

- In target therapy, drugs are given to hinder the tumor cell proliferation. On the other hand, in chemotherapy, cytotoxic drugs are given, which focus on killing cancerous cells.
- In target therapy, specific focus is given on the cancerous cell, but chemotherapy kills rapidly dividing entire cell (both normal and cancerous cells).

In this therapy, information related to an individual patient regarding his genes and associated proteins is derived to prescribe the medicine. Target therapy also has certain limitations because of drug resistance development. Therefore, target therapy is given in combination with several other drugs that counter the resistance offered. Several types of target therapies which are now being used include signal transduction inhibitors [27], gene expression modulators, apoptosis inducers and angiogenesis inhibitors.

a. Immunotherapy

It is a comparatively new approach that treats cancer by modulating the person's immune system [28] and it can be primarily done in two ways. First, through stimulating the immune system by targeting tumor-associated antigens (TAAs) to attack cancer cells, termed as activation immunotherapy. Secondly, by introducing the immune system components (for example man-made immune system proteins) and is called as suppression immunotherapy. Main types of immunotherapy include the following: (i). Monoclonal antibodies that deliver toxic molecules, (ii). Immune check point inhibitors, (iii). Cancer vaccines, (iv). Non-specific immunotherapies.

b. Hormone Therapy

In hormone therapy, a manipulation in endocrine system is usually done to administer a hormone exogenously. These are usually steroid hormones, which are needed for the gene expression in some cancer cells, and hence their blocking may be effective in stopping cancer cell growth. Hormone therapy is used for hormone sensitive or hormone-dependent cancers like breast, prostate, ovarian, womb and kidney [29]. In females, estrogen and progesterone hormones affect some breast cancers and hence they are stopped during the hormone treatment of cancer. Prostate cancer depends on the male hormone testosterone for its growth and hormone therapy is used to reduce or stop the production of testosterone by the body, thus aiming to slow down or stop the cancerous growth [30].

(iv). Chemotherapy

In 1940's, the use of nitrogen mustard during World War II had begun the era of chemotherapy, which led to a war on cancer through chemotherapy [31, 32]. Chemotherapy is considered as a cornerstone, which utilizes chemical substances known as anti-cancer drugs, to kill the metastasized cancerous cells, moreover it also targets main cancer; unlike surgery and radiation therapy, which target cancer cells at a localized area. Recent focus of chemotherapy is to postpone the appearance of disease-related symptoms and/ or to lengthen the survival, which is best illustrated in randomized controlled trials (RCTs) comparing chemotherapy with supportive care in patients with gastric cancer [33, 34]. On the basis of cancer size, its area of occurrence, stage of cancer and patient's personal health, an oncologist decides drugs in certain combinations, and this therapy is then termed as combination therapy. The chances of becoming resistant to any one chemo drug by cancer cells are minimized by combination therapy. The chemotherapy is usually given at a regular interval of time with certain gaps in between. The gap period is necessary for the normal cells to recover from the drug's side effects. The number of cycles is decided prior to even start of treatment, but in some cases, it may be flexible.

There are three major goals of chemotherapy:

1. **Cure:** If cancer is detected in the primary stage, then it has maximum chances of cure and even of complete eradication from the body, with no scope of reoccurrence. Also, eradication of circulation of cancer cells to prevent and treat metastasis is one of the major goals of chemotherapy.
2. **Control:** If cancer is spreading fast and curing it completely is not possible, then chemotherapy helps in controlling its fast growth.
3. **Palliation:** When cancer reaches to the advanced stage, chemo is used by the doctors to give a better quality life to the patient, as cancer can neither be controlled nor be cured at that stage.

Chemotherapy may be used as the only cancer treatment sometimes, but more often, it is used in combination with surgery, radiation therapy, or biological therapy (uses vaccines or bacteria to stimulate the body's immune system for fighting against cancer). For making a cancer smaller in size, chemotherapy can be done before surgery or radiation therapy, which is then called as neo-adjuvant chemotherapy. It also helps radiation and biological therapy to perform better by destroying recurrent cancer cells (cells that have come back) or spreading of them to other parts of the body. Chemotherapy may be given through various modes such as *via* injection, intra-arterial, intraperitoneal, intravenous, topically and orally. Common side effects of chemotherapy include diarrhea, constipation, nausea, vomiting, fatigue, pain, mouth & throat sores, blood disorders and effects on the nervous system.

Cytotoxic Chemotherapy

For last more than 60 years, cytotoxic chemotherapy remains as the main focus of clinical cancer therapy. It shows selective toxicity to cancer cells and most importantly, it quite selectively kills more cancer cells than normal cells [35]. It describes the chemical agents, which prevent cell growth and ultimately lead to cell death. Antitumor antibiotics, antimicrotubule agents, antimetabolites, and topoisomerase II inhibitors are few examples of drugs used in cytotoxic chemotherapy. Their mechanistic pathways involve inhibition of either microtubule or protein function, and DNA synthesis. The mechanism of cytotoxic chemotherapy may be cell-cycle dependent, including cell growth arrest at particular phases in the cell cycle. Drugs used in cytotoxic chemotherapy destroy proliferating cells generally by interfering with the process of mitosis. Cytotoxic chemotherapy has been proved useful in the treatment of hematological malignancies [36, 37]. It is usually offered in advanced stage cancer cases. Some of the cancers are highly proliferative as compared to normal cells, due to which they become selective for this treatment. As a result of cytotoxic chemotherapy, the proliferative cells cannot repopulate after the treatment, due to their inability to repair DNA damage [36]. Differential chemosensitivity may be a result of differential mitochondrial priming (mitochondrial readiness for apoptosis). Cytotoxic agents that selectively enhance the mitochondrial priming increases the response rate of cancers to chemotherapy, in spite of not causing cell death by themselves [35]. Patients undergo counseling even to use appropriate contraception during the course of chemotherapy, due to harmful effects of several cytotoxic drugs on the fetus. One of the main limitations of cytotoxic chemotherapy is the drug resistance to cancer cells.

OVERVIEW OF CHEMORESISTANCE

Several chemotherapeutic drugs such as alkylating agents, antimetabolites, and natural products have been developed for the successful treatment of different types of cancers [38]. Extensively active signaling pathways involved in cancer are responsible for susceptibility of cancerous cells to a broad range of anticancer drugs. The growth of signaling pathways or molecules, actively involved in gene expression or replication, are generally targeted by these drugs. Thus, treatment may sometimes cause more harm than the disease itself. Damage to kidney, immune system, lung, heart or reproductive system are few of the side effects caused by chemotherapy [7]. Drug resistance is considered as the most common cause for drug discontinuation [23, 39].

Drug response is highly dependent on the type of the cancer and the treatment chosen. It also depends on the patient's co-morbidities, due to which patient suffers from additional diseases. One such example is ovarian cancer, which is responsible for 6% of female deaths due to cancer [40]. It was reported that complete clinical response to first phase chemotherapy was achieved by 40-60% of patients [41], while relapse of cancer was observed in nearly 50% of these patients within 5 years. On the other hand, only 10% patients suffering from advanced stage cancer attain long-term remission [42]. It was reported that chemotherapy resistance is the main reason for high relapse rate of ovarian cancer [42]. Recent studies revealed that ovarian cancers show resistance to platinum-based chemotherapy agents, due to modulation of the DNA repair pathways, which increases the levels of miR-622, a regulator of DNA repair. It was found that cancer cells containing overexpressed miR-622 are resistant to chemotherapeutic agents (cisplatin and carboplatin) and also to PARP inhibitors (veliparib and olaperib). The increased expression of miR-622 results in suppression of two proteins, Ku70 and Ku80, which play crucial role in deciding the dominative DNA repair pathway [43].

There are several mechanisms of chemoresistance including variation in DNA repair enzymes, apoptosis gene alterations, enhanced expression of β-tubulin III, and ABC transporters [44]. In order to overcome the problem of resistance, various ATP-binding cassette (ABC) transporter inhibitors have been studied. ABC transporters involve the transport of several substrates across cell membrane with the help of ATP-binding energy. Certain breast cancers have a specific type of cancer, known as 'triple negative breast cancer' that show resistance to chemotherapy through cellular fates such as senescence and autophagy or tumor microenvironment. Due to enriched ABC transporter proteins in breast cancer, a fraction of cells become drug-resistant, which remain inactive in the G-null phase, which is an inactive and non-cycling state of cell cycle. These inefficient cells further divide and give rise to progenitor cells over a period of time. Few of these progenitor cells segregate into new tumor cells, characterized by chemoresistance phenotypes. This type of acquired chemoresistance is observed in breast cancer. Recurrent cancers develop in patients at this stage, which do not respond further to chemotherapy treatment [45]. Highly expressed ABC transporter proteins are involved in the exclusion of the fluorescent dyes Hoechst 33342 and Rhodamine 123 in tumor stem cells. Side population (SP) is the term defining the cells that are able to efflux Hoechst 33342 [46]. *In vitro* studies have revealed that excessive population of SP cells is found in chemoresistant breast cancer cells, in comparison to chemosensitive cells [47].

Non-small cell lung cancer (NSCLC) accounts for almost 80% of all the lung cancers and is one of the most common cancers in the world. Patients suffering from advanced NSCLC are suitable candidates for systemic chemotherapy treatment. Chemotherapy helps in determining the survival rate and quality of life of the patients [48]. But the overall response rate is only 30%, which ultimately leads to the therapeutic failure in NSCLC [49]. Mostly, platinum-based drugs are used in the chemotherapy treatment for NSCLC, along with tubulin binding agent, and vinca alkaloids. In spite of a broad array of chemotherapy drugs and efficient therapeutic ways, these drugs develop resistance, which give a hard challenge in the treatment of human neoplasms [48]. One report suggested that resistance to cisplatin was found in 63% of the samples, out of 3042 NSCLC patients, while carboplatin resistance was observed in 68% of the samples. On the other hand, resistance to topotecan, paclitaxel, and doxorubicin was reported in 31, 40, and 75% of samples respectively [50]. Resistance to platinum-based agents involves several mechanisms including intracellular accumulation, detoxification, and enzymatic inactivation. Enzymes related to glutathione, for instance, glutathione transferase (GST) and metallothioneins (MT) have been implicated in the resistance related to enzymatic inactivation. Expression of metallothioneins was found to be enhanced in NSCLC patients, who were initially treated with cisplatin [51]. Resistance to cisplatin in lung cancer cell lines has been linked with the expression of π-isoform of GST [52]. The resistance of NSCLC cancer cells to tubulin binding agents includes microtubule structural change and alteration in the mitotic checkpoints regulating apoptosis [53]. A recent report reveals that hypoxia-induced autophagy might add to the resistance factors of NSCLC. This can be done by reducing the apoptotic potential and increasing the survival of autophagy under hypoxic conditions [54].

Human cancers related to chromosomal abnormalities also experience chemoresistance. Chronic myelogenous leukemia (CML) is one such example, in which a translocation mixes together a 3'-sequence of the *ABL*1 proto-oncogene present on chromosome 9, with the 5'-sequence of the *BCR* gene found on chromosome 22. The chemotherapeutic treatment of CML involve imatinib mesylate, which is a competitive inhibitor of the *BCR-ABL* tyrosine kinase protein [55]. Initially, the cytogenetic and haematologic response rates were high in the CML patients but resistance to imatinib was developed after few days of treatment. Almost 50% of the CML patients had an imatinib-resistance related point mutation in *BCR-ABL* protein, which hinders the binding of imatinib [56]. So, in this case, chromosomal aberrations, *BCR-ABL* gene amplification, overexpression at the protein & mRNA levels, and the reduced intracellular uptake of imatinib are believed to be involved in the imatinib-resistance [57, 58].

Over the last two decades, the curing rate of the most common malignancy in children, acute lymphoblastic leukemia (ALL) has made improvements to approx. 90% in some clinical trials. In spite of a huge success rate, few groups of patients had also experienced chemotherapeutic resistance [59]. Evidences show that the bone marrow mesenchymal cells (MSCs) have the potential to contribute to the drug resistance of leukemic cells. Various mechanisms including asparagine generation by MSCs, interaction of vascular cell adhesion molecule-1 to MSC and interaction between antigen-4 (VLA-4) integrin and leukemic cells have been proposed as the reasons of drug resistance. The mechanism involving antigen-4 (VLA-4) integrin shows resistance against etoposide and cytarabine cytotoxicity [60 - 62]. Recent reports have suggested that hERG1 (human ether-à-go-go-related gene 1) channels play an important role in the development of drug resistance and enhance the survival rate of leukemic cells [63].

Glucose metabolism in glycolysis and oxidative phosphorylation (OXPHOS) are very well studied phenomenon and are known as "the Warburg effect". It establishes a relationship between the energy requirements of cancer cells and glycolysis. It is now widely exploited for better determination of metabolic effects of cancer. The metabolic differences of cancer cells not only promote tumourigenesis but they also increase drug resistance. For the energy requirements, cancer cells are also dependent on glycolysis and OXPHOS. Different theories and mechanisms which associate a linkage between glycolysis and drug resistance are quite elegantly reviewed by Bhattacharya et.al. 2016 [64]. It has already been documented that alteration in glucose metabolism rates also leads to chemotherapy resistance in several types of cancers. Acute myeloid leukemia (AML) is one of the most common type of cancer in adults and is usually treated with anthracycline-based chemotherapy, but 40-60% of AML patients encounter relapse and around 20% patients do not respond to the chemotherapy. Increased expression of genes associated with glucose metabolism have shown increased resistance in leukemic cell lines [65, 66].

FACTORS RESPONSIBLE FOR CHEMORESISTANCE

There are various classes of chemotherapeutics, which include platins, antimetabolites, antimitotics, enzyme inhibitors and DNA-alkylating agents [67]. Cancer cells resist the chemotherapeutics by different mechanisms depending on the mode of action of drugs. For effective chemotherapy and high cure rates, multiple drugs with different cellular targets and different mechanisms of influx are used simultaneously. Cancer cells can offer resistance to structurally and mechanistically unrelated drugs and this phenomenon is known as Multiple Drug resistance (MDR) [68]. Genetic and epigenetic factors as well as tumor microenvironment are involved in chemoresistance, thereby limiting the therapeutic efficacy of anticancer drugs.

a. Genetic Factors

Genetic mechanisms include alterations in drug transport across the cell membrane, enhanced DNA repair mechanism, modification of drug targets, inhibition of apoptosis, and alteration in drug metabolism. These mechanisms are discussed in detail in the following sections and have also been illustrated in Fig. (**2**).

(i). Alteration in Drug Transport Across the Cell Membrane

An alteration in membrane transport represents one of the major obstacles in present cancer chemotherapy. It is one of the most common mechanisms by which cancer cells become resistant to anticancer agents. MDR in cancer cells can result either due to reduced uptake of drugs or due to increase in drug efflux from cancer cells [69]. A number of transporters or receptors have been identified that can transport anticancer drugs into cells. For example, volume regulated anion channel (VRAC) transporter, which regulates anion transport, is responsible for the uptake of cisplatin. In a similar manner, proton-coupled folate transporter

(PCFT; SLC46A1) as well as reduced folate carrier (RFC; SLC19A1) import the antifolates methotrexate and pemetrexed [70 - 72]. The mutations in these transporters reduce the entry of drugs into the cells. There are very few anticancer drugs such as immunotoxins that enter the cell by the process of endocytosis, which can be either receptor mediated or non-specific [73]. The receptors that mediate endocytosis are present on the surface of cells, bind to the anticancer drugs and ingest the drug by endocytosis. The mutation in these receptors leads to defective endocytosis and such cells become resistant to both toxins and immunotoxins [74].

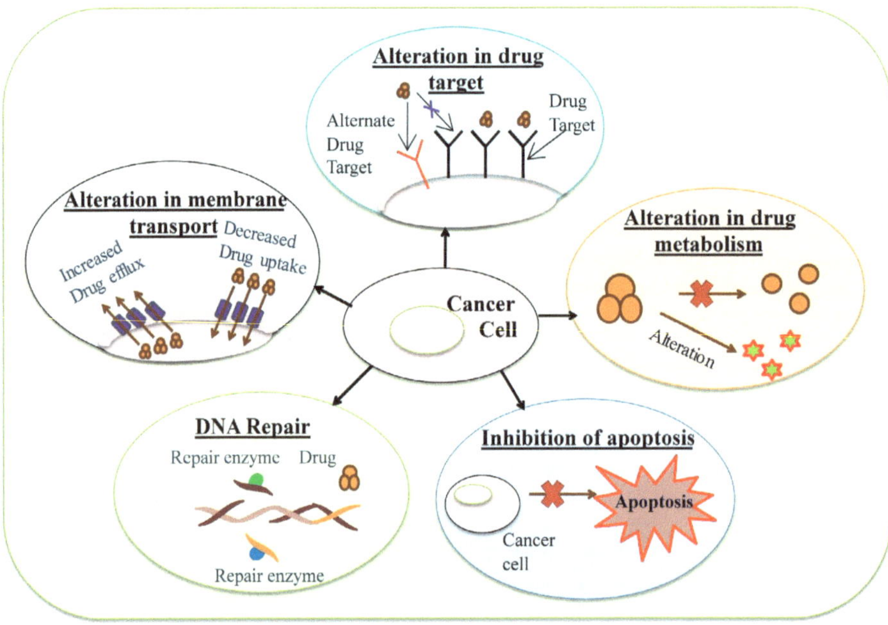

Fig. (2). Genetic factors affecting chemoresistance in cancerous cells.

A group of cell membrane transport proteins is also involved in resistance to chemotherapeutics, by promoting drug efflux and hence keeping the concentration of drug in the cell below the cell killing threshold. ABC transporter family is most significant among all the transmembrane proteins. The members of this family regulate the transport of many structurally and mechanistically unrelated chemotherapeutics across the plasma membrane. They abolish lipids and various hydrophobic compounds including major cancer chemotherapeutics such as topoisomerase inhibitors, platinum compounds, antimetabolites and microtubule inhibitors [5]. There are 49 members of ABC transporter family, of which three are selectively investigated in relation to MDR. These three members are multi-drug resistance protein 1 (MDR1 or P-glycoprotein and ABCB1), MDR-associated protein 1 (MRP1 or ABCC1) and breast cancer resistance protein (BCRP or ABCG2) [75].

MDR1 is a membrane-bound glycoprotein that mediates resistance to multiple hydrophobic drugs with a positively charged or neutral polyaromatic backbone. It contributes to intrinsic drug resistance by its overexpression in many cancers including lung, kidney and colon. In Leukemias and lymphomas, the expression of MDR1 increases as a result of chemotherapy.

*MDR*1 gene encodes P-glycoprotein (P-gp; MDR1, ABCB1), which is a phosphorylated and glycosylated protein of 1, 280 amino acids with the molecular weight of 170 kDa [76]. It is the most common efflux pump in the plasma membrane. The polypeptide P-gp consists of two homologous parts, which are separated by an intracellular linker region. Each part contains a hydrophobic transmembrane domain (TMD), which contributes to the drug specificity by forming channels with transmembrane helices and nucleotide-binding domain (NBD) that involve in ATP binding and hydrolysis [77]. The P-gp drug efflux process is associated with dynamic conformational changes of the ABC drug transporter. According to the most common hypothesis, the binding of the drug to a particular site of P-gp leads to the activation of one of the ATP-binding domains and then the hydrolysis of ATP results into alteration in the shape of P-gp. This alteration in shape of P-gp causes the drug efflux [78]. To restore the original state of the transporter, the hydrolysis of second ATP molecule takes place so that cycle of drug binding and release can be repeated [79]. P-gp can bind to many hydrophobic cancer chemotherapeutics like taxol, doxorubicin, paclitaxel, vinblastine, and daunorubicin and release them in extracellular space to prevent cell death.

MDR associated proteins come into play, when there is MDR in cancer cells in the absence of overexpression of P-gp. One of these proteins MRP1/ ABCC1 is widely expressed in many human tissues as well as cancers and has a broad spectrum of anticancer drug transport activity. Unlike MDR1, MRP1 transports hydrophobic anionic conjugates (negatively charged natural-product drugs), in addition to drugs modified by glutathione (GSH), glucuronic acid or sulfate and extrudes uncharged hydrophobic drugs [79, 80].

BCRP/ ABCG2 with the molecular weight approx. 75 kDa is the third ABC transporter, which is involved in MDR. Unlike P-gp and MRP1, BCRP contains one region with six TMD and one NBD. Overexpression of this protein in the cell results into resistance to chemotherapeutics.

(ii). Enhanced DNA Repair Mechanism

It is another important mechanism, by which cancer cells may become chemoresistant. Mode of action of many cytotoxic drugs like platinum drugs, alkylating agents and topoisomerase inhibitors involves direct or indirect DNA damage, which can cause death of cancer cells. The favored results are either due to massive DNA damage, causing direct inhibition of the metabolism of cancer cells or induction of apoptosis by tumor suppressor genes like *p*53 or *p*21. The inhibition of the repair of damaged DNA is the essential requirement for proper functioning of DNA damage causing agents.

In normal cells, there are six pathways: nucleotide excision repair (NER), base excision repair (BER), mismatch repair (MMR), homologous recombination (HR), inter-strand cross-link repair (ICL) and the non-homologous end-joining (NHEJ), by which a damaged DNA can be repaired [81]. But in cancer cells, there would be an impairment of at least one of these pathways, so that an alternative pathway comes into play, which is generally not present in normal cells. The inhibition of this alternative pathway is an effective strategy to stop repairing of damaged DNA. For therapeutic purpose, certain molecular targeted agents have been developed, which specifically target the components of cells that are involved in repair of damaged DNA. For example, poly (ADP-ribose) polymerase 1 (*PARP*1), a single strand break DNA repair enzyme is an important drug target for the cancer cells with mutated *BRCA*1 and *BRCA*2 genes [82]. *BRCA*1 and *BRCA*2 genes are necessary for homologous recombination DNA repair pathways and their mutation leads to the loss of the same. An alternative pathway that is *PARP* function dependent single strand repair mechanism compensates the inhibition of homologous recombination, which arises due to mutated *BRCA*1 and *BRCA*2 genes. Therefore, block of *PARP*l leads to inhibition of this alternate DNA damage repair pathway, which results in genetic instability and ultimately causes the death of cancer cell. It has been reported that during the treatment of *BRCA*2 gene mutated breast cancer, *PARP* inhibitors suffer drug resistance due to some *BRCA*2 mutations, which restore the function of *PARP* [83].

(iii). Modification of Drug Target

The mechanisms of cell death by several drugs involve interactions between the drug and an essential intracellular enzyme. These interactions result in modification or inhibition of normal functions of these enzymes. Alterations in the drug target, such as changes in expression level or mutations are also important factors which affect drug response and resistance. Cytotoxic drugs target the component of cells which have significant cellular influence, so that their mutation or blockage leads to death of cancer cells. But sometimes, modification of drug targets occurs or their level decreases during the course of reaction, so that their blockage becomes insignificant. For example, androgen receptors are the drug targets in prostate cancer and AR antagonists such as bicalutamide or testosterone lowering drugs such as leuprolide are used for therapeutic purposes. The effectiveness of these drugs decreases during the course of reaction because of increased androgen receptor expression [5].

The progression and initiation of these hormone related cancers like breast and prostate cancer is mediated by certain steroid hormone receptors. These receptors include progesterone receptor (PR), estrogen receptor (ER), and androgen receptor (AR). The hormone therapy drugs target the blocking of these receptors, in order to cease the function of disease causing hormones. But the over expression of these hormone receptors or the induction of alternative pathways for the functioning of hormones result in hormonal therapy drug resistance [84]. In many cancer cells, activation of certain alternative signaling pathways act as a general negative profactor for responses to chemotherapy. For example, tumor necrosis factor NFκB is involved in immune system activation, therefore its expression facilitates the action of drugs that use immune system for discarding cancer cells. But in most of the cancer cells, it activates pro-survival signaling pathways to inhibit cell apoptosis and hence actively contribute to chemoresistance [14].

Another common example is the endocrine resistance during the treatment of breast cancer with antiestrogens like tamoxifen. In resistant cancer cells, the estrogen receptors are apparently lost, which lead to unresponsiveness to tamoxifen treatment. However, estrogen synthesis inhibitors such as aromatase inhibitors stop the growth of cancer and then it does not respond to any endocrine manipulation. The growth of cancer cells is then assumed to be independent of estrogen and then, strategy of blocking estrogen receptors is no longer useful [7].

Generally, chemotherapeutics targets the essential cell components and disable their functions, so that cells cannot survive. Chemotherapy resistant cells contain a mutated gene, which encodes for a protein retaining its function but cannot be targeted by a drug. Hence the action of drug gets interrupted due to unavailability of target and cell becomes drug resistant. For example, certain cytotoxic drugs target topoisomerase enzyme, which play a key role in DNA replication. Cancer cells become resistant to topoisomerase II inhibitors like doxorubicin and etoposide, in case of mutation within this protein. Mutation in topoisomerase protein alters its localization, and hence absence of target molecules leads to inactivity of corresponding drug [7].

(iv). Inhibition of Apoptosis

Apoptosis is one of the balances and checks built into the cell cycle. In cells, apoptosis is an important regulatory event that causes cell death. Induction of cell death *via* apoptosis is one of the major strategies that have been used to treat cancer. There are two important activation mechanisms, intrinsic and extrinsic that regulate the initiation of cell apoptosis. Mitochondria forms the basis of intrinsic apoptotic pathway, which is mediated by caspase 9, BCL-2 protein family *etc* [5]. On the other hand, extrinsic apoptotic pathway is mediated by Tumor necrosis factor alpha and FAS ligand-receptor interactions [85]. Apoptosis can also be caused by downstream caspase 3 activation, which is the combination of intrinsic and extrinsic pathways [8].

When an appropriate amount of drug enters the cell and causes DNA damage, then a downstream reaction occurs, which results in cell death *via* apoptosis. The *p*53 gene plays an important role in regulation of apoptosis, as it transfers DNA damage signals to the *BCL*-2 family of pro-apoptotic proteins. Anti-apoptotic proteins of the same family may inhibit the process. Their overexpression in cancer cells lead to inhibition of intrinsic mitochondria-associated pathway of apoptosis and may enable cancer cells to survive therapy. Genetic mutations or alterations that block the apoptosis also make the cancer cells resistant to anti-cancer drugs [75].

Apoptosis involve a number of events including DNA fragmentation, chromosome condensation, nuclear cleavage and cellular shrinkage, which are controlled by a group of genes and proteins. The phosphatase and tensin homolog (*PTEN*) and *p*53 are reported tumor suppressor genes, which organize apoptotic events in cancer cells. *p*53 remains inactive until cell encounters stress conditions like DNA damage, lack of oncogenic signaling *etc*. In case of stress conditions, the activation of *p*53 occurs, which elevates intracellular *p*53 protein levels that trigger apoptosis. It also causes cell cycle arrest at certain phases for the repairing of damaged DNA by the cell and provides apoptotic signals in case of failure. In several kinds of cancer, cells develop anticancer drug resistance due to mutation in *p*53 gene. The damaged DNA cannot be repaired due to abnormality in gene function, which results into deactivation of apoptotic signaling pathways and uncontrolled cell proliferation. *PTEN* gene obstructs certain cancer promoting pathways like PI3K/ Akt signaling pathway, in case of different cancers. The abnormalities in the function of *PTEN* gene lead to the activation of these signaling pathways and development of anticancer drug resistance in cancer cells. On the other hand, *Bcl*-2 family of genes include oncogenes as well as tumor suppressor genes. Genes encoding proteins like *Bax*, *Bad*, *Bim* trigger apoptosis in cancer cells, while genes encoding *Bcl-XL* and *Bcl*-2 proteins enhance cancer cell proliferation and suppresses apoptosis. The effect of these encoded proteins on apoptosis is governed by the type of dimerization (homo or hetero) between the members of this family [78].

(v). Alteration in Drug Metabolism

Many anti-cancer drugs require metabolic activation in order to exert their cytotoxic effects. Cancer cells can resist the treatment *via* these drugs through decreased activation of the drug. For example, a nucleoside drug cytarabine (AraC), which is used in the treatment of AML needs to be activated first by series of phosphorylations. Initial phosphorylation is done by deoxycytidine kinase and cytarabine gets converted into cytarabine-monophosphate, which is further converted into cytarabine triphosphate by subsequent phosphorylations [86]. But if some mutations or alterations occur in the enzymes involved in these metabolic pathways, then activation of the drug does not take place and cancer cells develop drug resistance. Other important example of drug inactivation includes conjugation of antioxidant glutathione-*S*-transferase (GST) with drugs like cisplatin, which make these drugs substrate for ABC transporters and enhance drug efflux [8]. Rabinoside is administrated as inactive prodrug, which needs to undergo catalytic conversion to an active form by the action of an enzyme, deoxycytidine kinase. The mutation or under-expression of this enzyme reduces the efficacy of drug [7].

The mechanisms of drug activation involve their interaction with certain proteins, which lead to modification, partial degradation or active complex formation of the drug. There are certain drug metabolizing enzymes present in cells that can play a key role in drug resistance. The reduced expression or over expression of these enzymes alters the drug metabolism pathways, resulting in either activation or deactivation of drug. For example, Cytochrome P-450 is involved in activation of potential chemotherapeutic cyclophosphamide. Therefore, altered expression level of this enzyme can also contribute to the potential mechanism of drug resistance [8].

There are certain carrier molecules like glutathione or ubiquitin, which are overly expressed in cancer cells as a physiological response to drug's presence. These molecules form conjugates with the drugs that can be easily excreted from the cancer cells and render inactivity of the drugs. For example, over expression of dihydrodiol dehydrogenase in ovarian carcinoma cells make the cell resistant to drugs like cisplatin [7].

It is also reported that phase I drug metabolizing enzyme, Cytochrome P-450 is also involved in deactivation of the topoisomerase I inhibitor, Irinotecan, which is used in the treatment of colon cancer. The binding of cisplatin drug with a small cysteine-rich protein, metallothionein (MT) renders its inactivity [86].

b. Epigenetic Factors

The term 'epigenetics' refers to the study of all the molecular pathways that change the behavior of a person's genes that are mitotically and/ or meiotically heritable, without entailing a change in DNA sequence [87]. It is impossible for a multicellular organism to develop and function in the absence of the interaction between genome, epigenome and proteome. Epigenome consists of specific covalent modifications, resulting in silencing or gene activation *via* mechanisms like DNA methylation, histone modification and microRNA regulation, which are crucial for normal development and maintenance of cell type-specific responses. However, dysregulated silencing or activation of genes are recurrent events during the initiation and progression of cancer [88]. In addition, they play an important role in acquired drug resistance to the chemotherapeutics and are independent of genetic mutations, thereby forming a parallel pathway in drug resistance.

(i). DNA Methylation

Transfer of a methyl (CH_3) group from the endogenous co-factor S-adenosyl methionine (SAM or AdoMet) to the C5 position of cytosine ring, facilitated by enzyme DNA methyl-transferases (DNMT) results in to a covalent modification, which is known as DNA methylation [89]. The methylation at different DNA regions is associated with different functions. For example, methylation at transcribed regions aid transcription elongation, whereas that of the promoter region is linked with gene silencing. Generally, frequent hyper-methylation of guanine-rich and cytosine-rich DNA sequences in cancer cells prevail in order to silence the tumor suppressor genes. Several studies revealed that certain nucleoside analogs like 5-azacytidine can inhibit DNA methyl-transferases, which grants therapeutic effects by re-expression of silenced genes. The resistance to these nucleoside analogs is associated with abnormal transcription of protein transporters involved in drug efflux, enzymes involved in DNA-repair and apoptotic factors in cancer cells [90]. The drug resistance in AML and myelodysplastic syndrome (MDS) patients is attributed to the somatic DNMT3A mutations, which leads to the generation of cells with drug-resistant epigenome.

(ii). Histone Modification

The nucleosomes which are the basic unit of chromatin, consist of DNA and an octamer of small basic histone proteins (H2A, H2B, H3 and H4). Nucleosomes participate in the regulation of a diverse range of biological processes, main among them being DNA packaging and gene transcription, DNA repair *etc*. The N-terminal tail modified histone proteins extruded from the core of nucleosome control the interaction between DNA and histone proteins. The modification of histone can be done by acetylation, methylation, phosphorylation, butyrylation and N-formylation *etc*. Various enzyme families like lysine or histone acetyl transferases (KATs or HATs), protein arginine methyl-transferases (PRMTs) and lysine methyl-transferases (KMTs) catalyze the modification of lysine or arginine residues present at histone tails. It has been found that DNA and histone modifications play an important role in driving certain mechanisms like increased drug efflux, activation of parallel signal transduction pathway, disturbed DNA repair, inactivation of pro-apoptotic genes, and secondary mutations in drug targets.

The cell adhesion mediated drug resistance (CAM-DR) is involved in the epigenetic modifications. The sensitivity of proteasome inhibitors is determined by Class I histone deacetylases (HDACs). At the time of acquisition of CAM-DR by myeloma cells, the histone methyl-transferase, EZH2 regulates the transcription of anti-apoptotic genes. Another histone methyl-transferase, MMSET assists DNA repair of myeloma cells and thereby confers drug resistance [13].

(iii). MicroRNA Regulation

MicroRNAs, non-coding RNAs of short nucleotide lengths help in regulating gene expression at the post-transcriptional level. They also assist in regulating several biological processes involving sensitivity of cancer cells towards anticancer drugs. Various cellular processes such as cell proliferation, regulation of cell cycle, stress response, metabolism, apoptosis, angiogenesis, and differentiation get affected by miRNAs, which in turn, affect human diseases, such as cancer. Some of the different targets of miRNAs involved with various cancers and the drugs that either experience chemoresistance or chemosensitivity due to miRNAs involvement have been summarized in Fig. (**3**). miRNA molecule can be transferred from one cancerous cell to another through gap junction, including bone marrow stroma and breast cancer cells due to their small size and stability [91]. Expression of genes associated with drug metabolism and transport can also be modified by miRNAs [92]. MicroRNAs also play an important role in the progression of cancer by controlling the cell proliferation genes such as *E2F*1 and *c-myc*. Evidences show that the levels of miR-143 and miR-145 are found to be reduced in colon cancers, chronic lymphocytic leukemia, Burkitt lymphoma cell lines, and B-cell lymphoma [93]. Specific types of anaplastic thyroid cancers (ATCs) are largely regulated by miR-17-92 and thus, this miRNA can be a novel target for the treatment of ATC, whereas miR-21 plays a crucial role in the oncogenesis of neck and head cancer. Inhibition of cell growth and phosphorylation of Akt protein kinase occurs in the hepatoma cells due to miR-

125 [94].

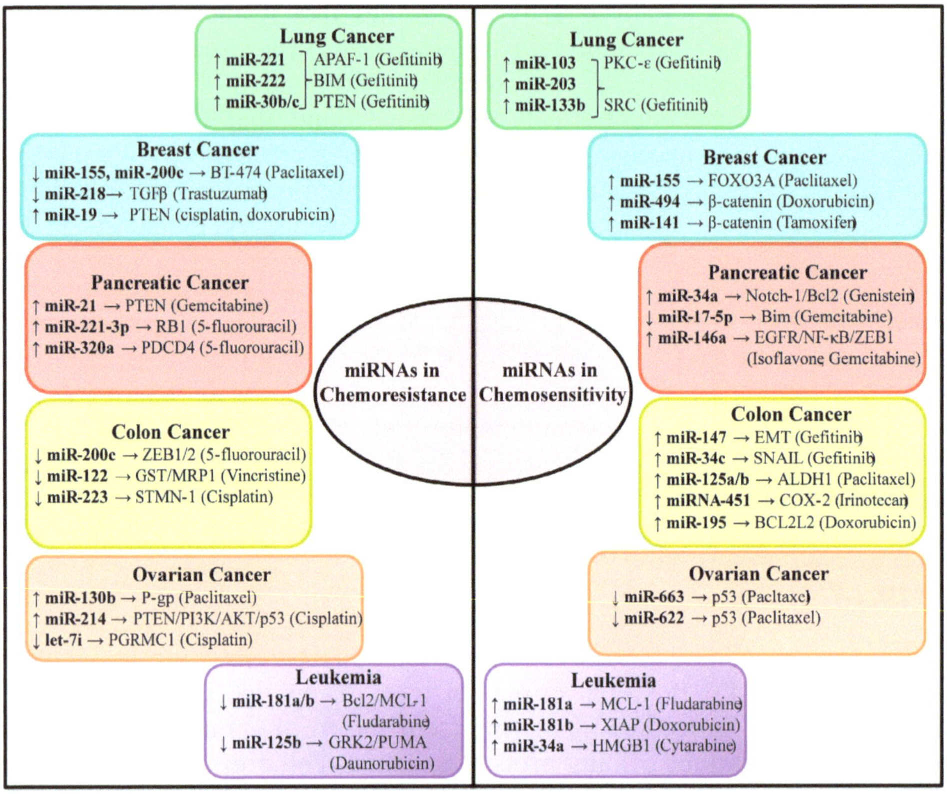

Fig. (3). MicroRNAs involved in modifying the expression of important targets in chemoresistance as well as chemosensitivity in different cancers for various drugs. (Arrows describe the up-regulation or down-regulation of specified miRNA) [92, 95 - 103].

Numerous studies describe the role of miRNAs in chemoresistance of different types of cancers, which may result due to dysregulation of miRNA expression. In women, use of tamoxifen is the known treatment of breast cancer, but 30% of breast cancer patients show resistance to tamoxifen. Recent reports show that the eight microRNAs are highly expressed, while seven microRNAs are down-regulated in the tamoxifen-resistant breast cancer cell lines. They also show an association between increased resistance to tamoxifen and reduction of p27 (Kip1) due to increased expression of miR-221/ 222 [104]. On the other hand, doxorubicin-resistant breast cancer cells also involve the de-regulation of miRNA expression as well as altered expression of miRNA-processing enzymes. The expression of an important factor of drug resistance, multi drug resistance 1 (*MDR*1) gene is regulated by miR-451 [105]. The up-regulation of miR-214 increases the survival rate of ovarian cancer cells, by carrying out resistance to cisplatin. Some studies have shown the down-regulation of miRNAs in independent drug-resistant cell lines, thus demonstrating a relationship between some miRNAs and specific drug resistance. The miR-335, miR-130a, and miR-30c are involved in chemoresistance, as these were down-regulated in almost all the resistant cell lines. Further report suggests that the down-regulation of miR-130a is associated with the activation of *M-CSF* gene, which is already known as a resistant factor for ovarian cancer [94]. X-linked inhibitor of apoptosis (XIAP) plays a vital role in the cisplatin-chemoresistance in ovarian cancer cell line A2780. The mRNA and protein levels of XIAP are extraordinarily decreased as a result of increased expression of miR-130a [106]. In ovarian cancer, silencing of miR-199b has also been linked with the acquired chemoresistance through the activation of JAG-Notch1 signaling pathway [107, 108].

In NSCLC, few of the miRNAs including *has*-miR-145, *has*-miR-182, *has*-miR-21, *has*-miR-210, *has*-miR-126, and *has*-miR-183 are differentially expressed. A fraction of human NSCLC is found to be resistant to *TNF*-α related apoptosis-inducing ligand (TRAIL). Five miRNAs (miR-100, miR-222, miR-221, miR-15b, and miR-125b) are found to be up-regulated in the TRAIL-resistant and TRAIL-semiresistant cell lines A459 and A549 [109]. The expression of miR-34a was significantly reduced in prostate cancer cells, which can further decrease the mRNA and protein levels of *SIRT*1 gene, leading to growth inhibition and cell cycle arrest. This further induces chemoresistance to camptothecin, an anticancer drug through apoptosis [110]. miR-508-5p, miR-106a, and miR-27a were found to be involved in the development of multidrug resistance in gastric cancer [111].

Breast cancer cells have the ability to transfer resistance capacity from affected cells to sensitive cells *via* releasing of exosomes, which take place through some intercellular transfer of particular miRNAs. Exosomal miR-21 is secreted by neuroblastoma cells, which then interacts with Toll-like receptor 8 (TLR8) in tumor-associated macrophages (TAMs), where it activates the NF-κB pathway, finally leading to the up-regulation of miR-155 in tumor-associated macrophages. A telomerase inhibitor, *TERF*1 is targeted by miR-155, which is transported back to the neuroblastoma cells. As a result, *TERF*1 is down-regulated, resulting into increased chemotherapy resistance in neuroblastoma [11]. miRNAs can also act as tumor-suppressors, by inhibiting cancer progression through post-transcriptional targeting of oncogenes. miRNA-17 was found to suppress cancerous cell growth in breast cancer, by reducing the expression of AIB1, while it also enhances the development of tumor in B-cell lymphoma. miRNAs such as miR-21, miR-19, and miR-34a are found to be highly expressed in chemoresistant cell lines and are also believed to play an important role in adaptation of cancer cells to chemotherapy. On the other hand, decreased expression of few miRNAs is related with the up-regulation of multidrug resistance proteins, resulting into chemoresistance [112, 95].

c. Tumor Microenvironment (TME)

In Cancer, transformation of non-maligned cells into maligned cells takes place, due to interactions between malignant and non-transformed cells, which create TME [113]. TME contains the immune system cells, tumor vasculature and lymphatics, as well as fibroblasts, pericytes, extracellular matrix, exosomes and sometimes adipocytes, along with the cancerous cells [114, 115]. The dynamics of this adaptive and heterogeneous system supplement the mutated genes with cardinal information to exert their cancer-promoting function at all the four definable stages of carcinogenesis: initiation, promotion, progression and malignant conversion.

Chemoresistance in cancer cells materializes through a variety of mechanisms in both primary and metastatic sites, entailing both intrinsic and extrinsic factors. However, the extrinsic factors such as TME often remain unnoticed. Each and every section of the TME are not mere onlookers but regulate all the four definable stages of carcinogenesis and more importantly, therapeutic response substantially. Decrease in drug penetration and providing anti-apoptotic as well as proliferative properties to surviving cells by TME expedite chemoresistance without causing genetic mutations and epigenetic changes. In this manner, it collectively modifies disease modality and contorts clinical indices [115]. Some important features of TME associated with chemoresistance are discussed below.

Cancer Cell Heterogeneity

TME facilitates innate as well as acquired resistance to cancer cells from chemotherapeutics. The innate resistance is expedited by TME through regular mutual interactions between cancer cells and neighboring TME components, even before the treatment using chemotherapeutics. It deviates from the inherited resistance, originating from the modifications at the genomic and/ or epigenomic levels of cancer cells. Acquired resistance imparted by the TME usually develops due to response to the long-term exposure to drug. Specifically, the chemoresistance triggered by TME worsens the prognosis as well as outlines the cancer progression route under varying treatments [115]. Fig. (**4**) demonstrates various mechanisms of drug resistance associated with TME.

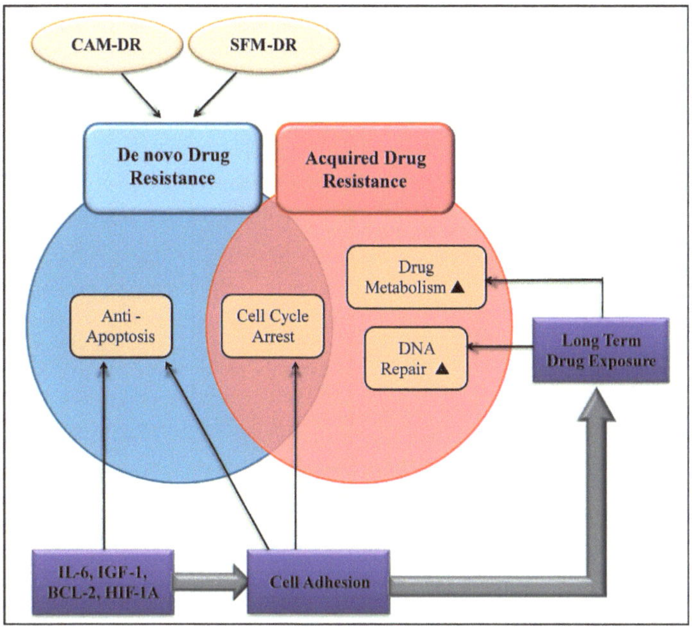

Fig. (4). Drug Resistance Mechanisms associated with tumor microenvironment.

Intimate relationship between cancer and TME leads to rapid development of drug resistance in some patients, who are initially responsive to chemotherapy. The multidimensional role of TME in cancer progression particularly therapeutic resistance involves cell adhesion, cell-extra cellular matrix (ECM) interactions *etc*. Furukawa and Kikuchi suggested two overlapping mechanisms, termed as soluble factor-mediated drug resistance (SFM-DR) and cell adhesion-mediated drug resistance (CAM-DR), with which myeloma cells develop drug resistance. SFM-DR and CAM-DR pathways are not independent of each other but rather are akin [13].

(i). Soluble Factor Mediated Drug Resistance (SFM-DR)

Stimulation of endothelial cells and their associated pericytes during the neovascularization are crucial for cancer growth, and are associated with a number of soluble factors present in the TME, such as vascular endothelial growth factor (VEGFs), fibroblast growth factors (FGFs), platelet-derived growth factors (PDGFs) and chemokines. Stimulation of angiogenesis results in sprouting of new vessels from the existing vasculature, due to hypoxic conditions in the TME or by receiving an angiogenic signal by a dormant blood vessel from cancerous cells. The cancer vasculature is similar to that in normal one, with the only difference in the abnormality of the structure and function of blood vessels [116]. Contrary to the highly ordered network of blood vessels in normal vasculator, cancer vasculature consists of heterogeneous leaky blood vessels with chaotic branching structures and an uneven vessel lumen [116, 114]. Unevenness of blood flow, oxygenation, nutrients and drug distribution in the TME are caused by the interstitial fluid pressure induced by the leaky vessel, thereby resulting in increased hypoxia, which further facilitates metastasis. VEGFs present in TME are most potent angiogenesis inducers and are produced by both malignant cells and inflammatory leukocytes. However, a diverse range of other angiogenic factors are produced by advanced cancers that can substitute for VEGFs [117]. VEGF can also induce the expression of anti-apoptotic proteins by modifying the apoptotic signaling pathway in endothelial cells and thus activating the survival pathway. Its overexpression can induce resistance to chemotherapeutic drugs in soft tissue sarcoma. In addition to the release of proangiogenic factors such as FGF or VEGF and matrix metalloproteinases (MMP9), mechanisms of cancer promotion also involve the release of immunosuppressive cytokines like interleukin-10 [118].

Transcriptional changes carried out by the soluble factors such as interleukin-6 (IL-6) activate signal transduction pathways leading to drug resistance in myeloma cells [13, 119]. Interleukin-6 (IL-6) secreted by adipocytes activates Chk1 protein in breast cancer cells, which results in an enhanced resistance to radiotherapy [118].

(ii). Cell Adhesion Mediated Drug Resistance (CAM-DR)

More than three decades ago, Durand and Sutherland demonstrated that multicellular spheroids of cancer cells were markedly more resistant to radiation and distinct doses of Adriamycin, than the same cells cultured in monolayer (Fig. **5**). In addition, it was also observed that this drug resistance was not associated with the drug penetration ability, as it was not compromised in case of spheroids [120,121]. In addition, down-regulation of E-cadherin or anti-integrin antibodies which act as "anti-adhesive" therapies, restore drug sensitivity *in vitro* [119]. This clearly suggests that the involvement of direct adhesive interactions between same cell types impart drug resistance.

Cell adhesion of bone marrow stromal cells leads to up-regulation of cell cycle inhibitors, anti-apoptotic members of the *Bcl*-2 family (*Bcl*-2 and *Mcl*-1) and ABC-family drug transporters in myeloma cells that transduce signals to modulate the expression of cell survival associated genes [13].

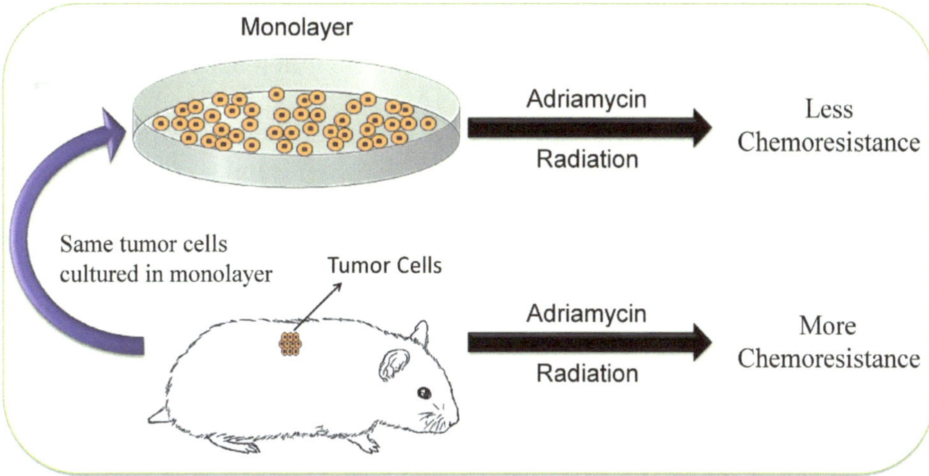

Fig. (5). Cell adhesion mediated drug resistance.

MOLECULAR MECHANISMS OF DRUG RESISTANCE

On the basis of structures and mode of actions, chemotherapy drugs are divided into various classes like alkylating agents, nucleoside analogs, alkaloids, antimetabolites, topoisomerase inhibitors, mitotic inhibitors *etc*. Cytotoxic nucleoside analogues include a variety of abnormal purine and pyrimidine nucleoside derivatives. Antimetabolites and nucleoside analogs either imitate nucleotides of DNA/ RNA or inhibit the enzymes responsible for biosynthesis of nucleotides, which ultimately leads to death of cancer cells.

Anticancer drugs, which act *via* inhibition of certain cellular processes like DNA replication and transcription by modifying the nitrogenous bases of DNA, referred as alkylating agents. These drugs (like cisplatin, temozolomide and cyclophosphamide) crosslink certain atoms to DNA in order to prevent strand separation and also induce base mis-pairing between strands. Enzyme inhibitors block enzymes like topoisomerases, which play key roles in DNA replication. Drugs like etoposide, topotecan and irinotecan are the examples of topoisomerase inhibitors. These enzyme inhibitors block the ligation step of the cell cycle, which leads to generation of single and double stranded breaks, subsequently rendering to apoptosis and cell death. Antimetabolites act *via* disabling the DNA replication machinery, either by inhibiting the supply of deoxy-oligonucleotides required for DNA replication and cell proliferation or by incorporation of chemically altered nucleotides. For example, Methotrexate, an antimetabolite used to treat acute lymphocytic leukaemia, inhibits an enzyme dihydrofolate reductase (DHFR) involved in nucleotide metabolism and reduction of the tetrahydrofolate precursors, *i.e.* folic acid and dihydrofolic acid. Pyrimidine and purine nucleoside analogues, also acts as antimetabolites, which block the synthesis of pyrimidine and purine bases respectively. They also mimic the nucleosides and incorporate themselves into newly synthesized DNA, resulting into premature chain termination. Acyclovir, 6-thioguanine and 6-mercaptopurine are examples of purine nucleosides analogues, whereas 5-fluorouracil, gemcitabine and arabinosyl-cytosine are examples of pyrimidine nucleosides [7]. Alkaloids are also an important class of compounds that serve as anticancer agents. The mode of action of alkaloids is the modulation of multiple signaling pathways, which leads to the blocking of carcinogenesis, induction of cell cycle arrest and apoptosis. Berberine, evodiamine, matrine, piperine and sanguinarine are some of the examples of alkaloids isolated from natural herbs which act as chemotherapeutics [122]. Mitotic spindle inhibitors are the anticancer drugs, which inhibit mitosis or cell division by blocking the formation or the function of spindle microtubule fibers needed for chromosome alignment. The plant-derived vinca alkaloids, paclitaxel and the synthetic docetaxel are some of the examples of mitotic spindle inhibitors [7].

As mode of action of anticancer drugs is different, therefore their drug resistance mechanisms are also specific and some of these have recently been reviewed [123]. The cell-intrinsic resistance mechanisms to antimetabolites like ARA-C involve drug efflux by breast cancer resistance protein ABCG2 and by target modification *via* target up-regulation or target mutation. The mode of action of platinum drugs and alkylating agents is cross linking of drug to DNA, which damages the target DNA. In cancer cells, resistance to the platinum drugs arise due to DNA repair *via* homologous recombination, nucleotide excision repair, and DNA mismatch repair. The mechanism of resistance to the alkylating agents include DNA repair through base/ nucleotide excision or DNA mismatch and drug efflux by MDR1 transporter. The mode of action of topoisomerase inhibitors is the blocking of DNA replication in cancerous cells by inhibiting the enzymes topoisomerase I and II. Cancer cells generally become resistant to TopoI and TopoII inhibitors due to target down-regulation, target mutation, DNA repair and drug efflux by ABC transporter proteins like ABCC1 and ABCG2. On the other hand, mitotic inhibitors act by halting formation of spindle, resulting in cell death due to mitotic arrest and mechanism of their resistance in cancerous cells include drug efflux by ABC transporters and mutations in ß-tubulin [14].

Molecular mechanisms of chemoresistance to various ligands have been discussed in the following sections.

a. Platinum Based Drugs

Platinum (Pt) based drugs like cisplatin, carboplatin, oxaliplatin and their analogs form an important class of cytotoxic drugs that are able to form adducts with DNA. These platinum based complexes get activated in the cytosol and react with guanine nucleobase by forming inter as well as intra-strand DNA crosslinks. These cross-links induce DNA damage, which results into blocking of DNA synthesis, RNA transcription and finally leading to the death of cancer cells [124].

These Pt based drugs show limited utility, because of their drug resistance property and side effects. These limitations pave a way for the search of new metal-based compounds with less toxicity and more effectiveness. As ruthenium (Ru) complex demonstrates similar ligand exchange kinetics to those of Pt(II) with less toxicity, so Ru complex can act as an alternate for Pt antitumor drug. Ru seems to be less reactive in comparison to Pt compound, so the combination of both is preferred, where Ru and Pt bind to different strands of DNA, which ultimately renders cell apoptosis [125].

After successful clinical trials of cisplatin, many of its analogues have also been developed such as carboplatin, oxaliplatin, ormaplatin, and enloplatin [126]. Working mechanism of cisplatin as anticancer drug enables the researchers to modify its chemical structure for bringing out more significant results [127]. Once the cisplatin or its analogues enter the cell, they become activated. Inside the cytoplasm, the leaving group of cisplatin is replaced by water molecule. The hydrolyzed cisplatin is a compelling electrophile, which can react with the N7 position of purine base (nucleophile). This reaction results into DNA damage of cancerous cell and leads to cell apoptosis (Fig. **6**).

Although platinum drugs are very effective chemotherapeutics, yet it has been reported that many cancer cells offer resistance to these drugs. An increased DNA adduct repair capacity is the main strategy of cancer cells to resist platinum-based drugs. Inter- or intra-strand DNA crosslinks induced by cisplatin are mainly repaired by Nucleotide Excision Repair (NER). Mismatch Repair (MMR) deficiency has also been linked to cause resistance to some platinum based drugs. *In vitro* studies showed that deficiency of MMR proteins like MSH2 or MSH6 give rise to cisplatin and carboplatin resistance. This is because these MMR proteins are required to maintain DNA-cisplatin inter-strand cross-links (ICLs), which are essential for the sensitivity of cells towards drugs. Their deficiency leads to increase in inter-strand cross-links repair and cells become resistant to cisplatin and carboplatin [128]. Homologous Recombination (HR) is another DNA double-strand breaks (DSBs) repair pathway. In breast, ovarian and pancreatic cancers; two HR genes *BRCA*1 and *BRCA*2 are inactivated due to mutations. BRCA mutated cancers sometimes become cisplatin resistant by restoring HR *via* secondary mutations in genes, in order to re-activate the proteins [14].

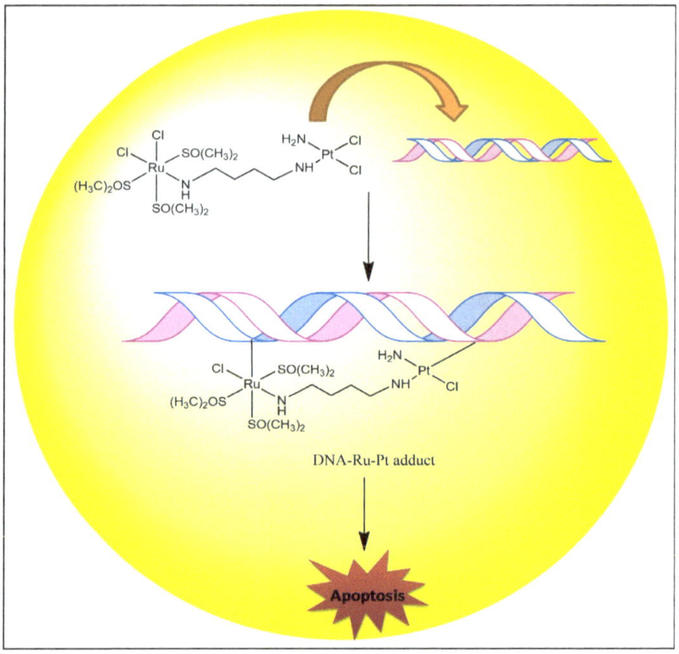

Fig. (6). Binding of Cis-platin with DNA leads to cell apoptosis.

Reduced intracellular accumulation of drug is an alternative mechanism for cisplatin drug resistance. Uptake of cisplatin may occur *via* passive membrane diffusion facilitated by subunits of volume-regulated anion channel (VRAC), therefore loss of these subunits account for reduced accumulation and drug resistance [69].

b. Antimetabolites

Antimetabolites are another class of anti-cancer drugs, which include folate, purine and pyrimidine antagonists. These are the chemical molecules, which mimic the endogenous metabolites and inhibit their functions. For example, antifolates like methotrexate inhibit the metabolism of folic acid. There are specific transporters like the proton-coupled folate transporter (PCFT/ SLC46A1), which are involved in import of anti folates into cell. Any qualitative or quantitative mutations in these transporters inhibit transport of antifolates and cell become resistant to them. Pyrimidine analog 5-fluorouracil inhibits thymidine synthesis by blocking thymidylate synthase. Hence, 5-fluorouracil selectively inhibits synthesis of DNA or RNA and this inhibition stops growth of cancer cells. But an enzyme dihydro-pyrimidine dehydrogenase (DPD) metabolizes 5-fluorouracil, and its high level allows cancer cell to resist 5-fluorouracil [129].

c. Alkylating Agents

Alkylating agents such as cyclophosphamide and ifosfamide also form another class of chemotherapeutics. Alkylating agents are initially pro-drugs which are bio-activated in the liver and then these agents interfere with replication and transcription of DNA by modifying the bases [14]. Alkylation of bases induce base mispairing, prevention of strand separation *etc*. The mode of action of alkylating agents is similar to cisplatin *i.e.* the formation of intra-strand/ inter-strand cross-links, which lead to DNA damage and subsequent cell death. Cells can repair the damaged DNA by removal of the alkylated guanine bases *via* certain pathways, like BER, NER or MMR. Aldehyde dehydrogenase ALDH1, glutathione has the tendency to deactivate alkylating agents and hence this contributes to drug resistance. ABC transporters like P-gp (ABCB1) or MRP2 (ABCC2) may be involved in removing the drugs from cells and decreasing the sensitivity of cells towards these drugs.

The covalent binders as drugs in chemotherapy try to attach a group with general formula C_nH_{2n+1} to the DNA, therefore they are also termed as alkylating agents [130]. The electron rich exposed portion of DNA groove is the favorable site for an alkylator to get attacked by S_N1 or S_N2 mechanism. The N7 and N3 position of guanine and adenine respectively, fulfill the criteria for an alkylator to attack. Some adenine specific alkylating agents are Duocarmycin and its analogues, Benz[e]indolones, analogues of KW-2189, Amino analogues, Bizelesin and other bis analogues, whereas Mitomycins, Carmethizole analogues and Pyrrolo-benzodiazepines are listed among the alkylating drugs, showing guanine specificity [131].

d. Enzyme Inhibitors

Enzyme inhibitors are the chemotherapy drugs that can inhibit the action of enzymes like topoisomerases and block DNA replication in cancer cells. There are two classes of topoisomerases, type I and II. Topoisomerase I mediate the breaking and re-ligation of single strand DNA, while its inhibitors like camptothecin and its analogs act *via* trapping, between DNA break and re-ligation. Therefore, Topoisomerase I inhibitors lead to single-strand DNA break by preventing re-ligation. Topoisomerase II is also involved in DNA replication/ transcription and catalyzes breaking/ re-ligation of two strands of DNA. Topo II poisons like etoposide and doxorubicin block the active form of Topo II at a stage after DNA double-strand break but before DNA ligation, yielding to DNA double-strand breaks. Similar to other cytotoxic drugs, major mechanisms of Topo I drug resistance are ineffective cellular uptake and/ or high efflux rate *via* the multidrug efflux transporter ABCG2 [132]. Camptothecin drugs specifically target Topo I enzymes, therefore reduced expression of Topo I leads to the drug resistance. Alteration in DNA damage repair pathway also contributes to the drug resistance due to increased repair of single-stranded DNA damage caused by Topo I inhibitor PARP1, XRCC1 and TDP1 [133], which are involved in DNA damage repair. It has also been reported that overexpression of XRCC1 promotes resistance to camptothecin.

The best studied resistance mechanism for Topoisomerase II poisons like etoposide and doxorubicin is drug efflux by the ABC transporter P-gp. Inhibition of transporter P-gp has not been proven to be much beneficial to increase sensitivity of cancer cells towards doxorubicin [134], which clearly indicates the prominence of alternate resistance mechanism. Due to down-regulation of topoisomerase II, sensitivity of cancer cells towards topoisomerase II also decreases [135].

e. Anti Microtubule Agents

Spindle poisons or anti-microtubule agent drugs generally bind to β-tubulin and are classified into two categories. First category includes drugs like taxanes and epothilones, which stabilize microtubules and other includes drugs like Vinca alkaloids, which destabilize the microtubules. These drugs interfere with microtubule dynamics and activate spindle checkpoints by preventing proper spindle formation, which leads to mitotic arrest. The major resistance mechanisms of these drugs involve drug efflux by ABC transporters and mutations in beta-tubulin. Alteration in microtubule regulating proteins (like stathmin) [136], overexpression of proteins like kinases aurora that are involved in the spindle assembly checkpoint, and modification in microtubule functioning, the HER2 signaling cascade have also emerged as potential mechanisms of resistance to spindle poisons [129].

OVERCOMING CHEMORESISTANCE

Ineffectiveness of chemotherapy due to chemoresistance adds to the troubles associated with targeting of cancer cells, hence it becomes extremely important to develop strategies to overcome drug resistance and also to make the treatment successful by employing the anticancer drugs in different combinations.

a. Combination Therapy

Clinical studies have focused on anti-resistant combinations of chemotherapy drugs to contribute towards the circumventing of drug resistance. Combination therapy is considered as the best method for cancer treatment because it helps in preventing the drug resistance and is more efficient than any one drug alone [8]. A mathematical model has been developed to exhibit the effectiveness of immunotherapy and combination therapy but the nature of effectiveness, whether it is additive or synergistic, remained unclear [137]. Generally, the most typical drug combinations developed till date are additive in nature, with only few synergistic combinations. Therapeutic synergism is the main aim of introducing combination therapy for which the ideal drug partners should be searched in order to enhance the cancer response and to inhibit the metastatic progression [138]. Breast cancer patients who receive the combination treatment utilizing doxorubicin plus an alkylating agent were more likely to experience the long-term survival. Clinical outcomes have also improved in metastatic breast cancer patients due to combination therapy. Gene amplification and/ or overexpression of human epidermal growth factor receptor 2 (HER2) are characteristics of 20% of all breast cancers, which result in poor prognosis and highly aggressive metastasis [139]. In addition to chemotherapy, an anti-HER2 humanized monoclonal antibody trastuzumab significantly enhances the overall survival rate among patients with HER2-positive metastatic breast cancer. Trastuzumab exercises its antitumor properties by binding to subdomain IV of the HER2 extracellular domain and blocking the cleavage of HER2. Trastuzumab further stimulates cytotoxicity, which inhibits the HER2-mediated mitogenic signaling [140,141,142]. Recently, more advanced combination therapies are developed, which employ pertuzumab, another humanized monoclonal antibody that binds HER2 at a different subdomain (subdomain II). Since pertuzumab and trastuzumab bind different subdomains and have complementary mechanisms of action, so their combination provides a more enhanced HER2 signaling blockade, while resulting into an increased antitumor activity. The combination of docetaxel, trastuzumab and pertuzumab significantly increases the survival of patients and no toxic side effects were observed during this combination therapy [143].

Capecitabine has been described as a potential chemotherapy partner for various types of cytotoxic drugs, due to its low potency for myelosuppression. Co-administration of taxanes and capecitabine increases the oncogenic activity of an enzyme (thymidine phosphorylase) involved in the activation of capecitabine. Apart from this, capecitabine, when combined with either nab-paclitaxel (nanoparticle albumin-bound paclitaxel) or docetaxel along with celecoxib result into enhanced synergistic antitumor activity against metastatic breast cancer [138,144]. Different mechanisms of action and partially non-overlapping tolerability profile of gemcitabine make it worthy candidate for combination therapy with taxanes. Other than this, gemcitabine plus paclitaxel have also been approved for the combination cancer treatment [138]. Several combinations including capecitabine plus ixabepilone, gemcitabine plus vinorelbine, eribulin-combined with other chemotherapeutic agents, lapatinib-containing combinations have either been approved in clinical trials or are undergoing different phases of clinical trials for the treatment of breast cancer [138].

Masui *et al.* 2013 have very elegantly reviewed a lot of aspects related to combination therapy, which include chemotherapy and targeted molecular therapy [145]. On detailed analysis, it was concluded that if drugs are chosen carefully for both therapies, then these can very effectively work complimentarily towards cancer treatment.

b. Nanoparticulate Drug Delivery System

Recently, nanocarriers have been paid a lot of attention because they have taken drug delivery to the next level from conventional chemotherapy. They can be formed by making use of a vast range of materials, which includes inorganic materials (*e.g.* metal nanoparticles, quantum dots), lipids (*e.g.* liposomes), polymers (*e.g.* biodegradable polymeric nanoparticles, dendrimers), and biological materials (*e.g.* viral and albumin nanoparticles) [146]. Some examples of these nanocarriers have been depicted in the Fig. (**7**). Nanocarriers can significantly target cancer cells, when they are used with a particular anticancer drug, in comparison to the drug alone, due to its ability to protect the drug from early degradation and intensifying its absorption into the specific tissues. It can also increase the pharmacokinetic profile of the drug molecule.

Multidrug resistance can also be prevented by using a nanoparticulate drug delivery system and is recently reviewed [147]. In this alternative method, drug molecules are encapsulated and conjugated with nanoparticles in order to provide a new alternative to the typical chemotherapeutics. The drug needed to be delivered is first dissolved, trapped, adsorbed and finally encapsulated into a nano-matrix. Nano-matrix can be of three types such as nanocapsules, nanoparticles and nanospheres. Nanocapsules are characterized by vesicular systems containing polymer membrane and a drug present in the cavity. On the other hand, nanoparticles and nanospheres comprise of matrix systems having physically and uniformly scattered drug. In order to attain disease-specificity in the localization of drug and controlled release of drugs, it is possible to tailor-made the polymeric nanoparticles by amending the surface chemistry as well as polymeric features. As soon as the nanoparticles are accumulated at the cancer site, they have the ability to act as a reservoir of an anticancer drug and deliver the continuous supply of encapsulated chemotherapeutic drug at the target site. This type of drug delivery system improves the bioavailability of drug and provides systematic delivery [148]. Various mechanisms of multidrug resistance and overcoming them through different ways has been recently reviewed by Minko et.al., 2013 [149].

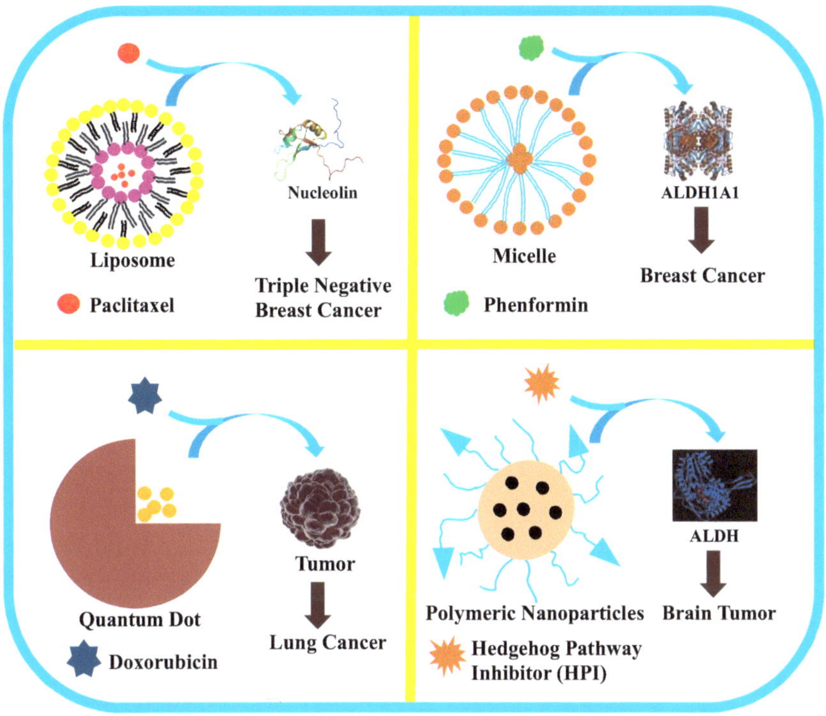

Fig. (7). Different Nanoparticles showing interaction with various drugs which acts on particular type of cancer or tumor. (150, 151); ALDH: Aldehyde Dehydrogenase, ALDH1A1: Aldehyde Dehydrogenase 1 family, member 1.

Nano-medicines play a vital role in target-specific drug delivery and early diagnosis of cancer, which make them superior in the field of cancer treatment. It is used in anticancer therapy to enrich the drug delivery, increasing the efficacy of the drug, reducing side effects and taking control over chemoresistance [16]. The targeted delivery of molecule into the cancer cell will enhance the formation of cancer biomarkers, which can be easily detectable. A very high resolution and accurate cancer imaging can be achieved with the help of nano-medicines [152]. Nano-medicines can easily be injected into the cancer cells, which destroy cancer cells upon activation. Molecular diagnostic tools utilizing nanotechnology are known as nano-diagnostics, which when combined with nano-therapeutics form the basic platform for personalized nano-medicines for patients. This is supposed to be more effective than the traditional chemotherapy [153].

The concentration and availability of anticancer drugs that are absorbed by the cells are the key factors that regulate the activity of these drugs. These cytotoxic internalized drugs are transferred to the outer environment of the cell with the help of various proteins including MRP (Multidrug resistance protein), P-gp (P-glycoprotein), LRP (Lung resistance-related protein), and BCRP (Breast cancer resistance protein) [154,155]. In order to increase the availability and function of these drugs in the cellular environment, it becomes the utmost priority to inhibit proteins involved in their transportation. Several inhibitors such as diltiazem, verapamil, quinidine, tariquidar, cyclosporine A, and doxorubicin-galiu--transferrin conjugate are employed to antagonize and inhibit the function of P-gp [156]. On the other hand, proteins inhibiting the MRP activity include itraconazole, erythromycin, rifampicin, NSAIDs, ofloxacin, difloxacin and many more [157], whereas estrone, genistein, fumitremorgin C were found to inhibit the activity of BCRP. Chemosensitizers such as nitazoxanide, derivatives of 6-mercaptohexanol, ethacrynic acid, benastatin A represent the class of GST inhibitors. Recently, an important class of selective and irreversible α-chloroacetamide inhibitors of glutathione S-transferase omega 1 (GSTO1) were reported to increase the sensitivity of cancer cells to cisplatin [158, 159]. These inhibitors can also block the target proteins present in the normal tissues, which lead to adverse side effects. That's why, it becomes extremely important to deliver these inhibitors specifically to the cancer sites, which can be done by encapsulating them in nanocarriers [6]. The inhibitors so delivered show specificity towards cancer cells and decrease the expression of responsible MDR proteins. Due to the down-regulation of MDR proteins, the anticancer drugs can be delivered along with the inhibitors or they can be delivered independently to the target site because of the presence of nanoparticles. It has been reported that due to the nanoparticle enclosed cyclosporine A and doxorubicin, the efficacy in doxorubicin-resistance leukemia cells increases two-fold [156]. Nanoparticles can also be prepared using liposomes that have the potential to carry drugs in either aqueous environment or in the lipid bilayer. Advances in the liposome-based nanoparticles have been made through the addition of polyethylene glycol, which lead to the enhanced circulation time and drug release [160].

Dendrimers containing hydrophobic inner cavities can interact with non-polar, uncharged molecules or drugs such as doxorubicin and paclitaxel *via* different types of interactions. The functional groups present at the outer layer have the capability to electrostatically interact with polar molecules. The outer layer also controls the drug delivery by carrying out the modifications, due to which the release of anticancer drug depends on certain pH and specific enzymes [161,162, 16]. Several clinical studies are currently in process using different types of modified dendrimers, which form micelles through self-assembly of dendrimers in order to deliver paclitaxel for the treatment of NSCLC, pancreatic and breast cancer [163 - 165].

c. Gene Delivery Through RNA Interference (RNAi)

Another important way to prevent drug resistance is gene delivery, which controls the function of a particular gene through RNAi. It is described as the defensive mechanism involving post-transcriptional gene silencing that is transmitted by small interfering RNAs (siRNA), constituting 20-25 nucleotides, resulting into the degradation of exogenous genes. Nucleic acids loaded with nanoparticles are transferred to the cancer site, which can further help to control the expression of *ABC* transporters. siRNA assists in the silencing of *MDR*-1, *Bcl*-2, and *MRP*1, due to which the expression of these proteins is decreased in cancer cells [156].

The road to cancer therapy involving siRNA silencing strategy is still not smooth. Under physiological conditions, siRNA is not stable, as it is easily digested by nucleases on passing through the blood. Intracellular siRNA is transported to the late endosomes, which are remarkably acidic. These endosomes are then transferred to lysosomes, which are even more acidified and constitute several nucleases that damage the siRNA [166]. The hydrophilicity and negative charge of siRNA inhibit them from crossing biological membranes. Thus, siRNAs are needed to be packed in vesicles for cellular entry. In addition, there are several other barriers in the siRNA-based cancer therapy. An efficient and safe delivery method is required to bring siRNA to the target site without harmful effects. Delivery systems should have features like serum stability, resistance to renal clearance, enhanced vascular permeability, smooth cell entry and less toxicity [167].

Chemically modified siRNA backbone, specially designed siRNAs, employment of nanoparticle-based synthetic lipid and polymer-based delivery systems are used to avoid these barriers [168]. siRNA and doxorubicin encapsulated in nanoparticles are delivered to the target site, where siRNA silences the P-gp expression, due to which the intracellular concentration of doxorubicin drug increases rapidly [169]. Likewise, modified liposomes comprising P-gp siRNA and doxorubicin were also developed for the treatment of drug-resistant cancers [170].

d. Novel Approaches to Overcome Chemoresistance

(i). Micro RNA Based Strategies

Restoration of down-regulated miRNAs can turn cancer cells more susceptible to apoptosis. Recent studies revealed that mir-221 inhibition leads to cell cycle arrest as well as induction of apoptosis and sensitization of pancreatic cancer cells to the effects of gemcitabine. Moreover, restoration of expression of the mir-200 family of miRNAs resulted in gemcitabine-resistant cells becoming sensitive to gemcitabine again. Down-regulation of mir-21 also results in increased sensitivity of these cells to gemcitabine [171].

(ii). Targeting Specific Transcription Factors

- *Nuclear factor (NF)-kappa B*: NF-κB binds the promoter of immunoglobulin kappa chain in B cells and is responsible for diverse cellular activities including DNA binding, immune response and inflammation. Many cancer cells which exhibit chemoresistance are characterized by constitutive activation of NF-κB, thereby, indicating it as a potent apoptosis suppressor. Its activation leads either to the up-regulation of anti-apoptotic genes or to the down-regulation of apoptotic genes, as well as, it mediates chemoresistance. NF-κB inhibition can be carried out by administration of natural compounds such as thymoquinone (TQ), sulforaphane (SFN), dihydroartemisinin (DHA) or 3,3-diindolylmethane (I3C/ DIM), which block both constitutive and anti-cancer drug induced NF-κB activity. It is successfully tested in preclinical experiments for sensitization of pancreatic ductal adenocarcinoma cells against chemotherapy. In addition, adenoviral delivery of a modified form of IκBα, a specific inhibitor of NF-κB, results in suppression of NF-κB sensitize chemoresistant cancers to the apoptotic potential of TNF-α leading to cancer regression [172, 173].
- *Tumor Necrosis Factor (TNF) alpha and its family members*: *TNF*-α is a master regulator of inflammation and is a key player in the cytokine network. In cancers, *TNF*-α can be produced by cancer cells, stromal fibroblasts, and inflammatory cells. *TNF*-α in the tumor microenvironment further stimulates the production of other cytokines and chemokines. As a consequence, primary cancer growth, metastasis, angiogenesis, and chemoresistance are enhanced, and the immune evasive tumor microenvironment is established. Inflammatory stimulating effects were well neutralized by anti-*TNF*-α treatments [173, 174].

(iii). Phytochemical Modulation of Apoptosis and Autophagy

Most cancer drug therapies act by inducing apoptosis in dividing cells but are ineffective in targeting quiescent leukemic stem cells (LSCs). Niches in the bone marrow, known as leukemic niches, behave as “sanctuaries”, where LSCs acquire drug resistance. Phytochemicals such as baicalein, chrysin and wogonin are used to overcome chemoresistance through the modulation of signaling pathways involved in the survival and death of LSCs, thereby inducing apoptosis in LSCs. Curcumin and resveratrol cause cell death by non-apoptotic pathway, that is autophagy. In order to be effective in eliminating LSCs, it is important to target signaling pathways (such as Wnt/ β-catenin, Notch, and Hedgehog). Resveratrol has been reported to induce apoptosis in leukemic cells through the inhibition of the Notch and Sonic hedgehog signaling pathways, therefore showing potential to affect LSCs [175].

(iv). Elimination of Drug-resistant Cell Types

Epithelial Mesenchymal Transition (EMT) is a renowned biological process that plays a significant role in cancer development and metastasis. Polarized sheets or several layers of cells attach to each other with the help of cellular junctions such as desmosomes and adhere junctions have been formed by epithelial cells. In addition, the basal surface of polarized epithelial cells interacts with basement membrane in order to achieve certain bio-chemical changes. On the other hand, the mesenchymal cells found inserted in extracellular matrix (ECM) does not form strong bond with neighboring cells. The changes in epithelial cells facilitate mesenchymal cell phenotype, which is characterized by enhanced production of ECM components, improved migratory aptitude, and elevated resistance to apoptosis. The epithelial cells transit into a mesenchymal state during specific embryonic morphogenesis process through the activation of epithelial-mesenchymal transition (EMT) [176,177]. In cancer, EMT enables cell invasion, which is required by cancer metastasis. The cancerous cells break through the basement membrane and enter the bloodstream with increased invasive properties. While exiting the bloodstream, these cancer cells form micro-metastases by undergoing mesenchymal-epithelial transition (MET) for clonal outgrowth at metastatic sites. The EMT and MET processes initiate as well as complete the invasion-metastasis cascade [178]. These EMT-related pathways could provide targets for chemotherapy. For instance, the growth of cells undergoing EMT process can be effectively prevented by Src kinase inhibitors. Pancreatic cancerous cells can be inhibited from undergoing EMT by hedgehog signal inhibitors [179].

Cancer Stem Cells (CSC) have the ability to self-renew, produce differentiated cells and drive continued expansion of the population of malignant cells that have invasive and metastatic properties and have critical role in chemoresistance [171, 180]. CSCs have been identified in pancreatic, breast as well as colon cancers. In cancers, the CSC ratio varies from 0.1 to 30% which depends on the type and the advancement of the cancer. Chemotherapy targets several cancer cells, but sometimes it is unable to eradicate the critical CSCs due to their protection by specific resistance mechanisms. The remaining CSCs lead to the production of new cancers and metastasis. The relapsed cancers gained more malignancy properties thereby, increasing the therapeutic failures [181]. CSC was initially recognized in human acute myeloid leukemia. By eradicating CSCs, cancer cure as well as cancer survival can be effectively improved. Surface markers which are used for identification and isolation of CSCs, signal generated in TME can be targeted in order to eliminate CSCs [182].

Elimination of EMT cells and cancer stem cells can be done through treatment with salinomycin or metformin to overcome chemoresistance. Quercetin is also reported to target pancreatic cancer stem cells and EMT-type cells [171].

RECENT INTERVENTIONS FOR APPROACHING CANCER BETTER

(a). Molecular Signatures of Cancer Using Bioinformatics and Systems Biology Approach

As our understanding of the molecular biology of cancers has advanced, it is imperative to take into account the contributions of modern genomics, proteomics, metabolomics and functional analytical techniques in determining the responsiveness of cancers to a particular drug treatment. The integration of bioinformatics and systems biology aids the interrogation of clinical samples in identifying molecular signatures that predict patient's response to specific therapeutics.

A recent review by Rahman *et al.* 2015 has quite elaborately discussed the cancer metabolism and drug resistance [183]. This review has focused upon the major findings associated with metabolic pathways in cancer cells, technologies used to study these pathways, proteins/ enzymes involved, genes associated with altered metabolism and anticancer agents used to target these pathways.

Recent advancements in systems biology approach which include high-throughput experimental platform technologies ranging from modern genomics, transcriptomic, proteomic and metabolomic profiling of the disease have significantly helped in the molecular-level characterization of cancer cells, which is extremely significant for deciding the line of cancer treatments. These computational modeling approaches actually help in understanding the pathways involved in cancer formation and progression, impact of particular mutation on pathway operation and various other related biological events [184]. The studies on integration of bioinformatics analysis may eventually guide to an advanced level in the understanding of cancer, its early diagnosis and choice of medicines/ therapies, which may be used for the treatment. These therapies can even be customized based on combination of single nucleotide polymorphisms (SNPs) and gene expression [185]. A large number of databases and web resources (*e.g.* cBioPortal, canSAR, UCSC cancer genomic browser, Cancer Genome Work Bench) are now available, to explore various genomes, which allow us to visualize gene expression, mutation, and cancer genomic data [186].

Gene expression signatures can be used to categorize cancer into molecular subsets that are of clinical importance [187]. Perou and their co-workers have used microarray-based transcriptome profiling on breast cancer samples and had generated five molecular subsets with great biological relevance [188]. Majumdar *et al.* have prepared a CANScript platform by integrating cancer ecosystem and a novel machine learning algorithm, which is able to predict therapeutic efficacy of targeted and cytotoxic drug in patients with colorectal cancer (CRC) as well as head and neck squamous cell carcinoma (HNSCC) [189]. Proteomic foot printing can easily detect drug-specific changes in cancer cell-surface antigens, which can be considered in selecting anticancer vaccines for checking reoccurrence of cancer [190].

Signaling elements of the EMT pathways involved in development and progression of cancer can be identified to analyze their associated genetic, molecular and cellular mechanisms that are responsible for inducing neoplasia in order to improve diagnosis and prognosis accompanying the cancer chemotherapeutics. Caveolin-1 is a scaffold protein that directly interacts with signaling molecules, including extracellular growth factor signaling receptor kinases, which can suppress cancers by regulating a diverse range of processes such as invasion, angiogenesis, metastasis, ECM remodeling, cell-cell interactions at cadherin elements, cell survival, polarity and directional migration. It also resists anoikis and apoptosis induced by taxol, *via* interaction with inhibitor of differentiation/ DNA binding (Id-1) and consequent Id-1 and protein phosphatase 2A (PP2A)-mediated AKT activation. A number of human miRNAs such as miR-126 and miR-335, which demonstrate tumor suppressive functions by arresting the overall cancer growth, proliferation and metastasis, are highly repressed in relapsed cancers of breast patients, compared to cancers from patients that did not relapse [191].

(b). Chemoresistance Biomarkers and Drug Companion Diagnostics

Unpredictable clinical response, chemoresistance and cytotoxicity of chemotherapeutic agents are some of the factors that can cause harmful side effects and even death of a patient. Determining the drug response of chemotherapeutic agents become highly essential to a particular cancer at an early stage in order to overcome the chemoresistance and to decrease the cost of health care [192]. Predictive biomarkers were identified, which would help in investigating the therapy response to a chemotherapeutic regime. P-glycoprotein was determined as a biomarker for multidrug resistance by Juliano and coworkers [193]. Gel-based mass spectrometry such as 2D coupled to MALDI-TOF-MS and gel-free mass spectrometry including 1D or 2D liquid chromatography coupled to tandem MS techniques can be used to predict the chemoresistance biomarkers.

14-3-3 sigma (stratifin) protein was used as a biomarker for Adriamycin-resistance in breast cancer cells [194]. Recently, new biomarkers such as aldehyde dehydrogenase (ALDH), epithelial-specific antigen (ESA), C-X-C chemokine receptor type 4 (CXCR4) and CD133 (prominin-1) for breast cancer have been identified [195]. Previously, carcino-embryonic antigens such as CA 19-9, CA-72 and CA-50 were the only biomarkers used for gastric cancer, which lacks the specificity and sensitivity. Recent study shows that genetic markers such as metallothionein-IG, MRP4, dihydropyrimidine and heparin-binding epidermal growth factor-like growth factor (HBEGF), epigenetic markers miRNAs and long non-coding RNAs, protein markers such as alpha-1-microglobulin, regenerating gene family member 4 (REG4), fork head box M1 transcription factor (FOXM1) and HNRNPC protein located in cytoplasm and nucleus are useful and prognostic biomarkers for gastric cancer [196]. The TNF superfamily ligand, APRIL (TNFSF13) has been identified as an effective chemoresistance biomarker against colorectal adenocarcinoma by gene expression profiling method [197].

microRNAs can also be used as biomarkers in order to predict drug resistance in chemotherapy. Recent reports have shown that the up- or down-regulation of specific miRNAs can be directly linked to the response of patients to the chemotherapeutic drugs. These also furnish high specificity and selectivity as compared to other biomarkers. Over expression of miR-92a-2, miR-574-5p, and miR-147 act as possible biomarkers for chemoresistance in patients of small cell lung cancer [198]. Increased expression of miR-23b and miR-21 indicates the high resistance to sunitinib in NSCLC patients. On the other hand, decreased expression of miR-424 was indicative of increased resistance to vandetanib as well as erlotinib [199].

It is essential to develop different methods of companion diagnostics in order to predict the selective and beneficial targeted therapies before clinical trials. Combination of targeted therapies with biomarkers provide a very good platform for creating personalized medicines. In order to perform an efficient companion diagnostic test, three important requirements should be fulfilled. The substance suitable for testing should be easily available, a set of mutations should be defined that can help in the prediction of therapeutic response and the technique employed for identifying mutations should be cost-effective [200]. The most common techniques used for companion diagnostics include Immunohistochemistry, DNA sequencing, fluorescent *in situ* hybridization (FISH), mutation-specific assays, indirect mutation analysis, qPCR, and cytogenetics [201]. Immunohistochemistry (IHC) is a simple technique used in most labs which determines the cells with mutation among the large reservoir of normal cells [202]. DNA sequencing method is used to detect mutations present in genome. Direct sequencing helps in the identification of mutations in an entire gene, whereas sanger sequencing is unable to find low frequency mutations. Pyro sequencing recognizes mutations with low frequency and is more sensitive than sanger sequencing [203]. FISH and qPCR techniques are used for the determination of genetic variations such as amplification or translocation. Mutation-specific assays or indirect mutation analysis are used, when the large quantity of non-neoplastic cells are present in clinical samples. Cytogenetic technique is utilized to detect chromosomal abnormalities in mitotic cells [200].

For metastatic colorectal cancer, targeting the epidermal growth factor receptor (EGFR) by monoclonal antibodies was found to be a successful treatment. Some of the biomarkers such as microsatellite instability, detection of *KRAS* mutations, *Her*2 amplification, and *BRAF* mutation are used to select patients who might be benefited by EGFR inhibition [204]. Epigenetic age of a tumor tissue is different from its chronological age and the difference between these two, blood Δ_{age} is also used as a biomarker to predict incidence of cancer and its mortality [205].

OUTLOOK AND FUTURE DIRECTIONS

Irrespective of the fact that new drugs and therapies are continuously being incorporated into a large repertoire of already available strategies of clinical practice, there is still a long way to win the war against cancer. It's quite disturbing that almost 90% chemotherapy drugs fail in metastatic cancers, because of chemoresistance. Drug resistance for chemotherapeutic drugs hence, defeat the purpose of treatment and cause mortality by curbing the therapeutic efficacy of anticancer drugs. A number of elusive mechanisms operate in cancer cells simultaneously in order to survive cancer treatment. As the mode of action of several classical anticancer drugs is to induce apoptosis, therefore, the main focus of biochemists is to develop drugs which cause cell death of apoptotic resistant cancer cells by alternate non-apoptotic pathways.

Recent advances in understanding of tumor biology, tumor subtypes, genomic variations causing cancer and more efficient diagnostic tools should now provide an important platform for designing more intense and efficient treatment plan. Using details of molecular mechanisms, novel approaches to overcome chemoresistance are evolving everyday. A combination therapy with the use of more efficient tools of molecular diagnosis like nano-diagnostics, personalized medicines for cancer patients can be planned. To address the challenges related to cancer chemoresistance, there is an urgent need for a multidisciplinary approach, which should not only focus on more in-depth tracing and characterization of tumors and their environment, but it should also provide a combination of better, personalized and target selective therapeutic strategies that contribute towards circumventing the drug resistance.

CONSENT FOR PUBLICATION

Not applicable.

CONFLICT OF INTEREST

The authors confirm that they have no conflicts of interest to declare for this publication.

ACKNOWLEDGEMENT

MK and SK would like to extend their gratitude to the University of Delhi, Delhi, India for their research and development grant.

REFERENCES

[1] American Cancer Society. Cancer Facts & Figures 2017.https://www.cancer.org/research/cancer-facts-statistics/all-cancer-facts-figures/cancer-facts-figures-2017.html

[2] Steeg PS. Targeting metastasis. Nat Rev Cancer 2016; 16(4): 201-18. [http://dx.doi.org/10.1038/nrc.2016.25] [PMID: 27009393]

[3] Hait WN. Anticancer drug development: the grand challenges. Nat Rev Drug Discov 2010; 9(4): 253-4. [http://dx.doi.org/10.1038/nrd3144] [PMID: 20369394]

[4] Trevor AJ, Katzung BG, Kruidering-Hall M. Katzung & Trevor's Pharmacology: Examination & Board Review, 10e; 2015, McGraw-Hill Education, United States of America; Chapter 54. Cancer Chemotherapy.

[5] Holohan C, Van Schaeybroeck S, Longley DB, Johnston PG. Cancer drug resistance: an evolving paradigm. Nat Rev Cancer 2013; 13(10): 714-26. [http://dx.doi.org/10.1038/nrc3599] [PMID: 24060863]

[6] Liang XJ, Chen C, Zhao Y, Wang PC. Circumventing tumor resistance to chemotherapy by nanotechnology. Multi-Drug Resistance in Cancer 2010; pp. 467-88. [http://dx.doi.org/10.1007/978-1-60761-416-6_21]

[7] Luqmani YA. Mechanisms of drug resistance in cancer chemotherapy. Med Princ Pract 2005; 14 (Suppl. 1): 35-48. [http://dx.doi.org/10.1159/000086183] [PMID: 16103712]

[8] Housman G, Byler S, Heerboth S, *et al.* Drug resistance in cancer: an overview. Cancers (Basel) 2014; 6(3): 1769-92. [http://dx.doi.org/10.3390/cancers6031769] [PMID: 25198391]

[9] Kaushik M, Chaudhary S, Mahendru S, Ahmed S, Pathak AK, Kukreti S. MicroRNA: A Multi-Facet Biological Target for Cancer and Other Diseases. Clin Cancer Drugs 2017; 4(1): 2-9. [http://dx.doi.org/10.2174/2212697X04666170123122037]

[10] Iorio MV, Croce CM. MicroRNA dysregulation in cancer: diagnostics, monitoring and therapeutics. A comprehensive review. EMBO Mol Med 2012; 4(3): 143-59. [http://dx.doi.org/10.1002/emmm.201100209] [PMID: 22351564]

[11] Fanini F, Fabbri M. MicroRNAs and cancer resistance: A new molecular plot. Clin Pharmacol Ther 2016; 99(5): 485-93. [http://dx.doi.org/10.1002/cpt.353] [PMID: 26875151]

[12] Pattabiraman DR, Weinberg RA. Tackling the cancer stem cells - what challenges do they pose? Nat Rev Drug Discov 2014; 13(7): 497-512.
[http://dx.doi.org/10.1038/nrd4253] [PMID: 24981363]

[13] Furukawa Y, Kikuchi J. Epigenetic mechanisms of cell adhesion-mediated drug resistance in multiple myeloma. Int J Hematol 2016; 104(3): 281-92.
[http://dx.doi.org/10.1007/s12185-016-2048-5] [PMID: 27411688]

[14] Wijdeven RH, Pang B, Assaraf YG, Neefjes J. Old drugs, novel ways out: Drug resistance toward cytotoxic chemotherapeutics. Drug Resist Updat 2016; 28: 65-81.
[http://dx.doi.org/10.1016/j.drup.2016.07.001] [PMID: 27620955]

[15] Saunders NA, Simpson F, Thompson EW, *et al.* Role of intratumoural heterogeneity in cancer drug resistance: molecular and clinical perspectives. EMBO Mol Med 2012; 4(8): 675-84.
[http://dx.doi.org/10.1002/emmm.201101131] [PMID: 22733553]

[16] Markman JL, Rekechenetskiy A, Holler E, Ljubimova JY. Nanomedicine therapeutic approaches to overcome cancer drug resistance. Adv Drug Deliv Rev 2013; 65(13-14): 1866-79.
[http://dx.doi.org/10.1016/j.addr.2013.09.019] [PMID: 24120656]

[17] Kuper H, Adami HO, Boffetta P. Tobacco use, cancer causation and public health impact. J Intern Med 2002; 251(6): 455-66.
[http://dx.doi.org/10.1046/j.1365-2796.2002.00993.x] [PMID: 12028500]

[18] Kushi LH, Byers T, Doyle C, *et al.* American Cancer Society Guidelines on Nutrition and Physical Activity for cancer prevention: reducing the risk of cancer with healthy food choices and physical activity. CA Cancer J Clin 2006; 56(5): 254-81.
[http://dx.doi.org/10.3322/canjclin.56.5.254] [PMID: 17005596]

[19] Roukos DH. Genome-wide association studies: how predictable is a person's cancer risk? Expert Rev Anticancer Ther 2009; 9(4): 389-92.
[http://dx.doi.org/10.1586/era.09.12] [PMID: 19374592]

[20] Milholland B, Auton A, Suh Y, Vijg J. Age-related somatic mutations in the cancer genome. Oncotarget 2015; 6(28): 24627-35.
[http://dx.doi.org/10.18632/oncotarget.5685] [PMID: 26384365]

[21] Young C. Solar ultraviolet radiation and skin cancer. Occup Med (Lond) 2009; 59(2): 82-8.
[http://dx.doi.org/10.1093/occmed/kqn170] [PMID: 19233827]

[22] Liao JB. Viruses and human cancer. Yale J Biol Med 2006; 79(3-4): 115-22.
[PMID: 17940621]

[23] Siegel R, DeSantis C, Virgo K, *et al.* Cancer treatment and survivorship statistics, 2012. CA Cancer J Clin 2012; 62(4): 220-41.
[http://dx.doi.org/10.3322/caac.21149] [PMID: 22700443]

[24] Lawrence TS, Ten Haken RK, Giaccia A. Principles of radiation oncology Cancer: principles and practice of oncology. 8th ed., Philadelphia: Lippincott Williams and Wilkins 2008.

[25] Ying AK, Feeley TW, Porter ME. Value-based healthcare: implications for thyroid cancer. Int J Endocr Oncol 2016; 3(2): 115-29.
[http://dx.doi.org/10.2217/ije-2015-0005]

[26] Gerber DE. Targeted therapies: a new generation of cancer treatments. Am Fam Physician 2008; 77(3): 311-9.
[PMID: 18297955]

[27] Vignot S, Faivre S, Aguirre D, Raymond E. mTOR-targeted therapy of cancer with rapamycin derivatives. Ann Oncol 2005; 16(4): 525-37.
[http://dx.doi.org/10.1093/annonc/mdi113] [PMID: 15728109]

[28] Farkona S, Diamandis EP, Blasutig IM. Cancer immunotherapy: the beginning of the end of cancer?

BMC Med 2016; 14: 73.
[http://dx.doi.org/10.1186/s12916-016-0623-5] [PMID: 27151159]

[29] Loren AW, Mangu PB, Beck LN, *et al.* Fertility preservation for patients with cancer: American Society of Clinical Oncology clinical practice guideline update. J Clin Oncol 2013; 31(19): 2500-10. [http://dx.doi.org/10.1200/JCO.2013.49.2678] [PMID: 23715580]

[30] Byar DP, Corle DK. Hormone therapy for prostate cancer: results of the veterans administration cooperative urological research group studies. NCI monographs: a Publication of the National Cancer Institute. 1987 Dec; (7): 165-170.

[31] Dos Santos Guimarães I, Daltoé RD, *et al.* Cancer Treatment-Conventional and Innovative approaches. 2013; 1-34.

[32] Bruce A. Chabner, BA and Roberts Jr, TG. Chemotherapy and the war on cancer. Nat Rev Cancer 2005; 5: 65-72.
[http://dx.doi.org/10.1038/nrc1529] [PMID: 15630416]

[33] Carlson LE, Rouleau CR, Speca M, Robinson J, Bultz BD. Brief supportive-expressive group therapy for partners of men with early stage prostate cancer: lessons learned from a negative randomized controlled trial. Support Care Cancer 2017; 25(4): 1035-41.
[http://dx.doi.org/10.1007/s00520-016-3551-1] [PMID: 28064386]

[34] Enard KR, Dolan Mullen P, Kamath GR, Dixon NM, Volk RJ. Are cancer-related decision aids appropriate for socially disadvantaged patients? A systematic review of US randomized controlled trials. BMC Med Inform Decis Mak 2016; 16(1): 64.
[http://dx.doi.org/10.1186/s12911-016-0303-6] [PMID: 27267490]

[35] Ni Chonghaile T, Sarosiek KA, Vo TT, *et al.* Pretreatment mitochondrial priming correlates with clinical response to cytotoxic chemotherapy. Science 2011; 334(6059): 1129-33.
[http://dx.doi.org/10.1126/science.1206727] [PMID: 22033517]

[36] Corrie PG. Cytotoxic chemotherapy: clinical aspects. Medicine (Baltimore) 2011; 39(12): 717-22.
[http://dx.doi.org/10.1016/j.mpmed.2011.09.012]

[37] https://www.mycancergenome.org/content/molecular-medicine/pathways/cytotoxic-chemothera-y-mechanisms-of-action

[38] DeVita VT Jr, Chu E. A history of cancer chemotherapy. Cancer Res 2008; 68(21): 8643-53.
[http://dx.doi.org/10.1158/0008-5472.CAN-07-6611] [PMID: 18974103]

[39] Hurley LH. DNA and its associated processes as targets for cancer therapy. Nat Rev Cancer 2002; 2(3): 188-200.
[http://dx.doi.org/10.1038/nrc749] [PMID: 11990855]

[40] Lloyd KL, Cree IA, Savage RS. Prediction of resistance to chemotherapy in ovarian cancer: a systematic review. BMC Cancer 2015; 15(1): 117.
[http://dx.doi.org/10.1186/s12885-015-1101-8] [PMID: 25886033]

[41] Li M, Yin J, Mao N, Pan L. Upregulation of phosphorylated cofilin 1 correlates with taxol resistance in human ovarian cancer *in vitro* and *in vivo.* Oncol Rep 2013; 29(1): 58-66.
[http://dx.doi.org/10.3892/or.2012.2078] [PMID: 23064469]

[42] Armstrong DK. Relapsed ovarian cancer: challenges and management strategies for a chronic disease. Oncologist 2002; 7 (Suppl. 5): 20-8.
[http://dx.doi.org/10.1634/theoncologist.7-suppl_5-20] [PMID: 12324630]

[43] Choi YE, Meghani K, Brault ME, *et al.* Platinum and PARP inhibitor resistance due to overexpression of MicroRNA-622 in BRCA1-mutant ovarian cancer. Cell Reports 2016; 14(3): 429-39.
[http://dx.doi.org/10.1016/j.celrep.2015.12.046] [PMID: 26774475]

[44] Leonessa F, Clarke R. ATP binding cassette transporters and drug resistance in breast cancer. Endocr Relat Cancer 2003; 10(1): 43-73.

[http://dx.doi.org/10.1677/erc.0.0100043] [PMID: 12653670]

[45] Dean M, Fojo T, Bates S. Tumour stem cells and drug resistance. Nat Rev Cancer 2005; 5(4): 275-84. [http://dx.doi.org/10.1038/nrc1590] [PMID: 15803154]

[46] Challen GA, Little MH. A side order of stem cells: the SP phenotype. Stem Cells 2006; 24(1): 3-12. [http://dx.doi.org/10.1634/stemcells.2005-0116] [PMID: 16449630]

[47] Chuthapisith S, Eremin J, El-Sheemey M, Eremin O. Breast cancer chemoresistance: emerging importance of cancer stem cells. Surg Oncol 2010; 19(1): 27-32. [http://dx.doi.org/10.1016/j.suronc.2009.01.004] [PMID: 19251410]

[48] Chang A. Chemotherapy, chemoresistance and the changing treatment landscape for NSCLC. Lung Cancer 2011; 71(1): 3-10. [http://dx.doi.org/10.1016/j.lungcan.2010.08.022] [PMID: 20951465]

[49] Sève P, Dumontet C. Chemoresistance in non-small cell lung cancer. Curr Med Chem Anticancer Agents 2005; 5(1): 73-88. [http://dx.doi.org/10.2174/1568011053352604] [PMID: 15720263]

[50] d'Amato TA, Landreneau RJ, McKenna RJ, Santos RS, Parker RJ. Prevalence of *in vitro* extreme chemotherapy resistance in resected nonsmall-cell lung cancer. Ann Thorac Surg 2006; 81(2): 440-6. [http://dx.doi.org/10.1016/j.athoracsur.2005.08.037] [PMID: 16427828]

[51] Matsumoto Y, Oka M, Sakamoto A, *et al.* Enhanced expression of metallothionein in human non-small-cell lung carcinomas following chemotherapy. Anticancer Res 1997; 17(5B): 3777-80. [PMID: 9427779]

[52] Rose MC, Kostyanovskaya E, Huang RS. Pharmacogenomics of cisplatin sensitivity in non-small cell lung cancer. Genomics Proteomics Bioinformatics 2014; 12(5): 198-209. [http://dx.doi.org/10.1016/j.gpb.2014.10.003] [PMID: 25449594]

[53] Sève P, Dumontet C. Is class III β-tubulin a predictive factor in patients receiving tubulin-binding agents? Lancet Oncol 2008; 9(2): 168-75. [http://dx.doi.org/10.1016/S1470-2045(08)70029-9] [PMID: 18237851]

[54] Lee JG, Shin JH, Shim HS, *et al.* Autophagy contributes to the chemo-resistance of non-small cell lung cancer in hypoxic conditions. Respir Res 2015; 16(1): 138. [http://dx.doi.org/10.1186/s12931-015-0285-4] [PMID: 26553068]

[55] Melo JV, Barnes DJ. Chronic myeloid leukaemia as a model of disease evolution in human cancer. Nat Rev Cancer 2007; 7(6): 441-53. [http://dx.doi.org/10.1038/nrc2147] [PMID: 17522713]

[56] Raguz S, Yagüe E. Resistance to chemotherapy: new treatments and novel insights into an old problem. Br J Cancer 2008; 99(3): 387-91. [http://dx.doi.org/10.1038/sj.bjc.6604510] [PMID: 18665178]

[57] Nimmanapalli R, Bhalla K. Mechanisms of resistance to imatinib mesylate in *Bcr-Abl*-positive leukemias. Curr Opin Oncol 2002; 14(6): 616-20. [http://dx.doi.org/10.1097/00001622-200211000-00005] [PMID: 12409651]

[58] Hochhaus A, Kreil S, Corbin AS, *et al.* Molecular and chromosomal mechanisms of resistance to imatinib (STI571) therapy. Leukemia 2002; 16(11): 2190-6. [http://dx.doi.org/10.1038/sj.leu.2402741] [PMID: 12399961]

[59] Coustan-Smith E, Mullighan CG, Onciu M, *et al.* Early T-cell precursor leukaemia: a subtype of very high-risk acute lymphoblastic leukaemia. Lancet Oncol 2009; 10(2): 147-56. [http://dx.doi.org/10.1016/S1470-2045(08)70314-0] [PMID: 19147408]

[60] Houthuijzen JM, Daenen LG, Roodhart JM, Voest EE. The role of mesenchymal stem cells in anti-cancer drug resistance and tumour progression. Br J Cancer 2012; 106(12): 1901-6. [http://dx.doi.org/10.1038/bjc.2012.201] [PMID: 22596239]

[61] Iwamoto S, Mihara K, Downing JR, Pui CH, Campana D. Mesenchymal cells regulate the response of acute lymphoblastic leukemia cells to asparaginase. J Clin Invest 2007; 117(4): 1049-57. [http://dx.doi.org/10.1172/JCI30235] [PMID: 17380207]

[62] Mudry RE, Fortney JE, York T, Hall BM, Gibson LF. Stromal cells regulate survival of B-lineage leukemic cells during chemotherapy. Blood 2000; 96(5): 1926-32. [PMID: 10961896]

[63] Pillozzi S, Masselli M, De Lorenzo E, *et al.* Chemotherapy resistance in acute lymphoblastic leukemia requires hERG1 channels and is overcome by hERG1 blockers. Blood 2011; 117(3): 902-14. [http://dx.doi.org/10.1182/blood-2010-01-262691] [PMID: 21048156]

[64] Bhattacharya B, Mohd Omar MF, Soong R. The Warburg effect and drug resistance. Br J Pharmacol 2016; 173(6): 970-9. [http://dx.doi.org/10.1111/bph.13422] [PMID: 26750865]

[65] Herst PM, Howman RA, Neeson PJ, Berridge MV, Ritchie DS. The level of glycolytic metabolism in acute myeloid leukemia blasts at diagnosis is prognostic for clinical outcome. J Leukoc Biol 2011; 89(1): 51-5. [http://dx.doi.org/10.1189/jlb.0710417] [PMID: 20959411]

[66] Song K, Li M, Xu X, Xuan LI, Huang G, Liu Q. Resistance to chemotherapy is associated with altered glucose metabolism in acute myeloid leukemia. Oncol Lett 2016; 12(1): 334-42. [http://dx.doi.org/10.3892/ol.2016.4600] [PMID: 27347147]

[67] Gottesman MM, Ambudkar SV, Ni B, *et al.* Exploiting multidrug resistance to treat cancer. Cold Spring Harbor symposia on quantitative biology. Cold Spring Harbor Laboratory Press.1994 Jan 1; (59):677-683. [http://dx.doi.org/10.1101/SQB.1994.059.01.078]

[68] Gottesman MM. Mechanisms of cancer drug resistance. Annu Rev Med 2002; 53(1): 615-27. [http://dx.doi.org/10.1146/annurev.med.53.082901.103929] [PMID: 11818492]

[69] Planells-Cases R, Lutter D, Guyader C, *et al.* Subunit composition of VRAC channels determines substrate specificity and cellular resistance to Pt-based anti-cancer drugs. EMBO J 2015; 34(24): 2993-3008. [http://dx.doi.org/10.15252/embj.201592409] [PMID: 26530471]

[70] Qiu A, Jansen M, Sakaris A, *et al.* Identification of an intestinal folate transporter and the molecular basis for hereditary folate malabsorption. Cell 2006; 127(5): 917-28. [http://dx.doi.org/10.1016/j.cell.2006.09.041] [PMID: 17129779]

[71] Rothem L, Ifergan I, Kaufman Y, Priest DG, Jansen G, Assaraf YG. Resistance to multiple novel antifolates is mediated *via* defective drug transport resulting from clustered mutations in the reduced folate carrier gene in human leukaemia cell lines. Biochem J 2002; 367(Pt 3): 741-50. [http://dx.doi.org/10.1042/bj20020801] [PMID: 12139489]

[72] Taddia L, D'Arca D, Ferrari S, *et al.* Inside the biochemical pathways of thymidylate synthase perturbed by anticancer drugs: Novel strategies to overcome cancer chemoresistance. Drug Resist Updat 2015; 23: 20-54. [http://dx.doi.org/10.1016/j.drup.2015.10.003] [PMID: 26690339]

[73] Pastan I, Kreitman RJ. Immunotoxins for targeted cancer therapy. Adv Drug Deliv Rev 1998; 31(1-2): 53-88. [http://dx.doi.org/10.1016/S0169-409X(97)00094-X] [PMID: 10837618]

[74] Lyall RM, Hwang JL, Cardarelli C, *et al.* Isolation of human KB cell lines resistant to epidermal growth factor-Pseudomonas exotoxin conjugates. Cancer Res 1987; 47(11): 2961-6. [PMID: 3494506]

[75] Klopfleisch R, Kohn B, Gruber AD. Mechanisms of tumour resistance against chemotherapeutic agents in veterinary oncology. Vet J 2016; 207: 63-72.

[http://dx.doi.org/10.1016/j.tvjl.2015.06.015] [PMID: 26526523]

[76] Chen CJ, Chin JE, Ueda K, *et al.* Internal duplication and homology with bacterial transport proteins in the mdr1 (P-glycoprotein) gene from multidrug-resistant human cells. Cell 1986; 47(3): 381-9. [http://dx.doi.org/10.1016/0092-8674(86)90595-7] [PMID: 2876781]

[77] Prajapati R, Sangamwar AT. Translocation mechanism of P-glycoprotein and conformational changes occurring at drug-binding site: Insights from multi-targeted molecular dynamics. Biochim Biophys Acta 2014; 1838(11): 2882-98. [http://dx.doi.org/10.1016/j.bbamem.2014.07.018] [PMID: 25068895]

[78] Kartal-Yandim M, Adan-Gokbulut A, Baran Y. Molecular mechanisms of drug resistance and its reversal in cancer. Crit Rev Biotechnol 2016; 36(4): 716-26. [PMID: 25757878]

[79] Sauna ZE, Ambudkar SV. Evidence for a requirement for ATP hydrolysis at two distinct steps during a single turnover of the catalytic cycle of human P-glycoprotein. Proc Natl Acad Sci USA 2000; 97(6): 2515-20. [http://dx.doi.org/10.1073/pnas.97.6.2515] [PMID: 10716986]

[80] Szakács G, Paterson JK, Ludwig JA, Booth-Genthe C, Gottesman MM. Targeting multidrug resistance in cancer. Nat Rev Drug Discov 2006; 5(3): 219-34. [http://dx.doi.org/10.1038/nrd1984] [PMID: 16518375]

[81] Ferreira JA, Peixoto A, Neves M, *et al.* Mechanisms of cisplatin resistance and targeting of cancer stem cells: Adding glycosylation to the equation. Drug Resist Updat 2016; 24: 34-54. [http://dx.doi.org/10.1016/j.drup.2015.11.003] [PMID: 26830314]

[82] Farmer H, McCabe N, Lord CJ, *et al.* Targeting the DNA repair defect in BRCA mutant cells as a therapeutic strategy. Nature 2005; 434(7035): 917-21. [http://dx.doi.org/10.1038/nature03445] [PMID: 15829967]

[83] Sakai W, Swisher EM, Karlan BY, *et al.* Secondary mutations as a mechanism of cisplatin resistance in BRCA2-mutated cancers. Nature 2008; 451(7182): 1116-20. [http://dx.doi.org/10.1038/nature06633] [PMID: 18264087]

[84] Rau KM, Kang HY, Cha TL, Miller SA, Hung MC. The mechanisms and managements of hormone-therapy resistance in breast and prostate cancers. Endocr Relat Cancer 2005; 12(3): 511-32. [http://dx.doi.org/10.1677/erc.1.01026] [PMID: 16172190]

[85] Wajant H. The Fas signaling pathway: more than a paradigm. Science 2002; 296(5573): 1635-6. [http://dx.doi.org/10.1126/science.1071553] [PMID: 12040174]

[86] Zahreddine H, Borden KL. Mechanisms and insights into drug resistance in cancer. Front Pharmacol 2013; 4: 28. [http://dx.doi.org/10.3389/fphar.2013.00028] [PMID: 23504227]

[87] Dupont C, Armant DR, Brenner CA. Epigenetics: definition, mechanisms and clinical perspective. Seminars in reproductive medicine. Thieme Medical Publishers. 2009 Sep; 27(5):351-357. [http://dx.doi.org/10.1055/s-0029-1237423]

[88] Zahnow CA, Topper M, Stone M, *et al.* Inhibitors of DNA Methylation, Histone Deacetylation, and Histone Demethylation: A Perfect Combination for Cancer Therapy. Adv Cancer Res 2016; 130: 55-111. [http://dx.doi.org/10.1016/bs.acr.2016.01.007] [PMID: 27037751]

[89] Das PM, Singal R. DNA methylation and cancer. J Clin Oncol 2004; 22(22): 4632-42. [http://dx.doi.org/10.1200/JCO.2004.07.151] [PMID: 15542813]

[90] Wilting RH, Dannenberg JH. Epigenetic mechanisms in tumorigenesis, tumor cell heterogeneity and drug resistance. Drug Resist Updat 2012; 15(1-2): 21-38. [http://dx.doi.org/10.1016/j.drup.2012.01.008] [PMID: 22356866]

[91] Lim PK, Bliss SA, Patel SA, *et al.* Gap junction-mediated import of microRNA from bone marrow stromal cells can elicit cell cycle quiescence in breast cancer cells. Cancer Res 2011; 71(5): 1550-60. [http://dx.doi.org/10.1158/0008-5472.CAN-10-2372] [PMID: 21343399]

[92] Donzelli S, Mori F, Biagioni F, *et al.* MicroRNAs: short non-coding players in cancer chemoresistance. Mol Cell Ther 2014; 2(1): 16. [http://dx.doi.org/10.1186/2052-8426-2-16] [PMID: 26056584]

[93] Akao Y, Nakagawa Y, Kitade Y, Kinoshita T, Naoe T. Downregulation of microRNAs-143 and -145 in B-cell malignancies. Cancer Sci 2007; 98(12): 1914-20. [http://dx.doi.org/10.1111/j.1349-7006.2007.00618.x] [PMID: 17892514]

[94] Zheng T, Wang J, Chen X, Liu L. Role of microRNA in anticancer drug resistance. Int J Cancer 2010; 126(1): 2-10. [http://dx.doi.org/10.1002/ijc.24782] [PMID: 19634138]

[95] Magee P, Shi L, Garofalo M. Role of microRNAs in chemoresistance. Ann Transl Med 2015; 3(21): 332. [PMID: 26734642]

[96] Liang Z, Li Y, Huang K, Wagar N, Shim H. Regulation of miR-19 to breast cancer chemoresistance through targeting PTEN. Pharm Res 2011; 28(12): 3091-100. [http://dx.doi.org/10.1007/s11095-011-0570-y] [PMID: 21853360]

[97] Xu CX, Xu M, Tan L, *et al.* MicroRNA miR-214 regulates ovarian cancer cell stemness by targeting p53/Nanog. J Biol Chem 2012; 287(42): 34970-8. [http://dx.doi.org/10.1074/jbc.M112.374611] [PMID: 22927443]

[98] Qu J, Zhao L, Zhang P, *et al.* MicroRNA-195 chemosensitizes colon cancer cells to the chemotherapeutic drug doxorubicin by targeting the first binding site of BCL2L2 mRNA. J Cell Physiol 2015; 230(3): 535-45. [http://dx.doi.org/10.1002/jcp.24366] [PMID: 23526568]

[99] Dong WH, Li Q, Zhang XY, Guo Q, Li H, Wang TY. Deep sequencing identifies deregulation of microRNAs involved with vincristine drug-resistance of colon cancer cells. Int J Clin Exp Pathol 2015; 8(9): 11524-30. [PMID: 26617885]

[100] Zhao L, Zou D, Wei X, *et al.* MiRNA-221-3p desensitizes pancreatic cancer cells to 5-fluorouracil by targeting RB1. Tumour Biol 2016. [http://dx.doi.org/10.1007/s13277-016-5445-8] [PMID: 27726102]

[101] Hawa Z, Haque I, Ghosh A, Banerjee S, Harris L, Banerjee SK. The miRacle in pancreatic cancer by miRNAs: tiny angels or devils in disease progression. International Journal of Molecular Sciences. 2016 May 26;17(6):809, 1-22. [http://dx.doi.org/10.3390/ijms17060809]

[102] Wang W, Zhao L, Wei X, *et al.* MicroRNA-320a promotes 5-FU resistance in human pancreatic cancer cells. Sci Rep 2016; 6(27641): 27641. [http://dx.doi.org/10.1038/srep27641] [PMID: 27279541]

[103] Li Y, Sarkar FH. MicroRNA targeted therapeutic approach for pancreatic cancer. Int J Biol Sci 2016; 12(3): 326-37. [http://dx.doi.org/10.7150/ijbs.15017] [PMID: 26929739]

[104] Miller TE, Ghoshal K, Ramaswamy B, *et al.* MicroRNA-221/222 confers tamoxifen resistance in breast cancer by targeting p27Kip1. J Biol Chem 2008; 283(44): 29897-903. [http://dx.doi.org/10.1074/jbc.M804612200] [PMID: 18708351]

[105] Kovalchuk O, Filkowski J, Meservy J, *et al.* Involvement of microRNA-451 in resistance of the MCF-7 breast cancer cells to chemotherapeutic drug doxorubicin. Mol Cancer Ther 2008; 7(7): 2152-9. [http://dx.doi.org/10.1158/1535-7163.MCT-08-0021] [PMID: 18645025]

[106] Zhang X, Huang L, Zhao Y, Tan W. Downregulation of miR-130a contributes to cisplatin resistance in ovarian cancer cells by targeting X-linked inhibitor of apoptosis (XIAP) directly. Acta Biochim Biophys Sin (Shanghai) 2013; 45(12): 995-1001.
[http://dx.doi.org/10.1093/abbs/gmt113] [PMID: 24145606]

[107] Liu MX, Siu MK, Liu SS, Yam JW, Ngan HY, Chan DW. Epigenetic silencing of microRNA-199b-5p is associated with acquired chemoresistance *via* activation of JAG1-Notch1 signaling in ovarian cancer. Oncotarget 2014; 5(4): 944-58.
[http://dx.doi.org/10.18632/oncotarget.1458] [PMID: 24659709]

[108] Favreau AJ, McGlauflin RE, Duarte CW, Sathyanarayana P. miR-199b, a novel tumor suppressor miRNA in acute myeloid leukemia with prognostic implications. Exp Hematol Oncol 2016; 5(1): 4.
[http://dx.doi.org/10.1186/s40164-016-0033-6] [PMID: 26848406]

[109] Garofalo M, Quintavalle C, Di Leva G, *et al.* MicroRNA signatures of TRAIL resistance in human non-small cell lung cancer. Oncogene 2008; 27(27): 3845-55.
[http://dx.doi.org/10.1038/onc.2008.6] [PMID: 18246122]

[110] Fujita Y, Kojima K, Hamada N, *et al.* Effects of miR-34a on cell growth and chemoresistance in prostate cancer PC3 cells. Biochem Biophys Res Commun 2008; 377(1): 114-9.
[http://dx.doi.org/10.1016/j.bbrc.2008.09.086] [PMID: 18834855]

[111] Dehghanzadeh R, Jadidi-Niaragh F, Gharibi T, Yousefi M. MicroRNA-induced drug resistance in gastric cancer. Biomed Pharmacother 2015; 74: 191-9.
[http://dx.doi.org/10.1016/j.biopha.2015.08.009] [PMID: 26349984]

[112] Li H, Yang BB. Friend or foe: the role of microRNA in chemotherapy resistance. Acta Pharmacol Sin 2013; 34(7): 870-9.
[http://dx.doi.org/10.1038/aps.2013.35] [PMID: 23624759]

[113] Hanahan D, Coussens LM. Accessories to the crime: functions of cells recruited to the tumor microenvironment. Cancer Cell 2012; 21(3): 309-22.
[http://dx.doi.org/10.1016/j.ccr.2012.02.022] [PMID: 22439926]

[114] Senthebane DA, Rowe A, Thomford NE, *et al.* The Role of Tumor Microenvironment in Chemoresistance: To Survive, Keep Your Enemies Closer. Int J Mol Sci 2017; 18(7): 1586.
[http://dx.doi.org/10.3390/ijms18071586] [PMID: 28754000]

[115] Sun Y. Tumor microenvironment and cancer therapy resistance. Cancer Lett 2016; 380(1): 205-15.
[http://dx.doi.org/10.1016/j.canlet.2015.07.044] [PMID: 26272180]

[116] Lefebvre KJ, Assadian S, El-Assaad W, Teodoro JG. Regulation of Angiogenesis by Tumour Suppressor Pathways.Experimental and Clinical Metastasis. Springer New York 2013; pp. 79-99.
[http://dx.doi.org/10.1007/978-1-4614-3685-0_8]

[117] Carmeliet P, Jain RK. Molecular mechanisms and clinical applications of angiogenesis. Nature 2011; 473(7347): 298-307.
[http://dx.doi.org/10.1038/nature10144] [PMID: 21593862]

[118] Castells M, Thibault B, Delord JP, Couderc B. Implication of tumor microenvironment in chemoresistance: tumor-associated stromal cells protect tumor cells from cell death. Int J Mol Sci 2012; 13(8): 9545-71.
[http://dx.doi.org/10.3390/ijms13089545] [PMID: 22949815]

[119] Box C, Rogers SJ, Mendiola M, Eccles SA. Tumor micro-environmental interactions: paths to progression and targets for treatment. Seminars in cancer biology. Academic Press. 2010 Jun 30; 20(3): 128-138.

[120] Durand RE, Sutherland RM. Effects of intercellular contact on repair of radiation damage. Exp Cell Res 1972; 71(1): 75-80.
[http://dx.doi.org/10.1016/0014-4827(72)90265-0] [PMID: 5063574]

[121] Morin PJ. Drug resistance and the microenvironment: nature and nurture. Drug Resist Updat 2003; 6(4): 169-72.
[http://dx.doi.org/10.1016/S1368-7646(03)00059-1] [PMID: 12962682]

[122] Lu JJ, Bao JL, Chen XP, Huang M, Wang YT. Alkaloids isolated from natural herbs as the anticancer agents. Evidence-Based Complementary and Alternative Medicine. Article ID 2012; 485042: 1-12.

[123] Pan ST, Li ZL, He ZX, Qiu JX, Zhou SF. Molecular mechanisms for tumour resistance to chemotherapy. Clin Exp Pharmacol Physiol 2016; 43(8): 723-37.
[http://dx.doi.org/10.1111/1440-1681.12581] [PMID: 27097837]

[124] Galluzzi L, Senovilla L, Vitale I, *et al.* Molecular mechanisms of cisplatin resistance. Oncogene 2012; 31(15): 1869-83.
[http://dx.doi.org/10.1038/onc.2011.384] [PMID: 21892204]

[125] Brabec V, Nováková O. DNA binding mode of ruthenium complexes and relationship to tumor cell toxicity. Drug Resist Updat 2006; 9(3): 111-22.
[http://dx.doi.org/10.1016/j.drup.2006.05.002] [PMID: 16790363]

[126] Dasari S, Tchounwou PB. Cisplatin in cancer therapy: molecular mechanisms of action. Eur J Pharmacol 2014; 740: 364-78.
[http://dx.doi.org/10.1016/j.ejphar.2014.07.025] [PMID: 25058905]

[127] Goodsell DS. The molecular perspective: cisplatin. The Oncologist. 2006 Mar 1; 11(3): 316-31Pan ST, Li ZL, He ZX, Qiu JX, Zhou SF, Molecular mechanisms for tumour resistance to chemotherapy. Clin Exp Pharmacol Physiol 2016; 43(8): 723-37.
[PMID: 27097837]

[128] Sawant A, Kothandapani A, Zhitkovich A, Sobol RW, Patrick SM. Role of mismatch repair proteins in the processing of cisplatin interstrand cross-links. DNA Repair (Amst) 2015; 35: 126-36.
[http://dx.doi.org/10.1016/j.dnarep.2015.10.003] [PMID: 26519826]

[129] Rebucci M, Michiels C. Molecular aspects of cancer cell resistance to chemotherapy. Biochem Pharmacol 2013; 85(9): 1219-26.
[http://dx.doi.org/10.1016/j.bcp.2013.02.017] [PMID: 23435357]

[130] Sirajuddin M, Ali S, Badshah A. Drug-DNA interactions and their study by UV-Visible, fluorescence spectroscopies and cyclic voltametry. J Photochem Photobiol B 2013; 124: 1-19.
[http://dx.doi.org/10.1016/j.jphotobiol.2013.03.013] [PMID: 23648795]

[131] Giddens AC, Lee HH, Lu GL, *et al.* Analogues of DNA minor groove cross-linking agents incorporating aminoCBI, an amino derivative of the duocarmycins: Synthesis, cytotoxicity, and potential as payloads for antibody-drug conjugates. Bioorg Med Chem 2016; 24(22): 6075-81.
[http://dx.doi.org/10.1016/j.bmc.2016.09.068] [PMID: 27745990]

[132] Tiwari AK, Zhang R, Gallo JM. Overlapping functions of ABC transporters in topotecan disposition as determined in gene knockout mouse models. Mol Cancer Ther 2013; 12(7): 1343-55.
[http://dx.doi.org/10.1158/1535-7163.MCT-13-0100] [PMID: 23635651]

[133] Xu Y, Her C. Inhibition of topoisomerase (DNA) I (TOP1): DNA damage repair and anticancer therapy. Biomolecules 2015; 5(3): 1652-70.
[http://dx.doi.org/10.3390/biom5031652] [PMID: 26287259]

[134] Kathawala RJ, Gupta P, Ashby CR Jr, Chen ZS. The modulation of ABC transporter-mediated multidrug resistance in cancer: a review of the past decade. Drug Resist Updat 2015; 18: 1-17.
[http://dx.doi.org/10.1016/j.drup.2014.11.002] [PMID: 25554624]

[135] Pilati P, Nitti D, Mocellin S. Cancer resistance to type II topoisomerase inhibitors. Curr Med Chem 2012; 19(23): 3900-6.
[http://dx.doi.org/10.2174/092986712802002473] [PMID: 22788766]

[136] Ganguly A, Cabral F. New insights into mechanisms of resistance to microtubule inhibitors. Biochim

Biophys Acta 2011; 1816(2): 164-71.
[PMID: 21741453]

[137] Hu WY, Zhong WR, Wang FH, Li L, Shao YZ. In silico synergism and antagonism of an anti-tumour system intervened by coupling immunotherapy and chemotherapy: a mathematical modelling approach. Bull Math Biol 2012; 74(2): 434-52.
[http://dx.doi.org/10.1007/s11538-011-9693-x] [PMID: 21972030]

[138] Yardley DA. Drug resistance and the role of combination chemotherapy in improving patient outcomes. Int J Breast Cancer 2013; 2013: 137414.
[http://dx.doi.org/10.1155/2013/137414] [PMID: 23864953]

[139] Ross JS, Slodkowska EA, Symmans WF, Pusztai L, Ravdin PM, Hortobagyi GN. The HER-2 receptor and breast cancer: ten years of targeted anti-HER-2 therapy and personalized medicine. Oncologist 2009; 14(4): 320-68.
[http://dx.doi.org/10.1634/theoncologist.2008-0230] [PMID: 19346299]

[140] Marty M, Cognetti F, Maraninchi D, *et al.* Randomized phase II trial of the efficacy and safety of trastuzumab combined with docetaxel in patients with human epidermal growth factor receptor 2-positive metastatic breast cancer administered as first-line treatment: the M77001 study group. J Clin Oncol 2005; 23(19): 4265-74.
[http://dx.doi.org/10.1200/JCO.2005.04.173] [PMID: 15911866]

[141] Junttila TT, Akita RW, Parsons K, *et al.* Ligand-independent HER2/HER3/PI3K complex is disrupted by trastuzumab and is effectively inhibited by the PI3K inhibitor GDC-0941. Cancer Cell 2009; 15(5): 429-40.
[http://dx.doi.org/10.1016/j.ccr.2009.03.020] [PMID: 19411071]

[142] Molina MA, Codony-Servat J, Albanell J, Rojo F, Arribas J, Baselga J. Trastuzumab (herceptin), a humanized anti-Her2 receptor monoclonal antibody, inhibits basal and activated Her2 ectodomain cleavage in breast cancer cells. Cancer Res 2001; 61(12): 4744-9.
[PMID: 11406546]

[143] Baselga J, Cortés J, Kim SB, *et al.* Pertuzumab plus trastuzumab plus docetaxel for metastatic breast cancer. N Engl J Med 2012; 366(2): 109-19.
[http://dx.doi.org/10.1056/NEJMoa1113216] [PMID: 22149875]

[144] Young SD, Lafrenie RM, Clemons MJ. Phase ii trial of a metronomic schedule of docetaxel and capecitabine with concurrent celecoxib in patients with prior anthracycline exposure for metastatic breast cancer. Curr Oncol 2012; 19(2): e75-83.
[http://dx.doi.org/10.3747/co.19.879] [PMID: 22514500]

[145] Masui K, Gini B, Wykosky J, *et al.* A tale of two approaches: complementary mechanisms of cytotoxic and targeted therapy resistance may inform next-generation cancer treatments. Carcinogenesis 2013; 34(4): 725-38.
[http://dx.doi.org/10.1093/carcin/bgt086] [PMID: 23455378]

[146] Liu Y, Miyoshi H, Nakamura M. Nanomedicine for drug delivery and imaging: a promising avenue for cancer therapy and diagnosis using targeted functional nanoparticles. Int J Cancer 2007; 120(12): 2527-37.
[http://dx.doi.org/10.1002/ijc.22709] [PMID: 17390371]

[147] Kunjachan S, Rychlik B, Storm G, Kiessling F, Lammers T. Multidrug resistance: Physiological principles and nanomedical solutions. Adv Drug Deliv Rev 2013; 65(13-14): 1852-65.
[http://dx.doi.org/10.1016/j.addr.2013.09.018] [PMID: 24120954]

[148] Singh R, Lillard JW Jr. Nanoparticle-based targeted drug delivery. Exp Mol Pathol 2009; 86(3): 215-23.
[http://dx.doi.org/10.1016/j.yexmp.2008.12.004] [PMID: 19186176]

[149] Minko T, Rodriguez-Rodriguez L, Pozharov V. Nanotechnology approaches for personalized treatment of multidrug resistant cancers. Adv Drug Deliv Rev 2013; 65(13-14): 1880-95.

[http://dx.doi.org/10.1016/j.addr.2013.09.017] [PMID: 24120655]

[150] Lu B, Huang X, Mo J, Zhao W. Drug delivery using nanoparticles for cancer stem-like cell targeting. Frontiers in pharmacology. 2016; 7, Article 84, 1-12. [http://dx.doi.org/10.3389/fphar.2016.00084]

[151] Cai X, Luo Y, Zhang W, Du D, Lin Y. pH-Sensitive ZnO quantum dots–doxorubicin nanoparticles for lung cancer targeted drug delivery. ACS Appl Mater Interfaces 2016; 8(34): 22442-50. [http://dx.doi.org/10.1021/acsami.6b04933] [PMID: 27463610]

[152] Miller MA, Gadde S, Pfirschke C, *et al.* Predicting therapeutic nanomedicine efficacy using a companion magnetic resonance imaging nanoparticle. Sci Transl Med 2015; 7(314): 314ra183. [http://dx.doi.org/10.1126/scitranslmed.aac6522] [PMID: 26582898]

[153] Jain KK. Nanotechnology in cancer drug delivery and selective targeting. Clin Lab Med 2012; 32: 15-31. [http://dx.doi.org/10.1016/j.cll.2011.10.001] [PMID: 22340841]

[154] Callaghan R, Luk F, Bebawy M. Inhibition of the multidrug resistance P-glycoprotein: time for a change of strategy? Drug Metab Dispos 2014; 42(4): 623-31. [http://dx.doi.org/10.1124/dmd.113.056176] [PMID: 24492893]

[155] de Figueiredo-Pontes LL, Pintão MC, Oliveira LC, *et al.* Determination of P-glycoprotein, MDR-related protein 1, breast cancer resistance protein, and lung-resistance protein expression in leukemic stem cells of acute myeloid leukemia. Cytometry B Clin Cytom 2008; 74(3): 163-8. [http://dx.doi.org/10.1002/cyto.b.20403] [PMID: 18200595]

[156] Kibria G, Hatakeyama H, Harashima H. Cancer multidrug resistance: mechanisms involved and strategies for circumvention using a drug delivery system. Arch Pharm Res 2014; 37(1): 4-15. [http://dx.doi.org/10.1007/s12272-013-0276-2] [PMID: 24272889]

[157] Choi CH. ABC transporters as multidrug resistance mechanisms and the development of chemosensitizers for their reversal. Cancer Cell Int 2005; 5(1): 30. [http://dx.doi.org/10.1186/1475-2867-5-30] [PMID: 16202168]

[158] Tsuboi K, Bachovchin DA, Speers AE, *et al.* Potent and selective inhibitors of glutathione S-transferase omega 1 that impair cancer drug resistance. J Am Chem Soc 2011; 133(41): 16605-16. [http://dx.doi.org/10.1021/ja2066972] [PMID: 21899313]

[159] Singh S. Cytoprotective and regulatory functions of glutathione S-transferases in cancer cell proliferation and cell death. Cancer Chemother Pharmacol 2015; 75(1): 1-15. [http://dx.doi.org/10.1007/s00280-014-2566-x] [PMID: 25143300]

[160] Perche F, Torchilin VP. Recent trends in multifunctional liposomal nanocarriers for enhanced tumor targeting. J Drug Deliv 2013; 2013: 705265. [http://dx.doi.org/10.1155/2013/705265] [PMID: 23533772]

[161] Madaan K, Kumar S, Poonia N, Lather V, Pandita D. Dendrimers in drug delivery and targeting: Drug-dendrimer interactions and toxicity issues. J Pharm Bioallied Sci 2014; 6(3): 139-50. [http://dx.doi.org/10.4103/0975-7406.130965] [PMID: 25035633]

[162] Tripathy S, Das MK. Dendrimers and their Applications as Novel Drug Delivery Carriers. 2013 Sep; 3(9): 142-149.

[163] Ma P, Mumper RJ. Paclitaxel nano-delivery systems: a comprehensive review. J Nanomed Nanotechnol 2013; 4(2): 1000164. [http://dx.doi.org/10.4172/2157-7439.1000164] [PMID: 24163786]

[164] Lim J, Simanek EE. Triazine dendrimers as drug delivery systems: from synthesis to therapy. Adv Drug Deliv Rev 2012; 64(9): 826-35. [http://dx.doi.org/10.1016/j.addr.2012.03.008] [PMID: 22465784]

[165] Ghosh B, Paul R. Nanoscience and Targeted Systems for Cancer Therapy: A Review. International

Journal of Pharmaceutical & Biomedical Research 2015; 2(4): 1-2.

[166] Dominska M, Dykxhoorn DM. Breaking down the barriers: siRNA delivery and endosome escape. J Cell Sci 2010; 123(Pt 8): 1183-9.
[http://dx.doi.org/10.1242/jcs.066399] [PMID: 20356929]

[167] Kanasty RL, Whitehead KA, Vegas AJ, Anderson DG. Action and reaction: the biological response to siRNA and its delivery vehicles. Mol Ther 2012; 20(3): 513-24.
[http://dx.doi.org/10.1038/mt.2011.294] [PMID: 22252451]

[168] Xu CF, Wang J. Delivery systems for siRNA drug development in cancer therapy. Asian Journal of Pharmaceutical Sciences 2015; 10(1): 1-2.
[http://dx.doi.org/10.1016/j.ajps.2014.08.011]

[169] Navarro G, Sawant RR, Biswas S, Essex S, Tros de Ilarduya C, Torchilin VP. P-glycoprotein silencing with siRNA delivered by DOPE-modified PEI overcomes doxorubicin resistance in breast cancer cells. Nanomedicine (Lond) 2012; 7(1): 65-78.
[http://dx.doi.org/10.2217/nnm.11.93] [PMID: 22191778]

[170] Kapse-Mistry S, Govender T, Srivastava R, Yergeri M. Nanodrug delivery in reversing multidrug resistance in cancer cells. Front Pharmacol 2014; 5(159): 159.
[PMID: 25071577]

[171] Wang Z, Li Y, Ahmad A, *et al.* Pancreatic cancer: understanding and overcoming chemoresistance. Nat Rev Gastroenterol Hepatol 2011; 8(1): 27-33.
[http://dx.doi.org/10.1038/nrgastro.2010.188] [PMID: 21102532]

[172] Li F, Sethi G. Targeting transcription factor NF-κB to overcome chemoresistance and radio-resistance in cancer therapy. Biochimica et Biophysica Acta (BBA)-. Rev Can 2010; 1805(2): 167-80.
[PMID: 20079806]

[173] Arlt A, Müerköster SS, Schäfer H. Targeting apoptosis pathways in pancreatic cancer. Cancer Lett 2013; 332(2): 346-58.
[http://dx.doi.org/10.1016/j.canlet.2010.10.015] [PMID: 21078544]

[174] Zhao X, Fan W, Xu Z, *et al.* Inhibiting tumor necrosis factor-alpha diminishes desmoplasia and inflammation to overcome chemoresistance in pancreatic ductal adenocarcinoma. Oncotarget 2016; 7(49): 81110-22.
[http://dx.doi.org/10.18632/oncotarget.13212] [PMID: 27835602]

[175] Owen HC, Appiah S, Hasan N, Ghali L, Elayat G, Bell C. Phytochemical Modulation of Apoptosis and Autophagy: Strategies to Overcome Chemoresistance in Leukemic Stem Cells in the Bone Marrow Microenvironment. Int Rev Neurobiol 2017; 135: 249-78.
[http://dx.doi.org/10.1016/bs.irn.2017.02.012] [PMID: 28807161]

[176] Xiao D, He J. Epithelial mesenchymal transition and lung cancer. J Thorac Dis 2010; 2(3): 154-9.
[PMID: 22263037]

[177] Tsai JH, Yang J. Epithelial-mesenchymal plasticity in carcinoma metastasis. Genes Dev 2013; 27(20): 2192-206.
[http://dx.doi.org/10.1101/gad.225334.113] [PMID: 24142872]

[178] Chaffer CL, Weinberg RA. A perspective on cancer cell metastasis. Science 2011; 331(6024): 1559-64.
[http://dx.doi.org/10.1126/science.1203543] [PMID: 21436443]

[179] Iwatsuki M, Mimori K, Yokobori T, *et al.* Epithelial-mesenchymal transition in cancer development and its clinical significance. Cancer Sci 2010; 101(2): 293-9.
[http://dx.doi.org/10.1111/j.1349-7006.2009.01419.x] [PMID: 19961486]

[180] Wang A, Chen L, Li C, Zhu Y. Heterogeneity in cancer stem cells. Cancer Lett 2015; 357(1): 63-8.
[http://dx.doi.org/10.1016/j.canlet.2014.11.040] [PMID: 25444897]

[181] Vinogradov S, Wei X. Cancer stem cells and drug resistance: the potential of nanomedicine. Nanomedicine (Lond) 2012; 7(4): 597-615.
[http://dx.doi.org/10.2217/nnm.12.22] [PMID: 22471722]

[182] Dragu DL, Necula LG, Bleotu C, Diaconu CC, Chivu-Economescu M. Therapies targeting cancer stem cells: Current trends and future challenges. World J Stem Cells 2015; 7(9): 1185-201.
[PMID: 26516409]

[183] Rahman M, Hasan MR. Cancer metabolism and drug resistance. Metabolites 2015; 5(4): 571-600.
[http://dx.doi.org/10.3390/metabo5040571] [PMID: 26437434]

[184] Kreeger PK, Lauffenburger DA. Cancer systems biology: a network modeling perspective. Carcinogenesis 2010; 31(1): 2-8.
[http://dx.doi.org/10.1093/carcin/bgp261] [PMID: 19861649]

[185] Fortney K, Jurisica I. Integrative computational biology for cancer research. Hum Genet 2011; 130(4): 465-81.
[http://dx.doi.org/10.1007/s00439-011-0983-z] [PMID: 21691773]

[186] Yang Y, Dong X, Xie B, *et al.* Databases and web tools for cancer genomics study. Genomics Proteomics Bioinformatics 2015; 13(1): 46-50.
[http://dx.doi.org/10.1016/j.gpb.2015.01.005] [PMID: 25707591]

[187] Roychowdhury S, Chinnaiyan AM. Translating cancer genomes and transcriptomes for precision oncology. CA Cancer J Clin 2016; 66(1): 75-88.
[http://dx.doi.org/10.3322/caac.21329] [PMID: 26528881]

[188] Perou CM, Sørlie T, Eisen MB, *et al.* Molecular portraits of human breast tumours. Nature 2000; 406(6797): 747-52.
[http://dx.doi.org/10.1038/35021093] [PMID: 10963602]

[189] Majumder B, Baraneedharan U, Thiyagarajan S, *et al.* Predicting clinical response to anticancer drugs using an ex vivo platform that captures tumour heterogeneity. Nat Commun 2015; 6(6169): 6169.
[http://dx.doi.org/10.1038/ncomms7169] [PMID: 25721094]

[190] Balashova EE, Dashtiev MI, Lokhov PG. Proteomic footprinting of drug-treated cancer cells as a measure of cellular vaccine efficacy for the prevention of cancer recurrence. Mol Cell Proteomics 2012; 11(2): 014480.
[http://dx.doi.org/10.1074/mcp.M111.014480] [PMID: 22074704]

[191] Sabbah M, Emami S, Redeuilh G, *et al.* Molecular signature and therapeutic perspective of the epithelial-to-mesenchymal transitions in epithelial cancers. Drug Resist Updat 2008; 11(4-5): 123-51.
[http://dx.doi.org/10.1016/j.drup.2008.07.001] [PMID: 18718806]

[192] van Waart H, Stuiver MM, van Harten WH, Sonke GS, Aaronson NK. Design of the Physical exercise during Adjuvant Chemotherapy Effectiveness Study (PACES): a randomized controlled trial to evaluate effectiveness and cost-effectiveness of physical exercise in improving physical fitness and reducing fatigue. BMC cancer. 2010 Dec 7; 10(1): 673, 1-10.

[193] Longley DB, Johnston PG. Molecular mechanisms of drug resistance. J Pathol 2005; 205(2): 275-92.
[http://dx.doi.org/10.1002/path.1706] [PMID: 15641020]

[194] Rontogianni S. Proteomics and its application to discover chemoresistant biomarkers in human cancer (Master's thesis).

[195] Verwey M, Joubert AM, Visagie MH, Theron AE. Chemoresistance in breast cancer stem cells. Biomed Res 2016; 27(1): 16-23.

[196] Matboli M, El-Nakeep S, Hossam N, *et al.* Exploring the role of molecular biomarkers as a potential weapon against gastric cancer: A review of the literature. World J Gastroenterol 2016; 22(26): 5896-908.
[http://dx.doi.org/10.3748/wjg.v22.i26.5896] [PMID: 27468184]

[197] Petty RD, Samuel LM, Murray GI, *et al.* APRIL is a novel clinical chemo-resistance biomarker in colorectal adenocarcinoma identified by gene expression profiling. BMC Cancer. 2009 Dec 11;9 (1): 434, 1-11.
[http://dx.doi.org/10.1186/1471-2407-9-434]

[198] Ranade AR, Cherba D, Sridhar S, *et al.* MicroRNA 92a-2*: a biomarker predictive for chemoresistance and prognostic for survival in patients with small cell lung cancer. J Thorac Oncol 2010; 5(8): 1273-8.
[http://dx.doi.org/10.1097/JTO.0b013e3181dea6be] [PMID: 20548249]

[199] Allen KE, Weiss GJ. Resistance may not be futile: microRNA biomarkers for chemoresistance and potential therapeutics. Mol Cancer Ther 2010; 9(12): 3126-36.
[http://dx.doi.org/10.1158/1535-7163.MCT-10-0397] [PMID: 20940321]

[200] Papadopoulos N, Kinzler KW, Vogelstein B. The role of companion diagnostics in the development and use of mutation-targeted cancer therapies. Nat Biotechnol 2006; 24(8): 985-95.
[http://dx.doi.org/10.1038/nbt1234] [PMID: 16900147]

[201] Deng H, Fang Y, Lahiri J, Ling S. Inventors; Corning Incorporated, assignee. Use of label-free biosensors to understand and identify treatment for cancer. United States Patent Application US 13/767, 142. 2013 Feb 14.

[202] Shia J. Immunohistochemistry *versus* microsatellite instability testing for screening colorectal cancer patients at risk for hereditary nonpolyposis colorectal cancer syndrome. Part I. The utility of immunohistochemistry. J Mol Diagn 2008; 10(4): 293-300.
[http://dx.doi.org/10.2353/jmoldx.2008.080031] [PMID: 18556767]

[203] Jancik S, Drabek J, Berkovcova J, *et al.* A comparison of Direct sequencing, Pyro-sequencing, High resolution melting analysis, Thera Screen DxS, and the K-ras Strip Assay for detecting KRAS mutations in non small cell lung carcinomas. Journal of Experimental & Clinical Cancer Research. 2012 Sep 20; 31(1): 79, 1-13.

[204] Loree JM, Kopetz S, Raghav KP. Current companion diagnostics in advanced colorectal cancer; getting a bigger and better piece of the pie. J Gastrointest Oncol 2017; 8(1): 199-212.
[http://dx.doi.org/10.21037/jgo.2017.01.01] [PMID: 28280626]

[205] Zheng Y, Joyce BT, Colicino E, *et al.* Blood epigenetic age may predict cancer incidence and mortality. EBioMedicine 2016; 5: 68-73.
[http://dx.doi.org/10.1016/j.ebiom.2016.02.008] [PMID: 27077113]

CHAPTER 3

Perspective on Trends in Drug Discovery: Deciphering GPCRs through Integration of Systems and Synthetic Biology

Aman Chandra Kaushik[1,2] and **Shakti Sahi**[1,*]

[1] *School of Biotechnology, Gautam Buddha University, Greater Noida, India*

[2] *The Shraga Segal Dept. of Microbiology, Immunology and Genetics, Faculty of Health Sciences, Ben-Gurion University of the Negev, Beer-Sheva, Israel*

Abstract: Human genome sequencing project has tremendously contributed towards the identification of GPCRs and orphan GPCRs. Currently, more than 800 GPCRs and orphan GPCRs are known. The exploration of the mechanism of the signalling pathways is imperative to understand various roles of these GPCRs, particularly orphan GPCRs. The characterization of physiological functions of orphan GPCRs through development of knockout mice has been a successful method. However, designing and development of novel molecules as agonist or antagonist, having higher efficacy, potency and better safety profile, requires consideration of a more dynamic, multipath way system that a drug can take to exert an effect. Most recently, synthetic circuits have become an important feature of electronic service delivery within the modern life science field, demanding close attention to their functionality, sustainability and usability. Some researchers have begun to examine the functionality of synthetic circuits but there is no relevant work done on software designing related to this specific field which will help to build a comprehensive model for regulatory system. The major objective of system biology is to identify all molecules and their interactions within a living cell. The knowledge about these molecules and the intermolecular interactions is critical to determine the function of human physiological system. The de novo engineering of genetic circuits, biological modules and synthetic pathways is a step towards understanding the biological machinery. The construction of genetic circuits involves proteins as input nodes (receptors), bowires (repressors and activators), or output nodes (enzymes and signaling) to the system. The intrinsic complexity of the proteins makes it difficult to correlate their sequence, structure, functions and interactions. Switching biology deals with redesigning natural proteins or designing novel proteins. An insight into the role of GPCR, its multiple interactions with other proteins, gene and other biomolecules is essential in order to design and develop better therapeutic intervention. An integrated approach combining structure based drug design, systems biology and biological circuits can help in gaining synergy in drug design. The chapter highlights the role of systems biology and biological circuits in drug discovery.

[*] **Corresponding author Shakti Sahi:** School of Biotechnology, Gautam Buddha University, Greater Noida, India; Tel: +91-9971791897; E-mail: shaktis@gbu.ac.in

Atta-ur-Rahman & M. Iqbal Choudhary (Eds.)

Keywords: 7TM, Biological Circuit, Bowires, De-novo, Drug Discovery, Genetic Circuit, GPCR's, Switching Biology, Synthetic Pathways, Synthetic Biology, Systems Biology.

INTRODUCTION

Current Pharmacological and Non-Pharmacological Therapies for Diseases

More than 140 GPCRs, referred to as 'orphan GPCRs' for which endogenous ligands are unknown have been identified. De-oprhanization is a major task for such receptors. The three dimensional structures of native and complex state of GPCRs provide insights into their mechanism and aid in the development of new therapeutic interventions [1, 2]. In biomedical field, computer-aided drug design (CADD) is widely used to facilitate the process of hit identification, lead selection and determination of pharmacokinetic properties (ADMET) with an objective to have a lead/drug with better efficacy, potency and least side effects [3 - 6].

Pharmaceutical industries as well as research groups are involved in the development of computational tools to reduce the time of drug discovery process involved in identification of lead compounds with high efficacy, high potency, and minimal side effects and also to decrease the use of animals. A critical step in drug discovery is the identification of a good lead.

Pharmacophore Hypothesis

The structure based pharmacophore method comprises of the analysis of the 3D structure of the active site residues and their spatial relationships. Different software packages have been used to generate the structure based pharmacophore models like, FLAP (Fingerprints for Ligands and Proteins), GBPM (GRID-based pharmacophore model), HS-Pharm (Hot-spots-guided receptor-based pharmacophores) and LigandScout (3-D pharmacophores derived from protein-bound ligands). The protein-ligand interaction pattern can lead to a pharmacophore model and finally the protein-ligand complex structure could be refined with the help of molecular dynamics (MD) simulation [7].

3D Quantitative Structure Activity Relationships (3D QSAR)

This is analogue based method which deals with designing of new compounds on the basis of compounds/drugs with known biological effect; it analyzes the co-relation between biological activity and molecular properties of compounds. Some important features such as cross validated correlation coefficient (q2), non cross validated correlation coefficient (r20), standard deviation of the error of prediction (SDEP) and standard deviation on SDEP (SD on SDEP) are calculated to get accurate result for 3D-QSAR model [8].

Molecular Dynamics (MD) Simulation Studies

Molecular Dynamics deals with the dynamics of the native as well as protein-ligand binding model system in terms of dynamic properties of atoms, molecules, solvent effects and free energy. Molecular dynamics has found its application most prominently in drug discovery field [9]. It provides an insight in different systems like peptides, proteins, nucleic acids, solvents, gas phases, drug like molecules and membranes. Membrane proteins consist of 30% of all proteins and 70% of drug targets but crystal structure of just 2% of it is reported in the PDB. Therefore, molecular dynamics play an important role in the field of membrane proteins in case of lack of high resolution structural information. Various accurate force field parameters and proper choice of the simulation condition are required to obtain a high quality result during the molecular dynamics [10, 11]. MD simulations have been helpful in exploring the effect of cholesterol on the receptor [12]. Recently, GPCRs structure availability along with the advances in using multiple templates for modeling and MD simulations reveal structural information. MD simulations have appeared as an important method to analyze structural properties and function of GPCRs.

G protein-coupled Receptors

In various organisms, GPCRs are known cell surface receptors playing vital role during signal transduction and metabolic pathways. These are amongst the largest receptor family which transmit signals from extracellular space to intracellular environment. All GPCRs share characteristic structural feature in terms of possessing seven transmembrane helices with distinct length of intra and extra-cellular loops. Certain indels are accompanied by seven helical transmembranes due to irregularities in the form of bulges and constraints in their structure [13]. GPCRs are involved in all regulating activities whether immune related or at physiological level in an organism [14]. Through the human genome sequencing project, it has been reported that around 800 GPCRs exist in human. These GPCRs have effective physiological role in all peripheral organs as well as in central nervous system [15]. These also act as targets for multiple diseases. Ligands act as an extracellular stimulus for GPCRs to cause physiological changes in organisms. Various ligands (outside signals) such as small interacting molecules, peptides (opioids, angiotensins, bradykinins, somatostatin, melano-cortin), fatty acids, nitrogenous substances, photons, proteins, amino acids, carbohydrates and ions stimulate GPCRs [16]. Due to the presence of different GPCRs it is not easy to identify the corresponding natural ligand of it [17]. About

40-50% of GPRCRs are probable drug targets [18, 19]. Currently 30% of recently developed drugs in the market target GPCRs. Melanocortin (MC) receptors are members of the rhodopsin family of seven transmembrane GPCRs, and categorized into five known categories of the melanocortin receptor; MC1R responsible for pigmentation genetics, MC2R known as ACTH receptor or corticotropin receptor, MC3R, MC4R associated with autosomal dominant obesity and MC5R [20]. The lysophospholipid receptor (LPA) are members of the GPCRs family of integral membrane proteins which are important for lipid signaling. P2Y receptors are member of purinergic GPCRs, which stimulated by nucleotides (ATP, ADP, UTP, UDP and UDP-glucose). Homology and studies of signaling transduction pathways suggest that P2Y5 is a member of a subgroup of LPA receptors, which also includes LPA4 and LPA5 [21]. The Muscarinic cholinergic receptors (CHRM3) are member of GPCRs; functional diversity of CHRM3 is binding of acetylcholine, cellular responses like adenylate cyclase inhibition, phosphoinositide degeneration, and potassium channel mediation [22].

Structure and function of the G protein-coupled Receptors

The super family of GPCRs share a common structural feature, that is, the presence of seven transmembrane helices sharing three intracellular and three extracellular loops with an intracellular carboxyl terminal and an extracellular amino terminal. Various structural rearrangements occur during binding to ligand, coupling with G protein and various protein-protein interactions. Rhodopsin is the first known crystal structure of GPCR. There are different sub families of GPCRs. The first family has large number of members including small peptide hormones, opsins, glycoprotein hormones, thyroid stimulating hormone, luteinizing hormone, odorant receptors, and receptors for chemokines, monokines and purines. Second family does not show any sequence homology with the first family within transmembrane helical regions. This family comprises of peptide hormone receptors like parathyroid hormone and related proteins. Third family includes members like receptors for pheromones, putative taste and calcium related extracellular signal sensors. An important feature of family third is that it forms dimer and this could be valid for other GPCRs too. GPCRs exist in equilibrium state between its active and inactive states. The difference in the state of GPCRs is due to the disposition of transmembrane helices and the G protein coupling at cytoplasmic domain. The activity of GPCR can be enhanced by certain agonists who provide stability to cell surface receptor. Agonists activate GPCR in a different way in both the families, in some first family receptors it directly binds with the transmembrane domain and alternates the helices disposition. When agonist binds to GPCR it alters intracellular secondary messengers and some ionic molecules that affect hormones, muscles and other physiological functions. All the G protein-coupled receptors are guanine nucleotide mediated as alpha unit binds to this nucleotide with higher affinity specificity and intrinsic GTPase activity. In the agonist bound state of GPCR (activated state) the guanine nucleotide binds to the alpha unit of heterotrimeric G protein and yield functional dimer, and then this leads to the activation of alpha subunit and its dissociation from the G protein beta gamma dimer. Alpha subunit possesses intrinsic GTPase activity where beta gamma units are involved specifically in functional dimer unit having no covalent interaction [23]. Alpha helical transmembrane protein of receptor is involved in different biological pathways such as signal transduction, transport of impermeable molecule across membrane and molecules and intercellular communication. Therefore, they act as prime targets of the currently known drugs in the market against several diseases. The detection of signal peptides is mainly done by topology prediction methods. Many other methods have been derived which are being used to predict re-entrant helices, kinks and coils within transmembrane helices, interfacial helices, tilted transmembrane helices other than their standard topological model. GPCRdb contains data related to GPCR crystal structures and the largest collections of receptor mutants, Crystal structure-based sequence alignments take into account helix bulges and constrictions, display statistics of amino acid conservation and have been assigned generic residue numbering for equivalent residues in different receptors, diagrams and web tools for G protein-coupled receptors (GPCRs) [24 - 27]. GPCRs are categorized into five main classes-

Class A GPCRs (rhodopsin-type),

Class B GPCRs (secretin-type),

Class C GPCRs (glutamate-type),

Class F GPCRs (frizzled-type) and

Unclassified GPCRs (orphan GPCRs, 7TM orphan receptors, Taste receptors, Vomeronasal receptors, Adiponectin receptors, Non signaling and 7TM receptors). GPCRs control most of physiological responses to hormones, neurotransmitters and environmental stimulants, and play as a potential therapeutic targets for leading diseases. GPCRs fascinating member in term of membrane protein structure and biology [28].

Mechanism of GPCRs

Low molecular weight ligands bind within the hydrophobic region of GPCRs whereas protein and peptide agonists bind to amino terminal and extracellular hydrophilic loops joined to the transmembrane domains. Second and third cytoplasmic loops and carboxyl terminal are known to be critical domains for the interaction with G protein [29]. Agonist is widely responsible for the activation of G protein activity and for that a ternary model for agonist activation has been developed. For most of the non-GPCR receptor families like tyrosine kinase receptor and growth hormone receptor family possessing single membrane spanning domains, agonist induced receptor dimerization occurs. Dimerization by certain GPCRs is very important for their activity and they also play a specific role for example in subtype specific receptor regulation [30]. In rhodopsin family the conserved DRY motif at the cytoplasmic side of TM3 remains highly conserved. The mutation in the third intracellular loop may be responsible for the increase in agonist independent receptor activity in several receptors. It could be an important mechanism for activation of rhodopsin by light. Within femtoseconds of photon activation cis retinal (inverse agonist) converts into trans retinal (full agonist) [31].

Models of Activation by Diffusible Agonists

The conformational selection model proposed by Koshland and Neet and extended ternary complex model are used to describe GPCRs activation. In the absence of agonist the transition between active state and inactive state occurs [32]. The interaction between the receptor and the agonist stabilizes the transmembrane domains. The inverse agonist can also be stabilized in similar way during the inactive state. Intermediate states are stabilized by the partial agonists by enhancing the spontaneous isomerization or stabilizing conformational states having low G protein binding affinity [33].

Classification of GPCR

In human genome about 800 GPCRs exist belonging to five families based on heptahelical structural frame and partial interaction with ligands- Adhesion, Glutamate, Secreting (GRAFS), Frizzled/taste2, and Rhodopsin [34].

The Adhesion Receptor Family

Adhesion family contains transmembrane spanning regions along with the fusion of single or many functional domains. These domains are fused with few motifs like EGF repeats, much in regions, cysteine rich motifs (few adhesive like motifs at N terminal). The N terminal region is rich in glycosylation sites and proline residue and has about 200-2800 amino acid residues. It possess very long N terminal possessing motifs that are involved in cell adhesion. The members of this family include secretin like receptor and latrotoxin receptor [35, 36].

The Secretin Receptor Family

Secretin family receptors bind to those peptide molecules which possess high similarity of amino acid sequences and are paracrine in nature. In A-F system this group belongs to the clan B and on N terminal contains about length of 60-80 amino acids and conserved cysteine residues bridges that acts as an important factor for the ligand binding. Various receptors have been found on N terminal like VIPR, PACAP that have ligand functional binding site. Few other reported members of this family are CALCR, CRHRs, GCGR, GIPR, GLPRs, GHRHR, PACAP, SCTR and VIPR [37].

The Glutamate Receptor Family

Glutamate receptor family contains eight metabotropic glutamate receptor (GRM), two GABA receptors, calcium sensing receptor (CASR) and five taste receptors (TAS 1). The two GABA receptors mainly are GAB-AbR1 (splice variants A and B) and GAB-AbR2. These receptors act as ion channels and belong to clan C. In the metabotropic glutamate recognition site for ligand is situated at N terminal. Glutamate binds at the N–terminal that forms two different lobes and is distinguished by a cavity where glutamate is close to the ligand around the lobes known as "Venus fly trap". The calcium sensing receptor (CASR) deals with the binding of Ca^{2+} and mediates its signal. It is quiet rich in cysteine residues at N-termini. GABA receptors do not have richness in cysteine residues at their N termini whereas TAS1 has long N terminus and has conserved cysteine residues at this terminal. These TAS1 receptors are found in tongue and mediate taste signals. The above two GABA receptors are present at basal region in the branch whereas CSAR lies with the TAS1 receptors. GRM2 and GRM3 are situated at 3p and 7q chromosomal region. GRM2 and GRM3 possess 67% of sequence similarity whereas GRM7 and GRM8 have 74% sequence identity. These regions are referred to as 1p/3p/7/22q paralog and provide evidence in favor of common ancestry [38].

The Frizzled/Taste2 Receptor Family

This receptor family consists of two different clusters that are frizzled and TAS2 receptors. It is clustered together based on their higher bootstrap value. Some consensus sequences were found to be located at different transmembrane like IFL in TMII, SFLL in TMV and SxKTL in TMVII. The TAS1 receptor of the glutamate receptor family is not similar to the TAS2 receptor. The TAS2 does not have ligand binding domain as it has a very short N terminus end, however, it is clearly known that in the hydrophobicity plot it acquires seven hydrophobic regions. The TAS2 receptor's role and function is not very clear. It is expressed in the palate epithelium and tongue. As per studies, 13 new TAS2 receptors have been identified in human database [39].

Rhodopsin family

It comprises of most of the current GPCRs as drug targets and is the largest amongst the five human GPCRs family with about 3000 members. It is activated by various ligands like biogenic amines, peptides, large proteins, nucleosides and nucleotides, lipids and eicosanoids and photons. As it is a vast family, complexity is there amongst the members of the ligands for these GPCRs. On the basis of phylogenetic classification and ligand based classification a new evolutionary tree of GPCR super family has been obtained. It is known that similar structures of proteins exhibit similar functions only when there is high identity in their sequences. So, as a step further, to identify the members in the rhodopsin family variant approaches have been defined to identify the natural ligands for the unidentified GPCRs. All new members have been identified on the basis of sequence and structural homology to known GPCRs and their ligands [40]. In the human rhodopsin family 900 olfactory receptors are found but two third of the genes for olfactory receptors are pseudo genes. Few receptors in this subgroup are ligand specific; however, the majority of olfactory receptors are ligand unspecific [41]. Various studies have been done to reanalyze the class A rhodopsin super family regarding phylogenetic relationship to characterize the ligands of this family. An evolutionary active set of GPCRs includes about 900 GPCRs out of which 534 are olfactory receptors [42, 43]. The class A rhodopsin receptor includes 36 different clusters and their subtypes.

Methodology

G protein-coupled receptors (GPCRs) are one of the largest and therapeutically important protein families. GPCRs are involved in many physiological processes and are effective drug targets for many diseases. In biomedical field computer-aided drug design (CADD) is widely used to facilitate the process of hit identification. Pharmacophore modeling is a good technique to identify leads belonging to chemically different structural classes. Various features which are involved in developing a pharmacophore model include H-bond donors, H-bond acceptors, positive, negative ionizing groups as well as aromatic rings and lipophilic regions. Biological network representations have been used to explain interaction as well as mechanism between entities. Biochemical mathematical models of the network help to investigate the pharmacokinetics mechanism for specific systems. Molecular dynamics (MD) simulation is a widely used technique which resolves Newton's equation of motion for a given system of atoms. It is a computational technique that provides information about the protein-ligand system as a function of time. Table **1** depicts the flow chart methodology to identify new or potential lead compounds for GPCRs.

Table 1. The workflow of modelling, Virtual screening, Pharmacophore generation, for potential lead compounds.

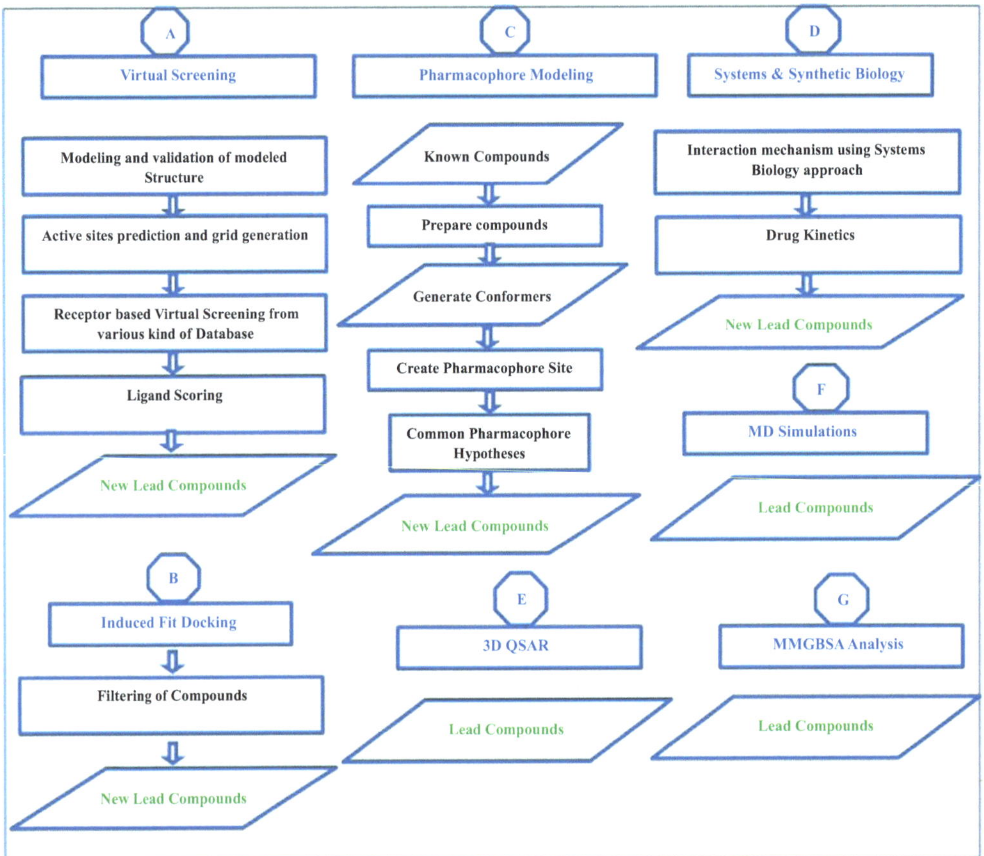

New trends of Drug Discovery of Orphan GPCRs through integration of Systems and Switching Biology

The human genome sequencing has led to the identification of about 140 orphan G protein-coupled receptors (O-GPCR). These orphan GPCRs do not have a known endogenous ligand and are of utmost significance as potential drug targets. The lack of identified ligands and the unique structural properties of the GPCRs are the major reasons for slow progress on orphan GPCRs research.

Systems Biology as a new trends for Drug Discovery of G protein-coupled Receptor

Biological networks and biochemical pathways are generally constructed using known interactions which are mined from literature and experimental data. The networks and pathways are based on identified interactions under many conditions and properties of entities. Statistical approaches help to investigate profile of genes, proteins and other network interactions. Bayesian network is based on probability for directed acyclic graph (DAG) while Gaussian approach is based on undirected graph structure (UGS) [44]. It is well known that the biological and chemical space comprises of a complex network of interacting molecules. These interactions can be dynamic, activate or suppress some molecular activity such as many proteins interact with specific as well as nonspecific ligands. Their interactions can be one to one or one to multiple. The biological network has wide application during drug designing to understand the system as a whole and finding therapeutic interventions which are more effective. The widely known example of cell surface receptor protein is GPCRs which are well known for their interactions with the large number of ligands possibly possessing many interactions at cellular as well as physiological level. These complex networks are dynamic, so it is required to design all interactions in a well-mannered way such that that a lead can be designed with better activity profile and least side effects. For the system level studies there is the requirement of description of the biological components, their relationship, dynamic behavior and interaction with other molecules [45]. Systems biology studies deal with the three areas: graph inference, graph analysis and dynamic network modeling. Various entities have been assigned for the graphical inference like graph nodes, genes, mRNA, proteins. The molecules (vertices) and edges can be directed as well as non directed (links). The edges can be positive for activation and negative for inhibition. In graph inference the notation of various entities are assigned, then in graph analysis the whole graph is studied with all the prediction results and the dynamic modeling represents all transitions occurring in the system's behavior [46]. The network formed between the various interacting molecules includes the relationship between the input and output state of a network [47]. The studies on the various models have been done to deal with different aspects of the system focusing on the ligand binding, receptor interaction and their activity at certain nodes. Through the most recent studies the signaling pathway of GPR142 is examined at network level involving systems biology. GPR142 is an important receptor acting as a drug target for Type 2 diabetes and has prominent role in pancreatic biology. It has wide interactions with many molecules such that they result in several cellular reactions, so to study them GPR142 signaling pathway was modeled followed by logical operations. Stimulus response may be analyzed through the kinetics such that the ups and downs of the activation level could be broadly studied by the help of these networking models [48]. Due to the complexity of the system, all unusual reactions, kinetics, allosteric behavior and other factors can directly affect the pharmacokinetics at system's level by influencing the ADME properties of a certain drug during its action. In a complex network target and ligand promiscuity leads to an interaction between the chemical and biological space. This complicated network in signal transduction pathway will regulate all the parameters at system level based on pharmacokinetics, drug kinetics and simulations. Further graph and data analysis can be done on the basis of the initial concentration values and order of the stoichiometric kinetic reactions.

The Role of Molecular and Cellular Biology in Regulatory Pathway Systems of G protein-coupled receptor using Systems Biology approach

Case Studies 1: GPR142 is a protein that in humans is encoded by the *GPR142* gene and is predominantly expressed in pancreatic cells [49 - 52]. GPR142 mRNA showed that it is expressed in the brain as well as in peripheral glands and organs. The high level of similarity between GPR142 and GPR139 results in development of a new sub branch within the phylogenetic tree and hypothesis that they may share a common ligand. GPR142 contains about two exons in mouse and three or four exons in humans. Both the receptors consist of single N-glycosylation site but lack cysteines residue in the extracellular loops EC1 and EC2. It is found that in HEK-293 cells GPR139 and GPR142 are present as monomers whereas in CHO-K1 cells GPR139 exists as a dimer unit. This lack of cysteine residues may destabilize the structure. It is expressed and preferentially acts in the mesodermal region of tissues and cells. So, GPR142 has a role in brain, bone and blood formation. The functions of both the receptors could be clarified only on the basis of their ligand identification. A ligand relaxin-3 belonging to insulin/relaxin super family acts as an endogenous ligand for GPR139; also it has an affinity and functional agonist properties for GPR142. On the basis of autoradiography it is reported that in the binding sites of rat central nervous system GPR142 acts as a pseudo gene, however it is a viable gene in the mouse. The nearest paralogs of this protein are GPR119 and GPR139. GPR142 and GPR119 both are G_q-coupled receptors expressed predominantly in pancreatic β-cells. The activation of both GPR142 and GPR119 receptors leads to signal transduction ultimately enhancing insulin secretion from pancreatic β-cells. The identity between GPR142 and GPR139 is approximately 50% [53].

Case Studies 2: GPR119 belongs to the rhodopsin family of GPCR and is mostly expressed in intestinal cells that lead to the formation and secretion of lowering blood glucose level hormone glucagon-like peptide-1(GLP-1) [54]. The literature survey revealed alternative names of GPR119 as SNORF25, RUP3, GPCR2,

19AJ, OSGPR116 and glucose dependent insulin tropic receptor. Similar forms of GPR119 are known in many mammalian species like chimpanzees, rhesus monkeys, cattle and dogs [55]. GPR119 forms a subgroup within the bioamine receptors. The nearest homolog of GPR119 is from cannabinoid receptors. Its expression in islet of β cells of pancreas plays a key role in modulating secretion of insulin. GPR119 raises the concentration of cAMP in β cells of pancreas that modulates GSIS (glucose stimulated insulin secretion) analogous to incretin hormone GLP-1 (glucagon like peptide-1) and GIP (gastric inhibitory polypeptide) that also acts as G-coupled β cell receptor. Agonist of GPR139 activates GLP-1 release and this makes it a good target for treatment of type 2 diabetes [56].

Disease for Case Studies 1 & 2: Type 2 diabetes is mainly associated with insulin resistance. FFA, TNFα and cellular stress induce insulin resistance by inhibiting functions of IRS. The various molecular mechanisms being stimulated by them include serine/threonine phosphorylation, SOCS interaction, modification of the cellular localization, regulation of expression and their degradation. Less involvement of β cell function leads to the development of type 2 diabetes and multiple defects in β cells can be induced by chronic hyperglycemia [57]. During hyperglycemia, rise in large amount of reactive oxygen species (ROS) in β cells result in ultimate damage to cellular component PDX-1. In regulation of insulin promoter activity, absence of PDX-1 is an important mechanism for β cell dysfunctioning. In insulin secretion and insulin signal transmission the two genes Kir6.2 and IRS play central role. Worldwide around 382 million people are affected by type 2 diabetes mellitus. With the increasing incidence of type 2 diabetes the therapies for treatment are essential [58]. Type 2 diabetes is distinguished by high level of glucose in blood occurring due to reduced insulin production coupled with insulin resistance from pancreatic β-cell [59 - 61]. Hyperglycemia, if uncontrolled can cause increase in the risk of cardiovascular complications in patients with diabetes that includes stroke, nephropathy, coronary heart disease, neuropathy, and retinopathy. So to prevent chronic diabetic complications it is important to have effective glycemic control [62, 63]. Presently, sulfonylureas and meglitinide are some of the insulin secretagogues used for treatment of moderate degree of β cell dysfunction [64]. These cause an insulin release independent of blood sugar levels and can thus cause hypoglycemia. Hence, the novel glucose stimulated insulin secretagogues such as 5 GLP-1 analogues, DPPIV inhibitors, and GPR142 and GPR119 agonists open new and alternative treatment of type 2 diabetes [65]. Fig. (**1**) depicts the role of GPR142 and GPR119 in physiological system.

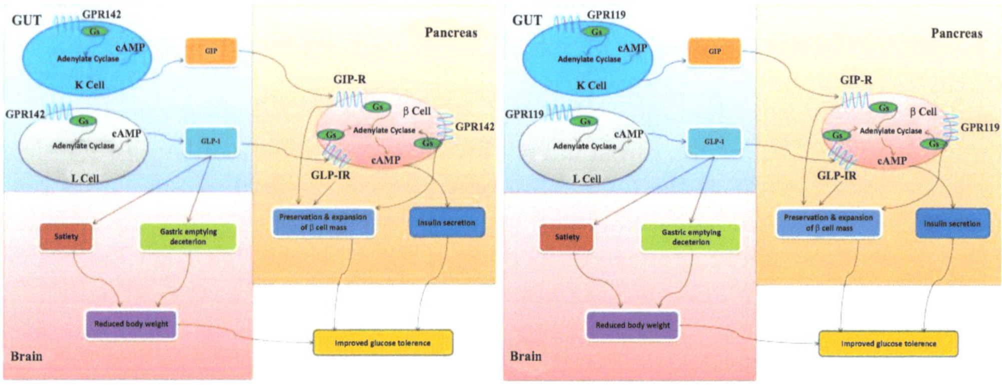

Fig. (1). Left panel represents the role of GPR142 in physiological system using systems biology approach and right panel represents the role of GPR119 in physiological system using systems biology approach.

Synthetic/switching biology as a new trends for Drug Discovery

The synthetic biology approach deals with the gene regulatory networks and demonstrates the basic principle governing the gene expression and networks [66]. Various computational tasks like memory storage and logical operations are being operated by gene circuits and may find applications in biotechnological and biomedical applications where they mimic cell behavior [67]. Non endogenous signals are used to design most of the networks as they respond to these signals and are directly coupled to the regulatory circuitry of the cell [68]. Various engineered gene networks can act as regulatory modules and link to the cell's genetic circuit [69]. They can then transmit specific biological signals [70]. Three distinct modules have been obtained on the basis of their design strategy:

A signaling pathway (biosensor module) is used to detect the appropriate signals and then linked with the regulatory signals.

An artificial genetic module (the regulatory circuit) is based on the engineered properties. It responds to signals transmitted by the biosensor module and directs output signals [71].

The signals are converted into biological response through an output interface. On the basis of the dynamical and logical properties of the regulatory module and the input and output signals the behavior of the programmed cell is determined [72].

Synthetic biology deals with two waves in terms of its understanding and establishment of cellular networks; First wave focuses on development of small modules and genetic devices whereas second wave combines these modules to develop a refined complex system.

First wave of synthetic biology

It involves the combination of basic known elements to form small modules with specific behaviours for promoters, ribosome binding sites and transcriptional repressors. Some recently known modules which are formed include cascades, pulse generators, switches, time delayed circuits, oscillators, logic formulas, partial patterning. The main role of modules is the regulation of expression of gene, protein function, metabolism and communication from one cell to the other.

Second wave of synthetic biology

System level circuitry has been created by the integration of basic parts and modules in the second wave of synthetic biology. Over the past few years various circuits have been published for controlling the various biological processes of the cell. It is possible to create a complex circuit of the concerned cell only if the regulatory gene of the cell is completely known. It focuses on development of new and effective bioengineering design principles. So the complexes synthetic pathways can be formed with the help of second wave which deals with the cellular behaviour by targeting multiple pathways, symptoms, and then further could be beneficial in creating therapeutic and environmental application [73].

Boolean signalling Networks- Switching biology is new stream of digital computing which uses DNA, protein, molecular biology and hardware as a substitute to established silicon based computer technologies. The "molectronics" term introduced by Leonard Adleman of the University of Southern California in 1994 [74] according to Leonard Adleman proposed theory DNA can be use in form of computation. In 1997, Mitsunori Ogihara and Animesh Ray introduced Boolean circuits and their implementation [75]. In 2002 Weizmann Institute of Science in Israel, introduced digital computing machine that composed of enzymes and DNA molecules in silicon microchips [76]. DNA computing approach is form of parallel computing using this implementation produce DNA based integrated circuits which able compute whole square roots.

Application- Switching biology accurate and robust approach to increasing genome sequence data and reduction of DNA synthesis costs. Switching biology has the ability to analyze circuit behavior and implement into developing control systems. It helpful to balance the states for efficient production when signal is "off" or "on"; the circuit becomes responsive in a log-linear fashion.

GPR142 and GPR119 biochemical pathway studies using Switching Biology- Switching Biology has ability to construct Boolean network that can traverse whole regulatory mechanism. According to recent article [77], biological circuit of GPR142 against type 2 diabetes proposed using Switching Biology approach, and analysed the GPR142 role in Type 2 diabetes [78]. The interaction of GPR142 with other interacting entities in form of logic GATE was analyzed [79, 80]. Fig. (**2**) depicts the biological circuit of GPR142 in Type 2 diabetes physiological system.

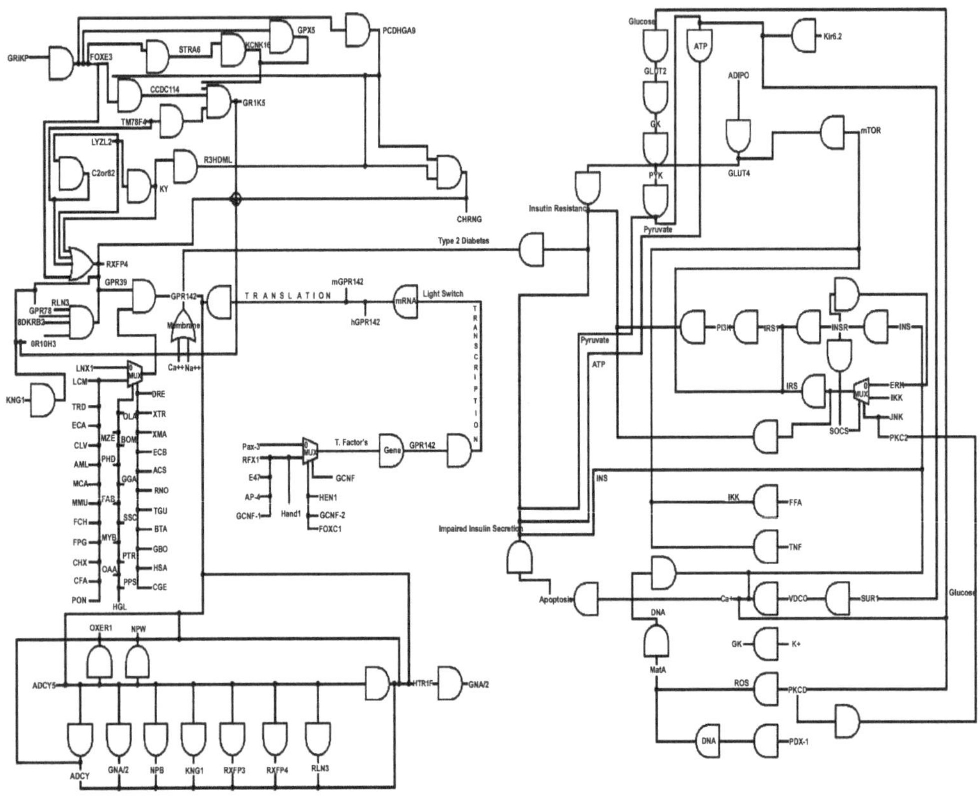

Fig. (2). Switching Biology of whole systems of GPR142 and GPR119 biochemical pathway for type 2 diabetes, and analysis of the effect of compound A on the pathway.

Systems biology and synthetic or switching biology both basically way of signifying biological entities interaction, how systems work and phase to phase involvement with other biological network and their effect in each single stage and investigate their entire biological pathway system and in form of true (1) and false (0). Fig. (**3**) shows the nomenclature and description used to define various interactions in systems biology and switching biology.

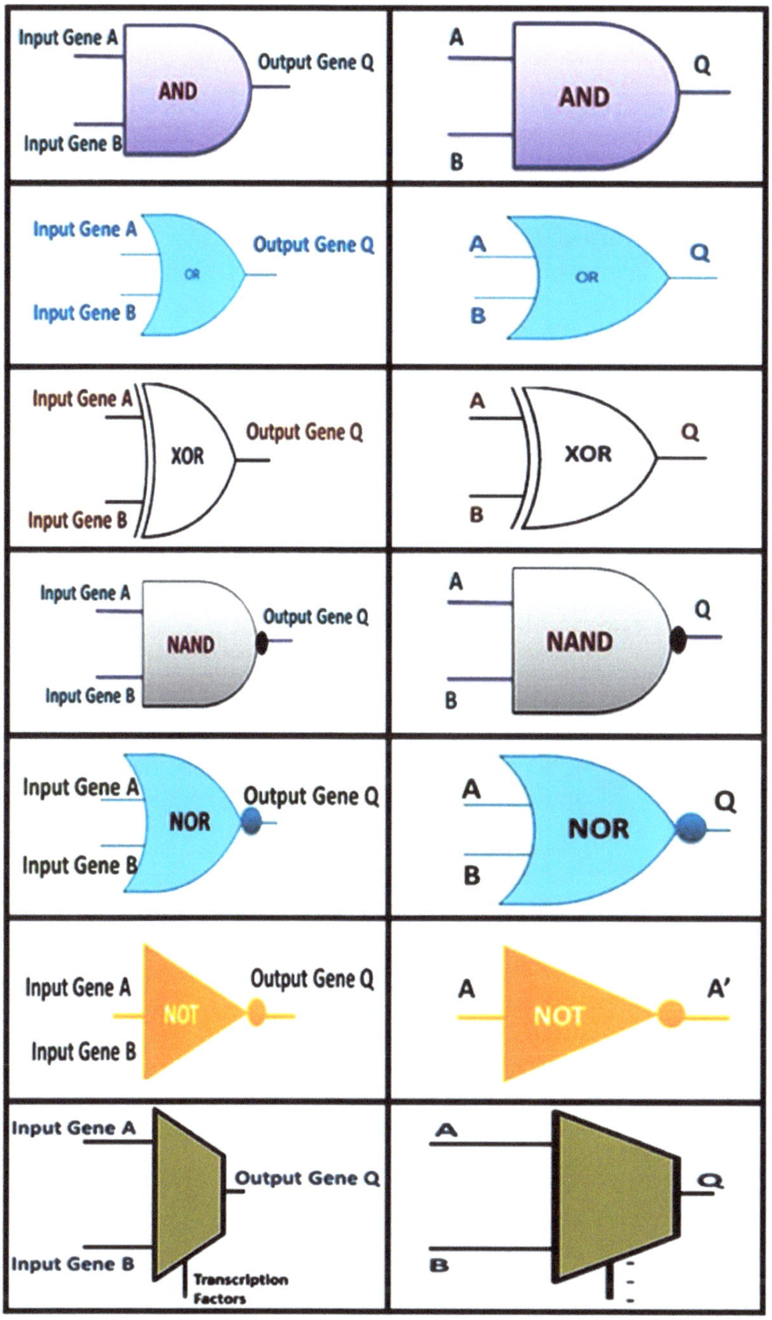

Fig. (3). Comparative analysis between Systems biology and Biological circuits.

CONCLUDING REMARKS

The ability to design drugs effectively depends on the fundamental understanding of the structures of molecules and the biochemical interaction networks. A paradigm shift has occurred in drug discovery with structure-based drug design playing a crucial role. The gap between the number of identified hits and actual drug candidates can be bridged by investigating the underlying interactions. G protein-coupled receptors (GPCRs) are one of the largest known and therapeutically significant protein families in human genome. Although a number of receptor classes exist in humans, GPCRs play a key role in various physiological processes *via* regulation of cellular functions. They are therapeutic targets for multiple ailments as reflected by the remarkable percentage of current drugs in the market and in developmental stages. Their key function is signal transduction which makes them critical in many physiological processes for instance neurotransmission, secretion, cellular metabolism, inflammatory and immune responses. GPCRs characteristic feature includes seven transmembrane (7TM) domains. Such new trends of Drug Discovery through integration of Systems and Switching Biology approach which can be helpful to develop application for predicting and validating the presence or absence of chemicals, genes or proteins for signalling pathway and converting it to an analogue output. This would facilitate in generating a pattern for expression of genes and proteins. Construction of pathways and circuits provide information regarding cell transduction, gene-gene, protein-protein and gene-protein interactions within a cell and their response to outer surface of cell in form of logical circuits.

CONSENT FOR PUBLICATION

Not applicable.

CONFLICT OF INTEREST

The author (editor) declares no conflict of interest, financial or otherwise.

ACKNOWLEDGEMENTS

Declared none.

REFERENCE

[1] Klabunde T, Hessler G. Drug design strategies for targeting G-protein-coupled receptors. ChemBioChem 2002; 3(10): 928-44. [http://dx.doi.org/10.1002/1439-7633(20021004)3:10<928::AID-CBIC928>3.0.CO;2-5] [PMID: 12362358]

[2] Flower DR. Modelling G protein-coupled receptors for drug design. Biochimica et Biophysica Acta (BBA)-. Reviews on Biomembranes 1999; 1422(3): 207-34.

[3] Congreve M, Marshall F. The impact of GPCR structures on pharmacology and structure-based drug design. Br J Pharmacol 2010; 159(5): 986-96. [http://dx.doi.org/10.1111/j.1476-5381.2009.00476.x] [PMID: 19912230]

[4] George SR, O'Dowd BF, Lee SP. G-protein-coupled receptor oligomerization and its potential for drug discovery. Nat Rev Drug Discov 2002; 1(10): 808-20. [http://dx.doi.org/10.1038/nrd913] [PMID: 12360258]

[5] Garland SL, Gloriam DE. A ligand's view of target similarity: chemogenomic binding site-directed techniques for drug discovery. Curr Top Med Chem 2011; 11(15): 1872-81. [http://dx.doi.org/10.2174/156802611796391276] [PMID: 21470171]

[6] Shi F, Shen JK, Chen D, *et al.* Discovery and SAR of a series of agonists at orphan G protein-coupled receptor 139. ACS Med Chem Lett 2011; 2(4): 303-6. [http://dx.doi.org/10.1021/ml100293q] [PMID: 24900311]

[7] Wieder M, Perricone U, Seidel T, Boresch S, Langer T. Comparing pharmacophore models derived from crystal structures and from molecular dynamics simulations. Monatsh Chem 2016; 147(3): 553-63.

[8] Qu Q, Tang X, Kuang B, Li S, Tu G. 3D-QSAR Studies, Molecular Dynamics Simulation and Free Energy Calculation of APN Inhibitors. Int J Pharmacol 2015; 11(8): 920-8. [http://dx.doi.org/10.3923/ijp.2015.920.928]

[9] Cordomí A, Caltabiano G, Pardo L. Membrane protein simulations using AMBER force field and Berger lipid parameters. J Chem Theory Comput 2012; 8(3): 948-58. [http://dx.doi.org/10.1021/ct200491c] [PMID: 26593357]

[10] Tautermann CS, Seeliger D, Kriegl JM. What can we learn from molecular dynamics simulations for GPCR drug design? Comput Struct Biotechnol J 2014; 13: 111-21. [http://dx.doi.org/10.1016/j.csbj.2014.12.002] [PMID: 25709761]

[11] Filizola M, Wang SX, Weinstein H. Dynamic models of G-protein coupled receptor dimers: indications of asymmetry in the rhodopsin dimer from molecular dynamics simulations in a POPC bilayer. J Comput Aided Mol Des 2006; 20(7-8): 405-16. [http://dx.doi.org/10.1007/s10822-006-9053-3] [PMID: 17089205]

[12] Stockert JA, Devi LA. Advancements in therapeutically targeting orphan GPCRs. Front Pharmacol 2015; 6: 100. [http://dx.doi.org/10.3389/fphar.2015.00100] [PMID: 26005419]

[13] Kalyaanamoorthy S, Chen YP. Modelling and enhanced molecular dynamics to steer structure-based drug discovery. Prog Biophys Mol Biol 2014; 114(3): 123-36. [http://dx.doi.org/10.1016/j.pbiomolbio.2013.06.004] [PMID: 23827463]

[14] Fisyuk AS, Kulakov IV, Goncharov DS, Nikitina OS, Bogza YP, Shatsauskas AL. Synthesis of 3-Aminopyridin-2 (1H)-Ones and 1H-Pyrido [2, 3-b][1, 4] Oxazin-2 (3H)-Ones. Chem Heterocycl Compd 2014; 50(2): 217-24. [http://dx.doi.org/10.1007/s10593-014-1464-9]

[15] Rosenbaum DM, Rasmussen SG, Kobilka BK. The structure and function of G-protein-coupled receptors. Nature 2009; 459(7245): 356-63. [http://dx.doi.org/10.1038/nature08144] [PMID: 19458711]

[16] Reaux A, De Mota N, Skultetyova I, *et al.* Physiological role of a novel neuropeptide, apelin, and its receptor in the rat brain. J Neurochem 2001; 77(4): 1085-96. [http://dx.doi.org/10.1046/j.1471-4159.2001.00320.x] [PMID: 11359874]

[17] Ahrén B. Islet G protein-coupled receptors as potential targets for treatment of type 2 diabetes. Nat Rev Drug Discov 2009; 8(5): 369-85. [http://dx.doi.org/10.1038/nrd2782] [PMID: 19365392]

[18] Bharate SB, Nemmani KV, Vishwakarma RA. Progress in the discovery and development of small-molecule modulators of G-protein-coupled receptor 40 (GPR40/FFA1/FFAR1): an emerging target for type 2 diabetes. Expert Opin Ther Pat 2009; 19(2): 237-64. [http://dx.doi.org/10.1517/13543770802665717] [PMID: 19441920]

[19] Swaminath G. Fatty acid binding receptors and their physiological role in type 2 diabetes. Arch Pharm (Weinheim) 2008; 341(12): 753-61. [http://dx.doi.org/10.1002/ardp.200800096] [PMID: 19009545]

[20] Kumar A, Bhandari A, Sinha R, Goyal P, Grapputo A. Spliceosomal intron insertions in genome compacted ray-finned fishes as evident from phylogeny of MC receptors, also supported by a few other GPCRs. PLoS One 2011; 6(8): e22046. [http://dx.doi.org/10.1371/journal.pone.0022046] [PMID: 21850219]

[21] Pasternack SM, von Kügelgen I, Al Aboud K, *et al.* G protein-coupled receptor P2Y5 and its ligand LPA are involved in maintenance of human hair growth. Nat Genet 2008; 40(3): 329-34. [http://dx.doi.org/10.1038/ng.84] [PMID: 18297070]

[22] Mahati E, Li P, Kurata Y, *et al.* M3 muscarinic receptor signaling stabilizes a novel mutant human ether-a-go-go-related gene channel protein *via* phosphorylation of heat shock factor 1 in transfected cells. Circ J 2016; 80(12): 2443-52. [http://dx.doi.org/10.1253/circj.CJ-16-0712] [PMID: 27803431]

[23] Spiegel AM, Weinstein LS. Inherited diseases involving g proteins and g protein-coupled receptors. Annu Rev Med 2004; 55: 27-39. [http://dx.doi.org/10.1146/annurev.med.55.091902.103843] [PMID: 14746508]

[24] Isberg V, Mordalski S, Munk C, *et al.* GPCRdb: an information system for G protein-coupled receptors. Nucleic Acids Res 2016; 44(D1): D356-64. [http://dx.doi.org/10.1093/nar/gkv1178] [PMID: 26582914]

[25] Takeda S, Kadowaki S, Haga T, Takaesu H, Mitaku S. Identification of G protein-coupled receptor genes from the human genome sequence. FEBS Lett 2002; 520(1-3): 97-101. [http://dx.doi.org/10.1016/S0014-5793(02)02775-8] [PMID: 12044878]

[26] Venter JC, Adams MD, Myers EW, *et al.* The sequence of the human genome. Science 2001; 291(5507): 1304-51.

[27] Buck L, Axel R. A novel multigene family may encode odorant receptors: a molecular basis for odor recognition. Cell 1991; 65(1): 175-87. [http://dx.doi.org/10.1016/0092-8674(91)90418-X] [PMID: 1840504]

[28] Rosenbaum DM, Rasmussen SG, Kobilka BK. The structure and function of G-protein-coupled receptors. Nature 2009; 459(7245): 356-63. [http://dx.doi.org/10.1038/nature08144] [PMID: 19458711]

[29] Nugent T, Jones DT. Membrane protein structural bioinformatics. J Struct Biol 2012; 179(3): 327-37. [http://dx.doi.org/10.1016/j.jsb.2011.10.008] [PMID: 22075226]

[30] Lebon G, Warne T, Edwards PC, *et al.* Agonist-bound adenosine A2A receptor structures reveal common features of GPCR activation. Nature 2011; 474(7352): 521-5. [http://dx.doi.org/10.1038/nature10136] [PMID: 21593763]

[31] von Zastrow M. Mechanisms regulating membrane trafficking of G protein-coupled receptors in the endocytic pathway. Life Sci 2003; 74(2-3): 217-24. [http://dx.doi.org/10.1016/j.lfs.2003.09.008] [PMID: 14607249]

[32] Levin ER. Cellular functions of plasma membrane estrogen receptors. Steroids 2002; 67(6): 471-5. [http://dx.doi.org/10.1016/S0039-128X(01)00179-9] [PMID: 11960623]

[33] Gether U, Kobilka BK. G protein-coupled receptors. II. Mechanism of agonist activation. J Biol Chem 1998; 273(29): 17979-82.

[34] Ghanouni P, Gryczynski Z, Steenhuis JJ, *et al.* Functionally different agonists induce distinct conformations in the G protein coupling domain of the β 2 adrenergic receptor. J Biol Chem 2001; 276(27): 24433-6.
[http://dx.doi.org/10.1074/jbc.C100162200] [PMID: 11320077]

[35] Gloriam DE, Schiöth HB, Fredriksson R. Nine new human Rhodopsin family G-protein coupled receptors: identification, sequence characterisation and evolutionary relationship. Biochim Biophys Acta 2005; 1722(3): 235-46.
[http://dx.doi.org/10.1016/j.bbagen.2004.12.001] [PMID: 15777626]

[36] Fredriksson R, Lagerström MC, Lundin LG, Schiöth HB. The G-protein-coupled receptors in the human genome form five main families. Phylogenetic analysis, paralogon groups, and fingerprints. Mol Pharmacol 2003; 63(6): 1256-72.
[http://dx.doi.org/10.1124/mol.63.6.1256] [PMID: 12761335]

[37] Bjarnadóttir TK, Fredriksson R, Höglund PJ, Gloriam DE, Lagerström MC, Schiöth HB. The human and mouse repertoire of the adhesion family of G-protein-coupled receptors. Genomics 2004; 84(1): 23-33.
[http://dx.doi.org/10.1016/j.ygeno.2003.12.004] [PMID: 15203201]

[38] Donnelly D. The arrangement of the transmembrane helices in the secretin receptor family of G-protein-coupled receptors. FEBS Lett 1997; 409(3): 431-6.
[http://dx.doi.org/10.1016/S0014-5793(97)00546-2] [PMID: 9224704]

[39] De Blasi A, Conn PJ, Pin J, Nicoletti F. Molecular determinants of metabotropic glutamate receptor signaling. Trends Pharmacol Sci 2001; 22(3): 114-20.
[http://dx.doi.org/10.1016/S0165-6147(00)01635-7] [PMID: 11239574]

[40] Karnik SS, Gogonea C, Patil S, Saad Y, Takezako T. Activation of G-protein-coupled receptors: a common molecular mechanism. Trends Endocrinol Metab 2003; 14(9): 431-7.
[http://dx.doi.org/10.1016/j.tem.2003.09.007] [PMID: 14580763]

[41] Cohen WM, Goto A, Nagata A, Nelson RR, Walsh JP R. Res Policy 2002; 31(8): 1349-67.
[http://dx.doi.org/10.1016/S0048-7333(02)00068-9]

[42] Kakarala KK, Jamil K. Sequence-structure based phylogeny of GPCR Class A Rhodopsin receptors. Mol Phylogenet Evol 2014; 74: 66-96.
[http://dx.doi.org/10.1016/j.ympev.2014.01.022] [PMID: 24503482]

[43] Okuno Y, Tamon A, Yabuuchi H, *et al.* GLIDA: GPCR--ligand database for chemical genomics drug discovery--database and tools update. Nucleic Acids Res 2008; 36(Database issue) (Suppl. 1): D907-12.

[44] Drton M, Perlman MD. A SINful approach to Gaussian graphical model selection. J Stat Plan Inference 2008; 138(4): 1179-200.
[http://dx.doi.org/10.1016/j.jspi.2007.05.035]

[45] Brown JB, Okuno Y. Systems biology and systems chemistry: new directions for drug discovery. Chem Biol 2012; 19(1): 23-8.
[http://dx.doi.org/10.1016/j.chembiol.2011.12.012] [PMID: 22284351]

[46] Albert R. Network inference, analysis, and modeling in systems biology. Plant Cell 2007; 19(11): 3327-38.
[http://dx.doi.org/10.1105/tpc.107.054700] [PMID: 18055607]

[47] Gianchandani EP, Papin JA, Price ND, Joyce AR, Palsson BO. Matrix formalism to describe functional states of transcriptional regulatory systems. PLOS Comput Biol 2006; 2(8): e101.
[http://dx.doi.org/10.1371/journal.pcbi.0020101] [PMID: 16895435]

[48] Samaga R, Saez-Rodriguez J, Alexopoulos LG, Sorger PK, Klamt S. The logic of EGFR/ErbB signaling: theoretical properties and analysis of high-throughput data. PLOS Comput Biol 2009; 5(8): e1000438.

[http://dx.doi.org/10.1371/journal.pcbi.1000438] [PMID: 19662154]

[49] Pin JP, Galvez T, Prézeau L. Evolution, structure, and activation mechanism of family 3/C G-protei--coupled receptors. Pharmacol Ther 2003; 98(3): 325-54. [http://dx.doi.org/10.1016/S0163-7258(03)00038-X] [PMID: 12782243]

[50] Hanyaloglu AC, von Zastrow M. Regulation of GPCRs by endocytic membrane trafficking and its potential implications. Annu Rev Pharmacol Toxicol 2008; 48: 537-68. [http://dx.doi.org/10.1146/annurev.pharmtox.48.113006.094830] [PMID: 18184106]

[51] Wilson JE, Kurukulasuriya R, Sinz C, *et al.* Discovery and development of benzo-[1,2,4]-triazo-o-[1,4]-oxazepine GPR142 agonists for the treatment of diabetes. Bioorg Med Chem Lett 2016; 26(12): 2947-51. [http://dx.doi.org/10.1016/j.bmcl.2016.04.018] [PMID: 27240550]

[52] Wang J, Carrillo JJ, Lin HV. GPR142 Agonists Stimulate Glucose-Dependent Insulin Secretion *via* Gq-Dependent Signaling. PLoS One 2016; 11(4): e0154452. [http://dx.doi.org/10.1371/journal.pone.0154452] [PMID: 27104960]

[53] Süsens U, Hermans-Borgmeyer I, Urny J, Schaller HC. Characterisation and differential expression of two very closely related G-protein-coupled receptors, GPR139 and GPR142, in mouse tissue and during mouse development. Neuropharmacology 2006; 50(4): 512-20. [http://dx.doi.org/10.1016/j.neuropharm.2005.11.003] [PMID: 16378626]

[54] Lauffer LM, Iakoubov R, Brubaker PL. GPR119 is essential for oleoylethanolamide-induced glucagon-like peptide-1 secretion from the intestinal enteroendocrine L-cell. Diabetes 2009; 58(5): 1058-66. [http://dx.doi.org/10.2337/db08-1237] [PMID: 19208912]

[55] Overton HA, Fyfe MC, Reynet C. GPR119, a novel G protein-coupled receptor target for the treatment of type 2 diabetes and obesity. Br J Pharmacol 2008; 153(S1) (Suppl. 1): S76-81. [http://dx.doi.org/10.1038/sj.bjp.0707529] [PMID: 18037923]

[56] Jones RM, Leonard JN, Buzard DJ, Lehmann J. GPR119 agonists for the treatment of type 2 diabetes. Expert Opin Ther Pat 2009; 19(10): 1339-59. [http://dx.doi.org/10.1517/13543770903153878] [PMID: 19780700]

[57] Intensive blood-glucose control with sulphonylureas or insulin compared with conventional treatment and risk of complications in patients with type 2 diabetes (UKPDS 33). Lancet 1998; 352(9131): 837-53. [http://dx.doi.org/10.1016/S0140-6736(98)07019-6] [PMID: 9742976]

[58] Modi P. Diabetes beyond insulin: review of new drugs for treatment of diabetes mellitus. Curr Drug Discov Technol 2007; 4(1): 39-47. [http://dx.doi.org/10.2174/157016307781115476] [PMID: 17630927]

[59] Rawal LB, Tapp RJ, Williams ED, Chan C, Yasin S, Oldenburg B. Prevention of type 2 diabetes and its complications in developing countries: a review. Int J Behav Med 2012; 19(2): 121-33. [http://dx.doi.org/10.1007/s12529-011-9162-9] [PMID: 21590464]

[60] Epidemiology of severe hypoglycemia in the diabetes control and complications trial. Am J Med 1991; 90(4): 450-9. [http://dx.doi.org/10.1016/0002-9343(91)80085-Z] [PMID: 2012085]

[61] Roth J, Glick SM, Yalow RS, Bersonsa . Hypoglycemia: a potent stimulus to secretion of growth hormone. Science 1963; 140(3570): 987-8. [http://dx.doi.org/10.1126/science.140.3570.987] [PMID: 13975313]

[62] American Diabetes Association. Diagnosis and classification of diabetes mellitus. Diabetes Care 2010; 33 (Suppl. 1): S62-9. [http://dx.doi.org/10.2337/dc10-S062] [PMID: 20042775]

[63] He CL, Soffer EE, Ferris CD, *et al.* Loss of interstitial cells of cajal and inhibitory innervation in

insulin-dependent diabetes. Gastroenterology 2001; 121(2): 427-34.
[http://dx.doi.org/10.1053/gast.2001.26264] [PMID: 11487552]

[64] National diabetes fact sheet: national estimates and general information on diabetes and prediabetes in the United States. 2011.

[65] Saltiel AR, Kahn CR. Insulin signalling and the regulation of glucose and lipid metabolism. Nature 2001; 414(6865): 799-806.
[http://dx.doi.org/10.1038/414799a] [PMID: 11742412]

[66] Andrianantoandro E, Basu S, Karig DK, Weiss R. Synthetic biology: new engineering rules for an emerging discipline. Mol Syst Biol 2006; 2(1): 0028.
[PMID: 16738572]

[67] Elbaz J, Wang F, Remacle F, Willner I. pH-programmable DNA logic arrays powered by modular DNAzyme libraries. Nano Lett 2012; 12(12): 6049-54.
[http://dx.doi.org/10.1021/nl300051g] [PMID: 22295948]

[68] Kobayashi H, Kaern M, Araki M, *et al.* Programmable cells: interfacing natural and engineered gene networks. Proc Natl Acad Sci USA 2004; 101(22): 8414-9.
[http://dx.doi.org/10.1073/pnas.0402940101] [PMID: 15159530]

[69] Guet CC, Elowitz MB, Hsing W, Leibler S. Combinatorial synthesis of genetic networks. Science 2002; 296(5572): 1466-70.
[http://dx.doi.org/10.1126/science.1067407] [PMID: 12029133]

[70] Hartwell LH, Hopfield JJ, Leibler S, Murray AW. From molecular to modular cell biology. Nature 1999; 402(6761) (Suppl.): C47-52.
[http://dx.doi.org/10.1038/35011540] [PMID: 10591225]

[71] Slusarczyk AL, Lin A, Weiss R. Foundations for the design and implementation of synthetic genetic circuits. Nat Rev Genet 2012; 13(6): 406-20.
[http://dx.doi.org/10.1038/nrg3227] [PMID: 22596318]

[72] Kobayashi H, Kaern M, Araki M, *et al.* Programmable cells: interfacing natural and engineered gene networks. Proc Natl Acad Sci USA 2004; 101(22): 8414-9.
[http://dx.doi.org/10.1073/pnas.0402940101] [PMID: 15159530]

[73] Purnick PE, Weiss R. The second wave of synthetic biology: from modules to systems. Nat Rev Mol Cell Biol 2009; 10(6): 410-22.
[http://dx.doi.org/10.1038/nrm2698] [PMID: 19461664]

[74] Adleman LM. Molecular computation of solutions to combinatorial problems. Science 1994; 266(5187): 1021-4.
[http://dx.doi.org/10.1126/science.7973651] [PMID: 7973651]

[75] Ogihara M, Ray A. Simulating Boolean circuits on a DNA computer. Algorithmica 1999; 25(2-3): 239-50.

[76] Lovgren S. Computer Made from DNA and Enzymes. National Geographic News 2003; p. 24.

[77] Kaushik AC, Sahi S. Boolean network model for GPR142 against Type 2 diabetes and relative dynamic change ratio analysis using systems and biological circuits approach. Syst Synth Biol 2015; 9(1-2): 45-54.
[http://dx.doi.org/10.1007/s11693-015-9163-0] [PMID: 25972988]

[78] Toda N, Hao X, Ogawa Y, *et al.* Potent and orally bioavailable GPR142 agonists as novel insulin secretagogues for the treatment of type 2 diabetes. ACS Med Chem Lett 2013; 4(8): 790-4.
[http://dx.doi.org/10.1021/ml400186z] [PMID: 24900747]

[79] Potter LA. Transcriptional profiling of pancreatic progenitor cells. Vanderbilt University 2011.

[80] Lin CL, Kuo TY, Chen YY, Bathula C, Singh S, Sen S. Systems and Synthetic Biology

CHAPTER 4

Ribavirin Against Viral, Neoplastic and Inflammatory Diseases: Focus on Mechanisms of Action

Danijela Savic[*], **Ivana Bjelobaba, Sanja Pekovic** and **Irena Lavrnja**

Department of Neurobiology, Institute for Biological Research "Sinisa Stankovic" University of Belgrade, Belgrade, Serbia

Abstract: Discovered 45 years ago, ribavirin still proves useful as a broad-spectrum anti-viral drug against different RNA and DNA viruses. Although many cellular and molecular mechanisms of ribavirin action have been proposed during several decades of the extensive research, the complete spectrum of its actions is still not fully known. The direct mechanisms of ribavirin anti-viral action involve RNA polymerase inhibition and lethal mutagenesis. The main intracellular target of ribavirin action is inosine 5'-monophosphate dehydrogenase (IMPDH), the rate-limiting enzyme in a *de novo* synthesis of guanine nucleotides. The inhibition of IMPDH activity leads to depletion of the guanine nucleotide pool and the consequent attenuation of GTP-dependent cellular processes, inducing cell cycle arrest, and the interruption of DNA and RNA synthesis. Since these processes are essential for the normal as well as mitogenic functions of all cells, the inhibition of IMPDH probably represents the central mechanism of ribavirin action, including its cytotoxic effect. Recent data suggest that ribavirin may also be effective as an anti-cancer drug. By inhibiting the eukaryotic translation initiation factor e4E (eIF4E), ribavirin impedes eIF4E-mediated oncogenic transformation. Additionally, immunomodulatory and immunosuppressive actions of ribavirin have been shown in different *in vivo* and *in vitro* models of neuroinflammation. Accumulating evidence points to important cell type-specific differences in response to ribavirin that may arise from cell specific variations in ribavirin metabolism, as well as the functional status of targeted cells. This chapter reviews the antiviral, anticancer and anti-inflammatory activities of ribavirin and its metabolites, and discusses the possible mechanisms of action.

Keywords: Anaemia, Astrocytes, Cancer, Cap structure, Encephalitis, Guanosine nucleotides, Hepatitis C, Inhibition, Immunomodulation, Immunosuppression, Leukaemia, Microglia, Multiple sclerosis, Mutagenesis, Nucleoside analogue, Nucleoside transporters, Oncogenes, Ribavirin, RNA viruses, T-cells.

[*] **Corresponding author: Danijela Savic:** Institute for Biological Research "Sinisa Stankovic ", Boulevard Despota Stefana 142, 11060 Belgrade, Serbia; Tel: +381 11 2078 340; Fax: + 381 11 2761 433; E-mail: danisto@ibiss.bg.ac.rs

Atta-ur-Rahman & M. Iqbal Choudhary (Eds.)

INTRODUCTION

The efficacy of nucleoside analogues has been proven in a variety of diseases, particularly in the treatment of viral infections and cancers. Due to the similarity to endogenous nucleotides, they can be incorporated into newly synthesised deoxyribonucleic acid (DNA), resulting in the inhibition of synthesis and chain termination. Several agents are able to inhibit crucial enzymes involved in the generation of purine nucleotides and ribonucleic acid (RNA) synthesis. The transport of nucleoside analogues into the cell is carried out *via* passive diffusion, or by active transport, through specific nucleoside transporters. Upon entry to the cells, they are phosphorylated to mono-, di- and triphosphate active forms, which are inhibited by the enzymatic action of soluble 5'-nucleotidase. Among the different analogues approved by the US Food and Drug Administration (FDA) which share mutual characteristics, each drug also has specific properties in terms of drug-target interactions and exerts differences in terms of efficacy against different diseases (for a review on nucleoside analogues see [1]).

This chapter summarises past and current knowledge of the FDA approved drug ribavirin. Ribavirin (also known as Rebetol, Copegus, Ribasphere and Virazole) is a guanosine analogue and strong synthetic antiviral agent [2, 3]. This drug is commonly used for the treatment of hepatitis C-infected patients [4]. In addition, it can be included in the therapy for severe respiratory syncytial virus disease or Lassa fever [5, 6]. Current studies indicate that it could be beneficial in the treatment of acute myeloid leukaemia, and could also be an anti-glioblastoma therapeutic. The main target of ribavirin in these disorders is the known potent oncogene eIF4E [7, 8]. Except for viruses and neoplastic cells, different cell types of the immune system, such as T cells, macrophages and dendritic cells, are sensitive to ribavirin. The immunomodulatory and immunosuppressive actions of ribavirin have been shown in models of *in vivo* and *in vitro* neuroinflammation. Specifically, the anti-inflammatory potential of ribavirin was shown in an animal model of multiple sclerosis - experimental autoimmune encephalomyelitis and experimental traumatic brain injury in rats [9, 10]. Furthermore, pleiotropic and cytotoxic effects were confirmed in activated microglia [11].

The well-known mechanism of ribavirin action is the inhibition of IMPDH (EC 1.1.1.205), which is known to be a principal enzyme in the *de novo* synthesis of guanine nucleotides (Fig. **1**) [12]. The inhibition of IMPDH leads to reduced levels of intracellular guanine, which serves as a precursor for RNA and DNA, as an energy source for protein synthesis, as a co-factor for G-proteins and signalling molecules, as a precursor for glycosylation and as an allosteric regulator [13]. IMPDH inhibitors reduce intracellular guanine nucleotide pools, and at the same time increase adenine nucleotide pools [13]. This induces a range of cellular effects, depending on the cell type, differentiation and cycle status. Cells that rapidly grow have a high demand for guanine nucleotides that cannot be provided by salvage pathways [14]. The direct relationship between biological activity and guanine nucleotide depletion was established *in vitro* by the reversibility of ribavirin effects with the repletion of guanine nucleotide pools [15].

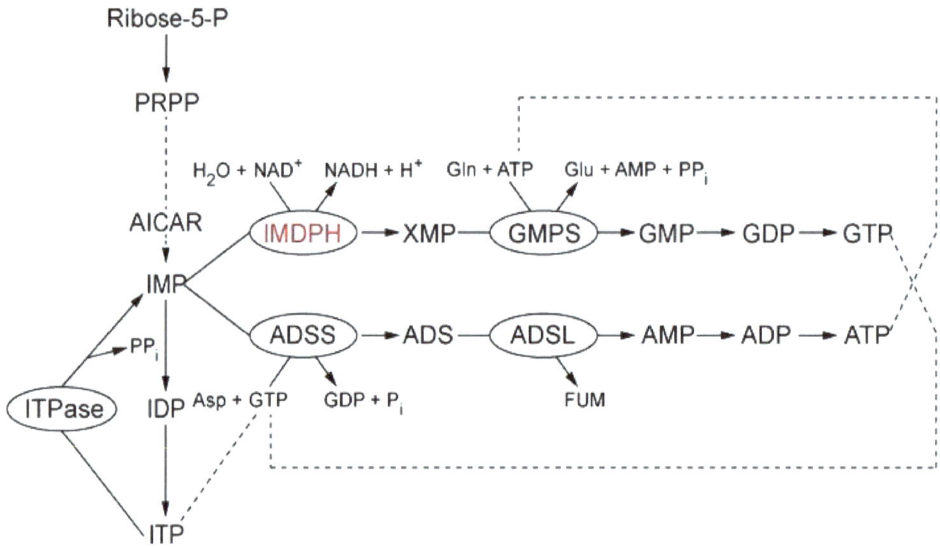

Fig. (1). De *novo* synthesis of purine nucleotides. A key regulatory step in *de novo* purine biosynthesis is the production of PRPP, followed by a complex multistep pathway to gain IMP, which is the first intermediate ribonucleoside monophosphate in purine metabolism to have whole purine ring. Both adenylate (AMP) and guanylate (GMP) are derived from IMP. IMPDH is the target of ribavirin action and represent the rate-limiting enzyme in the *de novo* synthesis of guanine nucleotides. In ITPase-deficient erythrocytes, ADSS could recognise ITP as a substrate instead of GTP, which is especially important when red blood cells are exposed to ribavirin and there is a lack of GTP molecules. ADS – adenylosuccinate; ADSL - adenylosuccinate lyase; ADSS - adenylosuccinate synthase; AICAR - 5-aminoimidazole-4-carboxamid--β-D-ribofuranoside; AMP, ADP, ATP – adenosine mono-, di- and triphosphate; Asp – aspartate; FUM – fumarate, Gln – glutamine; Glu – glutamate; GMP, GDP, GTP - guanosine mono-, di- and triphosphate; GMPS - guanosine monophosphate synthase; IMP, IDP, ITP – inosine mono-, di- and triphosphate, respectively; IMPDH - inosine 5'-monophosphate dehydrogenase; ITPase – inosine triphosphatase; NAD - Nicotinamide adenine dinucleotide; Pi - inorganic phosphate; PP_i – pyrophosphate; PRPP - 5-phosphoribosy--1-pyrophosphate; Ribosa-5-P – ribose-5-phosphate; XMP - xanthosine monophosphate.

This chapter will recapitulate the current knowledge of the mechanism of action, pharmacological properties, clinical usage and toxicity of ribavirin.

RIBAVIRIN AGAINST VIRUSES

In the ongoing fight against viruses, weapons that can be used include antiviral drugs and vaccines. Some viruses are sensitive to either one or both of these medical approaches, while defence strategies have yet to be developed for other some viral infections. Antiviral drugs are approved against nine human viral infections caused by: human immunodeficiency virus, hepatitis C virus, influenza virus, respiratory syncytial virus, hepatitis B virus, human papillomavirus, human cytomegalovirus, herpes simplex virus and varicella-zoster virus [16]. There are ninety approved antiviral drugs (including US, Japan and Europe) which can be separated into thirteen functional groups according to De Clercq and Li (see review for details [16]). At first, it was believed that the similarity between virus replication and DNA and RNA synthesis in the host (mammalian) cells is so close that the development of antiviral drugs is almost unattainable. However, precisely because of this similarity, *i.e.* the similarity to endogenous nucleosides, which are constituents of DNA and RNA, an entire class of antiviral drugs has been synthesised, belonging to the functional group of nucleoside analogues.

Ribavirin Chemistry

The structure of naturally occurring nucleosides consists of a pentose sugar moiety (ribose or 2-deoxyribose) linked to a heteroaromatic nucleobase through an N-glycosidic bond. Nucleobases are purine (two ring structures: adenine and guanine) or pyrimidine (one ring structures: thymine, cytosine and uracil) derivatives; therefore, the corresponding nucleosides are: adenosine/deoxyadenosine, guanosine/deoxyguanosine, thymidine, cytidine/deoxycytidine, and uridine. The phosphorylation of the sugar moiety of nucleosides leads to the formation of nucleoside monophosphates, which are also referred to as nucleotides, and are the main building blocks of DNA and RNA. Depending on the number of added phosphate groups, there are also nucleoside di- and triphosphates. In addition to the role in the synthesis of DNA and RNA, nucleosides and nucleotides are essential for a variety of cellular processes, such as energy and enzyme regulation, cell signalling and neurotransmission.

Nucleoside analogues are synthetic compounds, with a role in the aforementioned cellular processes, in place of their endogenous counterparts. They disrupt cellular metabolism, with the final goal of killing a virus or a cancer cell, without damaging the host/normal cells. Because of this, nucleoside analogues are often referred to as antimetabolites. They are a diverse group of compounds, thanks to the modification of sugar (*e.g.* by hydroxylation or dehydroxylation, saturation, ring opening, methylation, halogenation) and base moieties (*e.g.* by azotation, N-conjugation, halogenation) [1]. Another important feature of nucleoside analogues is that they are prodrugs, meaning that after the administration and cellular uptake *via* appropriate nucleoside transporters, they need to go through metabolic conversion into pharmacologically active forms. Thus, inside the cell they are phosphorylated into mono-, di- and triphosphate nucleosides by the activity of an appropriate kinase. The first nucleoside analogue approved as an antiviral drug, fifty four years ago, is still in use. The use of this drug, idoxuridine (5-iodo-2'-deoxyuridine, brand name Stoxil), was limited to herpes keratitis [16, 17]. Since then, scientists have been looking for a substance with a broad antiviral activity, which will have the same, or at least approximate success, as antibiotics have against bacteria [17]. Following that idea, in 1972, Sidwell *et al.* published a paper on the synthesis of ribavirin, a substance that displayed the aforesaid desirable traits [3].

Ribavirin (1-β-D-ribofuranosyl-1,2,4-triazole-3-carboxamide) consists of a ribose sugar moiety, β-D-ribofuranose, that has a modified base moiety on the C1-carbon atom represented by a heterocyclic compound: 1,2,4-triazole linked by an N-glycosidic bond, wherein triazole has a functional group of carboxamide (Fig. **2**). Based on a heterocyclic ring, analogy in the structure can be drawn between naturally occurring metabolites of ribavirin: 5-aminoimidazole-4-carboxamide-β-D-ribofuranoside (AICAR) and nicotinamide [2, 18]. AICAR is one of the participants in the formation of inosine monophosphate (IMP), which is crucial for *de novo* purine biosynthesis (Fig. **1**). Nicotinamide is a coenzyme which serves as: an oxidising agent - nicotinamide adenine dinucleotide (NAD^+), and as a reducing agent - nicotinamide adenine dinucleotide phosphate ($NADP^+$) (Fig. **1**). Since ribavirin has ribose as a sugar moiety, a parallel in the structure should also be seen among naturally occurring ribonucleosides, such as adenosine, guanosine, inosine, xanthosine, uridine, cytidine. In the literature, ribavirin is often referred to as a guanosine analogue, but except for having a ribose ring, there are no other obvious structural analogies (Fig. **2**). However, single crystal X-ray diffraction measurements have shown high similarity in the geometry of the carboxamide group between these two compounds. Therefore, ribavirin probably interferes with the biological processes that involve guanosine or its phosphorylated forms. [19, 20]. In addition, possessing a ribose sugar moiety indicates that ribavirin action is mainly directed to RNA-related metabolism [21].

The methodology of ribavirin production is based on three different approaches: (i) a chemical/synthetic technique, (ii) the use of enzymes and (iii) a fermentative method. The chemical procedure starts by gaining tetra-acetyl *β*-D-ribofuranose from endogenous nucleoside (*e.g.* guanosine, inosine) by acid-catalysed acetylation. Then, tetra-acetyl sugar forms are condensed with 1,2,4-triazolecarboxylic acid esters, that result in ribavirin following aminolysis in a methanolic solution (Fig. **3**) [2, 22].

The chemical synthesis of ribavirin is the most common in pharmaceutical industry, since it provides pure yield. However, this process is burdened by the high demand for the starting components and the amount of hazardous waste [23].

Therefore, alternative methods are required. The enzymatic synthesis of ribavirin employs purine nucleoside phosphorylase which catalyses the reaction between naturally occurring purine nucleosides (adenosine, guanosine, inosine) and 1,2,4-triazole-3-carboxamide [23]. However, the mentioned components are very expensive, which makes the whole process unattractive to businesses. In order to overcome this shortcoming, the fermentative method was developed, based on the addition of 1,2,4-triazole-3-carboxamide to the bacterial culture that naturally produces or overproduces nucleosides and purine nucleoside phosphorylase as the catalytic enzyme [23].

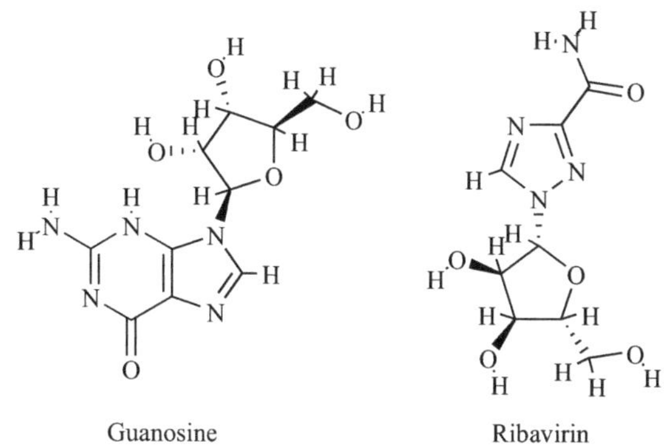

Fig. (2). Chemical structures of guanosine and ribavirin.

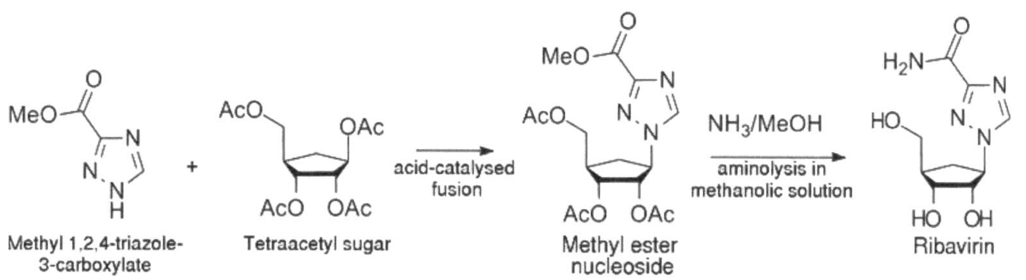

Fig. (3). Synthesis of ribavirin. This work "Figure 3. Synthesis of ribavirin" is a modification of "File:Ribavirin synthesis.png" by Nuklear, licensed under Creative Commons Attribution-Share Alike 4.0 International/Modification: Labelling of chemical compounds and chemical reactions by Danijela Savic.

Virus Sensitivity to Ribavirin

Early reports demonstrated the broad antiviral activity of ribavirin *in vitro* in cell cultures and *in vivo* in animal models of virus diseases. The research found that ribavirin was active against the following DNA viruses: herpes simplex virus type 1 and type 2, the pseudorabies virus, murine cytomegalovirus, myxoma virus, and adenovirus type 3; and RNA viruses: parainfluenza virus type 1 and type 3, influenza A and B virus, rhinovirus type 1A, type 13 and type 56, Coxsackie virus, poliovirus type 2 and vesicular stomatitis virus [2, 3]. *In vivo,* ribavirin was efficient for the treatment of rabbit keratitis induced by herpes simplex virus, mouse tail lesions induced by herpes simplex, vaccinia and vesicular stomatitis viruses, splenomegaly and hepatomegaly in mice induced by Friend leukaemia virus, and mouse pneumonia induced by influenza A0, A2 and B and parainfluenza 1 viruses [2, 3]. In addition, it was noted that the intraperitoneal or intravenous application of ribavirin was ineffective against mouse encephalitis induced by herpes simplex, vaccinia, Semliki Forest or Western equine encephalitis viruses, indicating a poor ability of the drug to cross the blood-brain barrier (BBB) [2, 3]. In line with this are the randomised clinical trials that do not support the use of oral ribavirin to treat encephalitis induced by La Crosse virus or the Japanese encephalitis virus [24, 25]. On the other hand, this drug may be neuroprotective if the route of administration is changed in order to achieve higher concentrations in the central nervous system (CNS). Accordingly, intraventricular ribavirin therapy may be helpful against Borna disease virus that cause multiphasic immune-mediated neurological syndrome [26], or as a treatment for subacute sclerosing panencephalitis triggered by persistent infection by the measles virus [27]. There are numerous reports that attest to ribavirin's wide antiviral activity *in vitro*. However, the focus of this study will be on real-life situations (*i.e.* clinical trials) and on the approved indication therapies. Although ribavirin is indeed used to inhibit more than one virus in clinical practice, its real use is far from the antiviral cure, with broad spectrum activity. Since ribavirin was highly active against influenza viruses under *in vitro* conditions, the hypothesis that ribavirin might be particularly useful as the treatment for infections caused by these viruses was reasonable [28]. Indeed, several clinical trials have addressed this issue, most of them with the encouraging results, but at the same time with a need for further confirmation in larger groups of subjects. A randomised, placebo-controlled, double-blind trial demonstrated that loading dose and larger maintenance doses of oral ribavirin significantly improved the symptoms of influenza A or B infection in 25 patients. However, based on the virus titre, there was no effect on elimination of the virus-positive status [29]. In addition, small pilot studies showed a beneficial effect of ribavirin in the treatment of immunocompromised patients with influenza A infection or patients with serious influenza and parainfluenza virus infection [30, 31]. Furthermore, ribavirin administered as an aerosol was efficient in the normalisation of temperature in college students infected with influenza and in young children hospitalised because of it [32, 33]. However, therapy with aerosolised ribavirin is costly, and these results were not persuasive enough to ensure that the treatment would be cost-effective. The transition to the clinical practice was also hindered by a well-performed clinical trial in which oral ribavirin failed to show any significant benefit in 97 patients with mild influenza

A infection [34]. Therefore, ribavirin never gained approval for the treatment of this indication. Clinical tests of ribavirin efficiency against HIV infection were also carried out, based on promising results obtained *in vitro* [35]. However, antagonism between zidovudine, which is the first nucleoside analogue approved for the treatment of AIDS, and ribavirin, discouraged the more thorough investigation of ribavirin in this field [36]. In addition, the results of several clinical studies conducted soon after HIV discovery did not support the use of ribavirin against AIDS, since this drug did not exhibit HIV-suppressive effects [37 - 40]. For the time being, ribavirin is primarily used as a treatment for infections caused by hepatitis C virus, and is effective only if it is used in combination with peg-interferon alpha-2b. In favour of ribavirin having little or no influence on HIV virus is the fact that the population of patients with HCV-HIV co-infection has been difficult to treat with this therapy [41]. Ribavirin is also approved for the treatment of severe infection caused by respiratory syncytial virus. Additionally, it may be used as an orphan drug to treat Lassa fever, since reports of its success rates are inconsistent. Before the pharmacokinetics of ribavirin and mechanisms of antiviral action are discussed, an overview of the basic facts about these three viruses will be presented.

Hepatitis C Virus

Hepatitis C virus (HCV) is a blood borne virus that was known as an inducer of non-A, non-B hepatitis before 1989. Host specificity is limited to humans, and so far there is no evidence of the zoonotic origin of HCV. However, recent findings of a non-primate hepacivirus in horses and dogs must be taken into account [42]. Once inside humans, HCV is primarily directed to the liver, where it infects hepatocytes and causes acute or chronic inflammation. In most infected individuals, acute hepatitis is subclinical and the disease often resolves by itself without medical treatment, usually within a period of six months [43]. However, 10–30% of individuals develop symptoms like fatigue, mild fever, muscle and joint pain, nausea, vomiting, abdominal pain and jaundice [44]. What makes HCV infection dangerous is the high probability of the disease becoming persistent and chronic; namely, 40–95% of acute HCV infections evolve into the chronic form, which is usually slowly progressive, and associated with severe hepatic impairment, such as compensated and decompensated cirrhosis and hepatocellular carcinoma [44]. As for the prevalence of HCV worldwide, based on anti-HCV antibody testing, it is estimated to be in the range of 1.3–2.1%, which is equivalent to 92–149 million individuals [45]. Recent data suggest that HCV can be classified into seven genotypes and 67 subtypes [46], and the presence of genotypes varies depending on the region. Thus, genotype 1 comprises 46% of all HCV infections, and is mainly found in North and South America, Europe, Australia, New Zealand, and Central and East Asia [41, 47]. Genotype 3 is the next most frequent on a global scale, accounting for 30% of all HCV cases, and is most common in India and Pakistan [41, 47]. Genotypes 2, 4, and 6 cause 23% of all HCV infections with the following regional distribution: genotype 2 prevails in West Africa; genotype 4 in Egypt, Central sub-Saharan Africa and parts of Scandinavia; and genotype 6 is found in South East Asia [41, 47]. The remaining 1% of all HCV cases have genotype 5 as a causative agent, and are primarily located in South Africa [41, 47]. So far, there has only been one case of HCV genotype 7 infection, which was registered in a person who emigrated from the Congo [46]. Disease severity and response to therapy are interrelated with HCV genotypes. When it comes to the structure of the HCV virion, like all members of the *Flaviviridae* family, it has a lipid bilayer envelope wrapped around the nucleocapsid. The nucleocapsid is composed of multiple copies of the core or capsid protein and contains the viral genome represented by a single-stranded, positive sense RNA [48, 49]. The HCV genome is around 9.6 kb in size, and encodes a single, long polyprotein of 3000 amino acids or more depending on the HCV genotype [49]. This polyprotein is proteolytically processed in the host cell, generating at least 10 proteins. Each component of the HCV life cycle could represent a potential target for future therapy. Hitherto, three different classes of direct-acting antiviral agents (DAAs) have been approved.

The Role of Ribavirin in the Treatment of HCV Infection

Before the role of ribavirin in the treatment of HCV infection is discussed, it is necessary to explain that the effect of any antiviral therapy of hepatitis C must be measured through sustained virological response (SVR) rates. SVR means that 12 or 24 weeks after the cessation of treatment, no HCV RNA is detected in the blood [41]. Before introducing ribavirin into the management of chronic hepatitis C, the mainstream approach was based on the administration of conventional interferon alpha, which resulted in the highest SVR rates of 29% for patients infected with HCV genotypes other than 1 and 7% for patients with HCV genotype 1 [50]. The therapy of chronic hepatitis C significantly evolved following the introduction of pegylated interferon (alpha 2a or 2b) together with ribavirin [50, 51]. During the decade from 2001 to 2011, the administration of pegylated interferon alpha once a week plus daily oral ribavirin for 24 to 48 weeks represented the standard of care for chronic hepatitis C [4, 41]. A successful SVR rate was achieved for patients infected with HCV genotype 2 or 3 (about 80-90%). However, in patients infected with the prevalent HCV genotype 1, the SVR rate was a modest 40% [52]. In addition, haemolytic anaemia, which is a well-known reversible and dose-dependent side effect of ribavirin, has led to the premature discontinuation of the treatment protocol in 27-36% of patients [53]. Hence, there was a constant pressure to find a more efficient medicine. A better understanding of the HCV life cycle has led to the breakthrough of new drugs, directed against the vital factors for virus survival (the above-mentioned DAAs). The currently approved DAAs are orally administered and well tolerated drugs that achieve SVR in 90-100% of patients, following 8-24 weeks of treatment, usually when two or three are combined [41, 54]. It can be said that ribavirin has achieved its zenith when it comes to the treatment of HCV infection, and that other players have stepped onto the scene. Nonetheless, under some circumstances, the inclusion of ribavirin is still helpful; for example, in the treatment of patients with HCV genotype 3 infection, which causes a more severe course of disease compared to other genotypes [54]. Moreover, ribavirin in combination with approved DAA interferon-free regimens reduces treatment time without affecting SVR (from 24 to 12 weeks for patients with HCV genotype 2 infection, or for cirrhotic patients with HCV genotype 1 or 4 infection) [54]. It seems that, at least for a while, ribavirin will play a supporting role in the treatment of chronic hepatitis C.

Respiratory Syncytial Virus

Respiratory syncytial virus (RSV) is a member of *Paramixoviridae* family, alongside parainfluenza 3 and 5, the mumps and measles viruses; however, it belongs to the other subfamily, the *Pneumovirus* genus [55]. RSV may be found with spherical or filamentous shapes that have an envelope represented by a lipid bilayer, which surrounds a helical ribonucleoprotein complex (*i.e.* the ribonucleocapsid). The envelope and capsid are connected through matrix proteins. The ribonucleocapsid encapsulates a 15 kb long viral genome, which is a single-stranded, negative sense RNA that encodes 11 proteins [56].

It was discovered in 1957, and has no other known hosts except for humans. This virus is globally present, displaying seasonal variation in outbreaks of infection. RSV attacks the respiratory system, usually causing mild upper respiratory tract inflammation that is resolved by self-care. However, in some high-risk groups (such as infants and young children; children with congenital heart disease, congenital lung malformations, chronic lung disease, and neuromuscular disease; elderly people ($\geq$ 65 years of age); and hematopoietic stem cell transplant recipients), RSV can induce severe complications related to lower respiratory tract infection, when hospitalisation might be necessary [57, 58]. A meta-analysis from 2005, showed that there are approximately 34 million cases of lower respiratory tract infections caused by RSV in children younger than 5 years worldwide, 10% of which needed clinical care, and about 2–6% of hospitalised cases are fatal [59]. Early RSV infection, even a mild one, represents a risk factor for recurrent wheezing and asthma development later in life [55, 57, 60].

Role of Ribavirin in the Treatment of RSV Infection

For the time being, complicated RSV infections that require medical attention still await specific antiviral therapy. The only available treatments are with ribavirin, or a prophylactic therapy with palivizumab. Palivizumab is a monoclonal antibody directed towards the envelope fusion protein. It is administered preventatively to children who are especially vulnerable to RSV infection, such as infants born prematurely or those with chronic diseases [61]. However, children who are at a high-risk of RSV infection represent only a small percentage of the total infant population hospitalised with RSV disease [61, 62]. Ribavirin, as the other option, has to be administered as an aerosol: this entails technical difficulty, since most of the patients are young children; is a matter of safety, because of the known drug toxicity and teratogenicity; and incurs high cost [5, 63]. Nevertheless, according to some randomised control trials, aerosolised ribavirin reduces the morbidity and mortality rate in infants and immunocompromised patients with RSV infection [5, 64]. However, there is a need for more convincing clinical evidence which will be drawn from larger cohorts [57, 58]. Although there is a rationale to the use of ribavirin in patients with mortality risk, it should not be routinely used against RSV infection, as recommended by the American Academy of Paediatrics [57].

Lassa Virus

Lassa virus (LASV) was identified in samples of patients who contracted the disease in the town bearing the same name in Nigeria, in 1969 [65]. It is a member of the *Arenaviridae* virus family, and belongs to the genus of *Arenavirus*. Like all arenaviruses, LASV has a lipid envelope derived from the plasma membrane of the host cell. An envelope surrounds the ribonucleocapsid, which contains the LASV genome, represented by two negative-sense, single-stranded RNAs (small ~3.4 kb and large ~7 kb) [66]. Each of the RNA segments encodes two genes in the opposite direction, in a so-called ambisense coding strategy [67].

Parts of West Africa, including Nigeria, Liberia, Sierra Leone and Guinea, represent the LASV endemic zone [68]. In addition, isolated LASV infections have been reported in Benin, Burkina Faso, Cote d'Ivoire, Ghana, Mali and Togo [68]. LASV has a zoonotic origin and comes from *Mastomys natalensis*, a multi-mammate rodent that is commonly found near human settlements across Africa [69]. Humans become infected by LASV if they come into contact with rat excretions (urine, faeces, respiratory secretions and blood), by the inhalation of aerosols saturated with infective particles of rat excretions, or by the consumption of rodent meat, since this rat is sometimes included in the local diet [68]. The route of transmission between humans includes direct contact with infected body fluids and excretions, and usually occurs in a hospital environment because of the lack of proper protection [70, 71]. LASV causes Lassa fever, an acute disease that in most cases is mild (flu like symptoms) or subclinical. However, in about 20% of infected people, it occurs as a severe disease (starts as a mild form and develops to abdominal pain, vomiting, diarrhoea, haemorrhage, respiratory and neurological problems, and facial oedema) which need clinical attention; the mortality rate is 15% of the hospitalised population [72]. The incubation period for Lassa fever ranges from 3 to 21 days, while recovery begins 8 to 10 days after onset. Nevertheless, approximately at that time in severe cases, the condition is rapidly deteriorating, with multiple organ failure and death in the third week [67]. Between 100,000 and 300,000 cases of Lassa fever are reported yearly, 5000 of which have a fatal outcome [68].

Role of Ribavirin in the Treatment of LASV Infection

A drug that would directly affect any of the components of the LASV life cycle has not yet been synthesised. Except for supportive care, the only available therapy is conducted with ribavirin. The most informative clinical trial of ribavirin effectiveness in the treatment of Lassa fever was published in 1986, and represented part of the program that started in 1977, which included the Ministry of health of Sierra Leone and the Centre for Disease Control [6]. First, the study determined two risk factors that were fatal: (i) the level of serum aspartate aminotransferase (AST) $\geq$ 150 IU/L and (ii) the level of viraemia $\geq 10^{3.6}$ TCID/ml. Ribavirin given 10 days intravenously significantly improved the survival of patients with increased levels of AST, from 45% to 95%, if the treatment started within the first 6 days after the disease began. The subsequent start of treatment was related to the lower survival rate of 74%. Regarding patients with increased levels of viraemia, intravenous ribavirin administered within the first 6 days after fever onset improved the survival rate from 24% to 91%. The later initiation of treatment achieved a survival rate of 53%. Oral ribavirin also improved the survival rate, but not as effectively as when administered intravenously [6]. Bausch *et al.* proposed guidelines for the use of ribavirin as a post-exposure prophylaxis for Lassa fever (see review for details [65]). They recommended that oral ribavirin should start immediately in situations of high risk exposure to LASV. Upon the appearance of any symptom of Lassa fever and confirmation of disease by PCR, the route of ribavirin administration should be changed to intravenous [65]. Ribavirin is currently the best option for the treatment of Lassa fever, a disease that threatens to break out of its endemic zone due to the increased level of travel into and out of Africa, and due to the annual outbreaks. According to WHO, the most recent cases of Lassa fever were recorded in June 2017 in Nigeria [73].

Pharmacokinetics of Ribavirin

The efficiency of ribavirin is positively correlated with its concentration in the serum. Nevertheless, the same applies to ribavirin toxicity. Therefore, interplay between the dose of ribavirin and its absorption, distribution, metabolism and elimination is very important for successful therapy. The route of ribavirin administration also has a considerable influence, which depends on the virus against which the drug is directed. Thus, ribavirin may be given orally in the treatment of HCV infection, as an aerosol against RSV or intravenously to treat LASV fever. The pharmacokinetics of oral ribavirin in humans was investigated in healthy volunteers, as well as in patients with HCV infection [74 - 76]. The recommended daily dosage of ribavirin taken orally for adult patients with chronic hepatitis C ranges between 800 and 1200 mg, depending on the body weight and HCV genotype [77]. Upon oral administration, ribavirin is absorbed to a level of about 85%. However it was estimated that the mean bioavailability for a single oral dose of ribavirin (400 mg) was ~50%, suggesting a significant role of gastrointestinal or hepatic metabolism [75]. Taking ribavirin with a high fat meal increases its bioavailability by 46% when compared to fasting or with a standard meal [76]. Following intestinal absorption, ribavirin enters the plasma, after which it is widely distributed, also reaching the cerebrospinal fluid. Data collected from patients infected with HCV gave insight into single and multiple dose pharmacokinetics of ribavirin. Thus, following a single dose of treatment (600 mg oral ribavirin plus interferon), the plasma concentration of ribavirin exhibits a three-phase pharmacokinetic profile: (i) a rapid absorption phase with the time to maximum concentration of approximately 2 h, and the maximum concentration of 1030 ng/ml; (ii) a rapid distribution phase; and (iii) a prolonged elimination phase (with a half-life in the range from 44-49 h) [74]. Multiple dosing treatments entailed the administration of ribavirin twice a day at a total dose of 1200 mg, plus interferon for 6 weeks [74]. Again, the rapid absorption and distribution phases were recorded. The mean time to maximum concentration was in the range of 2-3 h and the maximum concentration was 3483 ng/ml. The range of mean washout half-lives following multiple dosing was 274–298 h, due to the slow elimination of ribavirin from tissue compartments other than the plasma (so called "deep compartments"). Four weeks following treatment cessation, ribavirin was still detectable [74]. Elimination routes for ribavirin are metabolic and renal. Two enzymes, purine–nucleoside phosphorylase and adenosine deaminase, are involved in the degradation of ribavirin to 1,2,4-triazole-3-carboxamide and 1-β-d-ribofuranosyl-1,2,4-triazole-3-carboxylic acid, respectively. The amount of unchanged ribavirin and its main catabolite, triazole carboxamide, excreted in the urine ranges between 15-25% and 6% for intravenous doses, or 5-10% and 30% for oral doses, respectively [78]. These data, together with the above-mentioned modest bioavailability of ribavirin after oral administration, are in favour of the first-pass metabolism through the gastrointestinal tract, since it was shown that bioavailability is not affected by liver dysfunction [79].

Regarding the pharmacokinetics of ribavirin administered by aerosolisation, there are data obtained in animal studies or in preliminary clinical studies with a small number of children with suspected RSV infection [80 - 82]. Experiments performed on mice entailed ribavirin administered as a standard drug concentration for a long treatment schedule (20 mg/ml for 12 h) or a high dose short-duration treatment (60 mg/ml for 2 h), both for three consecutive days [81]. In the respiratory system, ribavirin rapidly reached peak concentrations of 65 nmol and 112 nmol per lung, for standard and high dose treatments, respectively. The drug clearance was also rapid and mediated through mucociliary activity and absorption into the blood, without any accumulation of the drug in the lungs. Peak concentrations in the serum were in the range from 20–30 μM, independent of dose regimen, with a significant clearance before the start of the next treatment [81]. Pharmacokinetic studies of ribavirin given as an aerosol to humans are scarce and data are obtained from a small group of subjects [80, 83]. Thus, nine patients (7 years old) with a suspected RSV infection received aerosol treatment with ribavirin at a high dose of 60 mg/ml for a short duration of 2-hour periods, three times daily for up to 5 days [83]. In respiratory secretions, ribavirin achieved a peak concentration of 1725 μmol/L with short half-life of about two hours. The maximum plasma level after the first dose was about 4 μmol/L, reaching a steady state of 5 to 10 μmol/L after multiple treatments. After 3 days of therapy, a reduction of viral load was almost complete without causing resistance [83].

Ribavirin Uptake

Purine and pyrimidine nucleosides move between cells and the extracellular environment, as well as between some intracellular compartments, by two classes of nucleoside transporters: equilibrative nucleoside transporters (ENT) and concentrative nucleoside transporters (CNT). ENT family transporters (in humans encoded by the SLC29 gene family) include: ENT1, ENT2, ENT3 and ENT4. They enable the bidirectional facilitated diffusion of nucleosides along their concentration gradients [84]. In humans, ENT1 and ENT2 have broad substrate specificity and wide tissue distribution, being mostly located on the plasma membrane, but can also be found on nuclear and mitochondrial membranes; ENT3 also displays broad selectivity, and is widely expressed, but typically inside cells on lysosomal and mitochondrial membranes; and ENT4 mediates the transport of adenosine and monoamines, and is expressed on the plasma membrane of heart, brain and skeletal muscle cells [85]. CNT family transporters (in humans encoded by the SLC28 gene family) include: CNT1, CNT2 and CNT3, all of which enable the sodium-dependent transport of nucleosides and nucleoside drugs, while CNT3 also being able to transfer substrates in a proton-coupled manner [84, 86]. Uridine is transported by all three CNTs, but each shows substrate selectivity [86, 87]. Thus, in humans, CNT1, which is located on the plasma membrane and is predominantly distributed in the kidney, liver and small intestine, has specificity for pyrimidine nucleosides; CNT2 is expressed on the plasma membrane of heart, skeletal muscle, liver, kidney, intestine, pancreas, placenta and brain cells and is selective for purine nucleosides; and CNT3 transports purine as well pyrimidine nucleosides, is broadly distributed with the highest concentration in mammary glands, the pancreas, bone marrow, the trachea and intestine, and can be found in plasma and intracellular membranes [85, 86].

Nucleoside analogues, many of which are antiviral and anticancer drugs, including ribavirin, are hydrophilic compounds transported across membranes *via* both classes of nucleoside transporters. *In vitro* experiments carried out on human choriocarcinoma cells and *Xenopus laevis* oocytes expressing human nucleoside transporters demonstrated that ribavirin is a substrate for CNT2, CNT3, ENT1 and ENT2, whereby CNTs have a higher affinity for ribavirin compared to ENTs [88]. Several studies have dealt with the issue of the role of nucleoside transporters on the availability, efficacy and toxicity of ribavirin. Thus, it was shown that in the mouse, the uptake of low concentrations of ribavirin (20 μM) from the intestinal lumen mostly relies on the sodium-dependent process mediated by CNT2, while exit from the enterocytes into the mesenteric blood is mediated by ENT1 [89]. Increases in the concentration of ribavirin within the intestinal lumen (200 or 5000 μM) lead to the saturation of these transporters, hence other mechanisms are involved in its absorption [89]. Since HCV is the clinically approved target of ribavirin, the uptake of this drug by hepatocytes has also been the focus of several studies. Thus, it was shown that ribavirin enters human hepatocytes mostly by ENT1, and that other nucleoside transporters have little or no significance in ribavirin uptake by these cells [90, 91]. Since haemolytic anaemia is a well-known and dose-limiting side effect of ribavirin, the entry of this drug into erythrocytes was also of interest. It was found that ENT1 plays a significant role in the erythrocyte distribution of ribavirin [92, 93].

Ribavirin Metabolism

Once inside the cell, ribavirin becomes a substrate for adenosine kinase (ADK), yielding the product ribavirin monophosphate (RMP). RMP is further phosphorylated by nucleoside mono- and di-phosphate kinases into ribavirin triphosphate (RTP), the major intracellular metabolite of ribavirin in mammalian cells [94 - 96]. However, ADK has a higher affinity for endogenous nucleoside adenosine, with *Km* values for ribavirin being greater than 1000 times that of adenosine [97 - 99]. Thus, the phosphorylation of ribavirin into RMP is a rate-limiting step of RTP production. *In vitro* experiments showed that the incubation of mouse lymphoma cells with pharmacologically relevant concentrations of ribavirin over the duration of a few hours results in excessive intracellular concentrations of RTP, which reach more than 100 μM [96]. Importantly, the phosphorylation of ribavirin is a reversible process in cells with a nucleus, and is irreversible in anucleate cells [100]. Thus, following the removal of ribavirin from the culture system, RTP is quickly dephosphorylated with a half-life of less than 2 h [100]. On the contrary, anucleate erythrocytes lacking cellular phosphatases (5′-nucleotidase and alkaline phosphatase) that degrade phosphorylated forms of ribavirin eliminate ribavirin rather slowly with a half-life of about 40 days [101].

The major side effect of ribavirin, a reversible haemolytic anaemia, is actually a consequence of the accumulation of RTP in erythrocytes. One of the explanations for this is that excessive RTP induces a relative deficiency of adenosine triphosphate (ATP), resulting in an interruption of metabolism in erythrocytes, which consists of ATP-dependent glycolysis and the pentose phosphate cycle [102, 103]. Consequently, the production of glucose-6-phosphate is reduced, leading to oxidative damage of the membrane. This, in turn, causes the premature elimination of injured erythrocytes, challenged by ribavirin from the circulation, reducing the lifespan of these cells [103]. Haemolytic anaemia occurs in particular in patients suffering from chronic HCV infection, due to the long-term exposure to ribavirin therapy. Although this undesirable effect of ribavirin is considered to be a mild one, since it is reversible, it may lead to dose modification (*i.e.* reduction) and even to the cancellation of ribavirin administration [104]. A positive correlation between ribavirin concentration and its adverse reaction is an aggravating circumstance. Thus, it is very important to maintain adequate dosages of ribavirin and to conduct a full length therapy in order to eradicate HCV virus from the body [104, 105]. There are some guidelines that are followed in order to maintain the balance between drug toxicity and efficacy. It was shown that a steady-state plasma concentration of ribavirin which exceeds 3.5 μg/mL is concurrently associated with a drop in haemoglobin level below 8.5 g/dL, and is predictive for the appearance of haemolytic anaemia, which would probably be in conflict with ribavirin therapy [104, 106, 107].

Results published in 2010, obtained from genome-wide association study, have shed light on ribavirin-induced anaemia. Fellay and colleagues demonstrated that genetic variants in the ITPA gene (which encodes inosine triphosphatase - ITPase), protects patients (European-Americans, African-Americans and Hispanics) infected with HCV genotype 1 from haemolytic anaemia caused by ribavirin treatment [108]. They identified a single nucleotide polymorphism (SNP), rs6051702, with a major A allele on chromosome 20 that was strongly associated with haemoglobin reduction, early in the therapy with pegylated interferon alpha and ribavirin. Furthermore, two polymorphisms in the ITPA gene (rs1127354 and rs7270101) were found almost exclusively on chromosomes that also carry the protective rs6501702 minor C allele [108]. These two functional variants are known to be responsible for the benign inherited condition of ITPase deficiency, characterised by the accumulation of the substrate of this enzyme, inosine triphosphate (ITP), in the erythrocytes. Thus, the question remains, how are hepatitis C patients with simultaneous ITPase deficiency protected against ribavirin-induced anaemia? Hitomi *et al.* provided evidence how excessive ITP in certain individuals, could have a protective role in the red blood cells that encountered ribavirin. [109]. Namely, one of the known effects of ribavirin is the depletion of the guanosine triphosphate (GTP) pool, due to the inhibition of IMPDH. GTP is the main source of energy for the conversion of inosine monophosphate (IMP) to adenylosuccinate (ADS), which marks the beginning of ATP biosynthesis (Fig. **1**). This reaction is catalysed by adenylosuccinate synthase (ADSS). It was shown that ITP could replace erythrocyte GTP as a substrate for ADSS, thus enabling the unhindered synthesis of ATP (Fig. **1**) [109]. As mentioned above, maintaining ATP at an optimal level is very important for avoiding oxidative damage in the erythrocytes. Furthermore, a protective effect of ITPase deficiency against ribavirin related anaemia was recorded not only early in the treatment (after 4 weeks), but also over the entire treatment period of 48 weeks [110]. In addition, ITPA variants were also protective against anaemia during the treatment of individuals with chronic genotype 2/3 hepatitis C [111]. In the Japanese population infected with HCV genotype 1, SNP rs7270101 was not polymorphic, while SNP rs1127354 was shown to be a predictive factor for ribavirin-induced anaemia [112]. The determination of ITPA polymorphisms was also assessed during the treatment of hepatitis C with a combination of DAA and ribavirin. For instance, SNPs rs1127354 and rs6051702 were shown to be helpful in the prognosis of anaemia during treatment with telaprevir-based triple therapy (containing pegylated interferon alpha and ribavirin) or in DAA interferon-free, ribavirin-containing regimens [113, 114]. Although the determination of ITP variants proved to be a valuable tool in the clinical management of ribavirin, the association of ITPA genotypes and sustained virological response were never documented [108, 110, 114, 115]. This is partly due to the fact that only a minority of HCV-infected individuals (about 30%) have moderate to severe ITPase deficiency, which is able to protect them against ribavirin-induced anaemia [110].

Nucleoside transporters, CNTs and ENTs have also emerged as factors that could influence ribavirin bioavailability *via* changes in function and/or number, and thus have an impact on anaemia and antiviral response. Analyses of the gene families SLC28 and SLC29, which encode CNTs and ENTs, respectively, revealed that a genetic variant (haplotype rs10868138G/rs56350726T) in the SLC28A3 gene, which encodes CNT3, was protective against haemolytic anaemia in patients with chronic genotype 1 hepatitis C, treated with pegylated interferon alpha and ribavirin [116]. It is supposed that this genetic variant encodes CNT3 with decreased transporter function, resulting in the hampered entry of ribavirin into erythrocytes and consequently to less of a haemolytic effect of the drug. At the same time, this genetic variant in SLC28A3 was protective against anaemia and did not affect SVR. After comparing the allelic frequency between SNPs in SLC28A3 and ITPA, it was concluded that the clinical importance of the genetic variants in SLC28A3 is lower than that of the variants in the ITPA gene [116]. In addition, Rau *et al.* analysed Swiss hepatitis C cohort treated with pegylated interferon alpha and ribavirin and found that SNP rs11854484 in SLC28A2 gene that encodes CNT2 (in charge for intestinal uptake of ribavirin) is associated with higher ribavirin levels and the development of anaemia [117]. Furthermore, SLC28A2 rs11854484 was a predictive marker of clinically-significant anaemia in patients with chronic genotype 1 hepatitis C treated with triple therapy (DAA – boceprevir, pegylated interferon alpha and ribavirin) [118]. Also, genetic analyses help to identify a subgroup of patients with a better tolerance of triple therapy. Those were found to carry a favourable haplotype of SLC28A2 rs11854484 and ITPA rs1127354, with no need for blood transfusion or erythropoietin administration [118].

Determining ITPA or the combination of ITPA/SLC28 polymorphisms prior to the ribavirin-containing therapy of HCV-infected patients could be a significant factor in clinical practice. It would enable individuals prone to the development of anaemia to be identified, who would need closer clinical monitoring and a more personalised approach to ribavirin dose management.

With regard to the interrelation between ribavirin and SVR, it was found that a higher concentrations of the drug in the plasma is associated with a desirable outcome in the treatment of HCV infection [104, 105]. This was particularly important in patients infected with HCV genotype 1 or 4, where therapeutic monitoring of ribavirin was highly recommended [105]. Thus, several clinical studies found that serum ribavirin concentrations equal to or higher than 3 μg/ml at 8 or 24 weeks after the initiation of combination therapy with IFN are related to higher rates of SVR [106, 119, 120]. However, bearing in mind that a ribavirin plasma concentration of 3.5 μg/mL measured in the same period represents the threshold for treatment cessation, it leaves a very narrow therapeutic gap (between 3 and 3.5 μg/ml), representing a seesaw between attaining SVR and inducing high toxicity [106]. In addition, Maynard *et al.* suggested that a concentration of ribavirin in plasma higher than 2 μg/mL measured at week 4 may be the cut-off level to achieving higher rates of SVR [121]. Although patients infected with HCV genotype 2 or 3 are highly responsive to ribavirin, there are data that suggest a certain threshold of drug plasma concentration leading to the high percent of SVR (*i.e.* 84% of SVR was achieved when ribavirin plasma level was higher than 2 μg/mL at day 29 of the combination treatment compared to 66% of SVR for ribavirin plasma level less than 2 μg/mL) [122].

Mechanisms of Antiviral Action of Ribavirin

Despite the fact it has been 45 years since the synthesis of ribavirin, it is still not exactly clear how this drug exerts its antiviral activity. There are several proposed mechanisms, each being supported with experimental evidence. However, none of them offer a full and exact understanding of the action of this drug against viruses. There is no universal explanation for each situation that implies the utilisation of this drug. Ribavirin has been tested against many viruses *in vitro* and a lot of information about its operating mechanisms has come from these investigations. However, in this section, attention will be paid to data related to viruses against which ribavirin is approved (*i.e.* HCV and RSV). Before we go further into this issue, we will emphasise one fact about ribavirin, which is that it is very important for the interpretation of the proposed mechanisms of action. Namely, ribavirin is mostly used to treat infections caused by HCV, and when given as a monotherapy, it induces a temporary reduction of alanine aminotransferase levels, but without any significant effect on virus clearance from the body [123, 124]. However, one clinical study on a small number of patients showed that ribavirin monotherapy induced the short-term reduction of viral load early in the treatment [125]. On the contrary, when given in combination with interferon, SVR rates were significantly improved when compared to interferon monotherapy [50, 51]. Dixit *et al.* presented a mathematical model of HCV RNA decay profile following interferon therapy, which consists of two phases: (i) first, a rapid one that happens due to the inhibition of viral replication (ii) and second, a much slower one which happens due to the elimination of infected hepatocytes by immune cells [101, 126]. This modelling revealed that ribavirin acts synergistically with interferon in some patients, by improving levels of viral decline of second slower phase, indicating that the origin of this synergy probably lies in ribavirin antiviral activity [101, 126].

RMP as an Inhibitor of IMPDH

We have already mentioned that ribavirin is a prodrug that achieves pharmacological activity upon phosphorylation into RMP, RDP and RTP. Here, known mechanisms of action of ribavirin will be presented with regard to its active metabolites.

The antiviral action of ribavirin that relies on RMP belongs to indirect mechanisms, as it implies an interaction with the enzyme of the host that will, in turn, result in the interruption of the virus life cycle (Fig. **4A**). Namely, RMP targets enzyme IMPDH in a competitive inhibition with its natural substrate – IMP [12, 127]. Actually, based on Ki values, RMP interacts at the ligand site of IMPDH with higher affinity than the natural substrates and products [18]. IMPDH catalyses the NAD-dependent conversion of IMP to xanthosine monophosphate (XMP) as a crucial step in the *de novo* synthesis of guanine nucleotide (Fig. **1**). XMP is a substrate for guanosine monophosphate synthase (GMPS); this catalyses the amination of XMP to guanosine monophosphate (GMP), which is then converted to guanine metabolites such as guanosine diphosphate (GDP), GTP, deoxyguanosine di- and tri-phosphate (dGDP, dGTP) (Fig. **1**). Thus, the inhibition of IMPDH results in prominent decline of guanilate pools. In turn, lack of GTP/dGTP could affect the virus in two ways: (i) since virus replication is GTP- and dGTP-dependent, one possible antiviral mechanism of ribavirin could be the deprivation of material for RNA or DNA synthesis, which will result in the inhibition of viral replication; and (ii) an imbalance in building elements for RNA/DNA synthesis could cause misincorporation by viral polymerase, resulting in mutations that could affect virus infectivity and survival. Cells exposed to ribavirin are characterised by GTP levels that are reduced up to 90% [18]. However, it is questionable to what extent the inhibition of IMPDH contributes to the antiviral activity of ribavirin. There are some *in vitro* studies that favour IMPDH inhibition as a predominant mechanism by which ribavirin exerts its antiviral activity against measles virus, yellow fever virus, human parainfluenza virus and GB virus B (GBV-B, member of the same genus like HCV) [12, 128, 129]. Evidence for this statement is grounded on the linear correlation between the antiviral activities of ribavirin (inhibition of viral RNA replication, reduction of viral infectivity) and level of GTP depletion, and the prevention of the inhibitory action of RMP by the exogenous administration of guanosine [12, 128, 129]. However, the antiviral activity of ribavirin against RSV, LASV or orthopoxviruses *in vitro* was not antagonised by the addition of guanosine [130 - 132]. It is interesting that orthopoxviruses are sensitive to mycophenolic acid (MPA), which is uncompetitive and well-established inhibitor of IMPDH, meaning that the reduction of GTP levels could produce an antiviral effect, at least when comes to orthopoxviruses [131]. On the other hand, when tamarin hepatocytes infected with GBV-B were treated with MPA, no antiviral effect was registered [128]. Furthermore, although ribavirin is effective against the same virus *in vitro*, in GBV-B-infected tamarins it has not shown any kind of antiviral activity at concentrations typically used *in vivo* [128]. Considering HCV, the situation is similar to that of its close relative GBV-B. In the HCV RNA replication system, ribavirin reduced HCV replicon colony-forming efficiency in a dose-dependent manner, in an effect that was even more pronounced by the addition of MPA and/or VX-497 (another potent IMPDH inhibitor) [133]. This action was markedly antagonised by the addition of exogenous guanosine [133]. However, founded on facts that MPA and VX-497 alone had marginal success in inducing an anti-replicon effect and that VX-497 or ribavirin as a monotherapy are not sufficient to produce SVR, it can be concluded that the inhibition of IMPDH may be required, but certainly is not primary or sufficient for the antiviral effect of ribavirin in HCV infection [123, 124, 133, 134].

Actions of RTP

RTP is the most abundant metabolite of ribavirin, with actions that are based on direct interactions with the virus, more precisely with the viral RNA polymerase that could result in: (i) the inhibition of virus replication due to the inhibition of the enzyme itself or due to the incorporation of RTP as an alternative substrate; and (ii) RNA mutagenesis due to the incorporation of RTP as an alternative substrate, but without causing chain termination (Fig. **4A**) [18]. There is an additional mechanism linked to this metabolite that may be functional in some RNA viruses that contain a 7-methylguanosine cap on the 5'-end, which is a structure that is necessary for RNA stability and translation (Fig. **4A**) [135].

The Inhibition of Virus Replication

Because concentrations of RTP in cells exposed to ribavirin exceed 0.1 mM and it was demonstrated that it can be recognised either as an analogue of GTP or ATP [21, 136, 137], which have intracellular concentrations of about 0.5 mM and 3 mM, respectively, then RTP has good potential to interfere in virus replication instead of natural nucleotides [138]. Thus, it is reasonable to suppose that viral polymerases could recognise RTP as a substrate, and that its misincorporation will cause chain termination or will prevent other nucleotides from binding, which will cause the inhibition of chain elongation. Indeed, it was shown *in vitro* that RTP is able to inhibit several viral RNA polymerases, such as those from influenza virus, vesicular stomatitis virus, La Crosse virus and Reo virus [28, 139, 140]. Furthermore, poliovirus polymerase and HCV polymerase can incorporate RTP opposite cytidine and uridine with equal efficiency [21, 136, 137]. While in the case of poliovirus this action results in the mutagenesis of the genome, in the case of HCV, the incorporation of RTP can also cause a significant block in RNA elongation [21, 136, 137]. However, the rate of RTP incorporation was 50-100 lower than that of the natural nucleotides, and concentrations of ribavirin necessary to induce the inhibition of polymerase activity were much higher than clinically applicable [137]. Furthermore, if we recall the fact that ribavirin given as a monotherapy in the patients with chronic hepatitis C exerts a transient antiviral effect that is significantly smaller when compared to the magnitude of the decline achieved with interferon, and that in combination therapy ribavirin does not affect the first rapid phase of the HCV RNA reduction, it is likely that ribavirin is a weak direct inhibitor of HCV viral polymerase in clinical conditions [101, 125, 126]. However, some claim that this weak antiviral activity of ribavirin becomes significant upon the decline of viral load caused by interferon, when ribavirin is able to eradicate the remaining low level of virus that could otherwise have caused a relapse [141].

Induction of Mutagenesis

With the release of the study of Crotty *et al.*, a new mechanism of action of ribavirin emerged [21]. This research started with the characteristics of RNA virus replication that imply high mutation rates, high yields, and short replication times in order to rapidly adapt to environmental challenges and ensure survival [142]. As a consequence, RNA viruses usually exist as viral quasispecies – a collection of related but genetically distinct variants that interact cooperatively on a functional level, and collectively contribute to the characteristics of the population [142, 143]. The quasispecies concept gave rise to an error threshold, *i.e.* a maximum mutation rate beyond which viral fitness is endangered [142, 143]. *In vitro* studies of Crotty *et al.* showed that poliovirus RNA polymerase incorporates RTP as a nucleotide analogue of either ATP or GTP, since it is equally paired with cytidine or uridine [21]. The result of this erroneous pairing was not chain termination, but the increase in mutation rate, particularly evident at higher concentrations of ribavirin (*i.e.* 2 point mutations/wild type poliovirus genome *vs.* 15 point mutations/poliovirus genome in the presence of ribavirin at a dosage 1000 μM). Furthermore, the mutagenic action of ribavirin correlated with the antiviral action of this drug, so the production of infectious poliovirus in cell culture was significantly reduced. The authors concluded that ribavirin forces the virus into lethal mutagenesis, aka 'error catastrophe'. Furthermore, the mutagenic activity of ribavirin could be enhanced by the inhibitory activity of RMP on IMPDH, since lowering GTP levels probably increase the chance of RTP being incorporated into poliovirus nascent RNA strands [21]. The infliction of lethal mutagenesis by ribavirin was shown to function in HCV *in vitro*; however, only at high ribavirin dosages, as mentioned above. Similar to the poliovirus RNA polymerases, it was demonstrated that HCV RNA polymerases recognise and equally base-pair RTP with cytidine and uridine [136, 137]. Vo *et al.* determined that under conditions of low GTP concentrations, HCV RNA polymerases extend the RNA chain past the incorporated RMP and that multiple RMP incorporations result in stalled elongation complexes with reduced levels of RNA synthesis [137]. Therefore, they concluded that the antiviral action of ribavirin is achieved through the induction of mutations and the inhibition of viral synthesis [137]. Clinical studies on this matter provide evidence for and against the increasing mutation rate during the treatment with ribavirin [144 - 147] (Table **1**).

Table 1. Clinical studies on ribavirin as HCV mutagen.

Clinical studies	No. of patients	Ribavirin dosage, duration	Sequence analysis of HCV non-structural (NS) proteins	Nucleotide mutations rate	Conclusion
Young *et al.* [147]	HCV genotype 1a, 5 patients	1200 mg/day for 36 weeks	NS5B	ribavirin *vs*. placebo group, 9.1 *vs*. 3.1 X 10^{-3}/site/year, (P=0.064)	"RBV could work as a weak mutagen for HCV RNA in HCV-infected patients. "
Asahina *et al.* [144]	HCV genotype 1b, 34 patients	600 or 800 mg/day for 4 weeks followed by a 24-week course of IFN/ribavirin therapy.	NS5A and NS5B	ribavirin *vs*. non-treatment observation period, 2.9 (NS5A) and 1.3 (NS5B) *vs*. 0.60 (NS5A) and 0.24 (NS5B) X 10^{-2}/site/year, (P=0.02)	"Ribavirin may act as a mutagen, and mutations occurring during ribavirin therapy correlate with the virological response to subsequent IFN/ribavirin combination therapy."
Chevaliez *et al.* [145]	HCV genotype 1b, 4 patients	1000 or 1200 mg/day for 24 weeks	NS3 and NS5A	See Table **5**. in Chevaliez *et al.* (error generation rate is shown individually for all patients in the study)	"Ribavirin monotherapy did not increase the rate of variation of the consensus sequence, the mutation frequency, the error generation rate, or the between-sample genetic distance."
Lutchman *et al.* [146]	HCV genotype 1, 31 patients	placebo-controlled trial of ribavirin for 48 weeks was crossed-over to open-label ribavirin 1200 mg/day for 48 weeks	NS5B	placebo that crossed over to ribavirin - after 4 weeks 2.1 *vs*. 4.4 X 10^{-3} per site/year, (P=0.02) - after 24 weeks 5.5 *vs*. 4.0 X 10^{-3} per site/year, (P=0.1) ribavirin *vs*. placebo in randomized phase after 24 weeks 6.6 *vs*. 4.3 X 10^{-3} per site/year, (P=0.4)	"Ribavirin therapy is associated with an early, transient increase in the mutation rate of HCV. Lethal mutagenesis and error catastrophe is unlikely to be the sole mechanism of action of ribavirin during therapy for CHC."

Inhibition of Capping Machinery

The capping machinery entails three enzymes: RNA triphosphatase, RNA guanylyltransferase, and RNA methyltransferase, each of which could represent a potential target for RTP, which will eventually result in the inhibition of the capping of RNA genomes. RTP could also be incorporated as a cap analogue, which will result in the interruption of RNA translation [148]. As early as 1979, Goswami *et al.* found that RTP is a potent competitive inhibitor of vaccinia virus mRNA guanylyltransferase [149]. Furthermore, Bougie and Bisallion showed that RTP can directly serve as a substrate for the vaccinia virus D1 protein, which displays activities like two RNA capping enzymes: triphosphatase and guanylyltranseferase [150]. In this case, vaccinia RNA transcripts are capped by ribavirin, which undermined the process of translation. The inhibitory activity of RTP considering capping machinery was also demonstrated in dengue virus. Precisely, RTP competes with GTP to bind the 2'-O-methyltransferase NS5 domain of the virus [151]. There is a possibility that the simultaneous inhibition of IMPDH could enhance this action by lowering the competing material (*i.e.* GTP) [151]. However this mechanism has noteworthy limitations, since some studies have demonstrated that ribavirin does not act as a functional mimic of the mRNA cap [152, 153]. Besides, ribavirin has a broad spectrum antiviral activity, whereas many viruses sensitive to it do not display a cap structure, HCV for example [148].

Additional Indirect Mechanisms of Ribavirin

There is some evidence in support of two more actions of ribavirin that could contribute to the antiviral effect achieved against HCV and RSV.

First, there is a possibility that ribavirin acts as an immunomodulator of host adaptive immune response, which plays a key role in the outcome of HCV infection, since well-coordinated cross-talk between all participants of humoral (antibodies) and cell-mediated (CD4+ T helper (Th) cells, and CD8+ T cytotoxic cells) immune responses is associated with viral clearance [154]. However, the main effector cells in anti-HCV immune response are virus-specific CD8+ T cells, which act by cytolytic or non-cytolytic mechanisms (*e.g.* killing of infected hepatocytes or secretion of antiviral cytokines such as interferon-gamma (IFN-γ) and tumour necrosis factor-alpha (TNF-α), respectively) (Fig. **4A**) [154]. Regarding the role of HCV-specific CD4+ Th cells, it is considered that they are essential in preventing viral escape from CD8+ T cell response [155]. As well as enhancing the activity of CD8+ T cells, they also produce cytokines as specific messengers to infected hepatocytes. Thus, there are two subpopulations of Th cells: Th1 cells that produce type 1 cytokines, such as IFN- γ, TNF- α, and IL-2, which promote cell-mediated immunity; and (ii) Th2 cells that produce type 2 cytokines, such as IL-4, IL-5, IL-6, and IL-10, in order to regulate Th1 cells and stimulate the humoral response [155]. Actually, disruption of the balance between Th1 and Th2 cells is probably responsible for the development of chronic HCV infection, provided that the data on the contribution of one or the other subpopulation are conflicting [155, 156]. Thus, it was determined that in patients with self-limited HCV infection, Th1 responses were predominant, while the activation of Th2 responses may accelerate transition to the chronic state [157]. In addition, elevated Th2 cytokine serum levels were detected in a proportion of patients with chronic HCV infection [158]. Furthermore, a decrease in HCV RNA levels as a consequence of interferon alpha treatment was associated with the drop in type 2 cytokine (IL-4 and IL-10) levels [159]. On the other hand, it was shown that the progressive liver injury in chronic hepatitis C correlates with the up-regulation of intrahepatic IL-2 mRNA and the down-regulation of IL-10 mRNA levels [160].

Several studies have demonstrated that ribavirin alters the T-helper cell phenotype from type 2 to type 1 (Fig. **4A**). Thus, Tam *et al.* showed that in isolated and activated human T-lymphocytes, ribavirin enhanced a type 1 (IL-2, IFN- γ, TNF-α) while suppressing a type 2 cytokine response (IL-4, IL-5 and IL-10), at the level of both protein and mRNA expression [161]. In mice immunised with hepatitis B virus (HBV) core- or e-antigens and treated with ribavirin, core antigen- or e antigen-specific IL-2 and IFN-γ levels were more than 25-fold higher compared to untreated mice [162]. In addition, changes in the Th1/Th2 balance in favour of Th1 were noted in ribavirin-treated HBV e-antigen transgenic mice [162]. Furthermore, when peripheral blood mononuclear cells (PMBCs) from patients with chronic hepatitis were co-cultured with HCV antigens, ribavirin treatment (at a dosage of 2-5 μM) induced *de novo* or enhanced T-cell proliferation to HCV antigens with the concomitant suppression of IL-10 [163]. Cultured PMBCs from chronic hepatitis C patients were also treated with a combination of ribavirin and interferon alpha. Together, they had an opposite effect on the production of IL-2 and IL-12, and enhanced the secretion of IFN-γ and IL-10 [164]. On the other hand, in similar experiments, it was shown that the synergistic action of combination therapy may rely on ribavirin ability to induce an additional decrease of IL-10 secretion compared to interferon alpha alone [165].

Another possible mechanism of ribavirin action is based on the interaction with its partner in combination therapy - interferon alpha. A significant share of the antiviral activity of interferon alpha is achieved through the activation of interferon-stimulated genes (ISG), the products of which exert antiviral effector functions (Fig. **4A**) [166, 167]. Thus, liver biopsies from chronic hepatitis C patients responsive to interferon alpha therapy revealed that ISG levels were low before, but increased significantly after treatment. On the contrary, non-responders had high expression levels of ISGs before therapy that remained similar after treatment [168]. Ribavirin has been suggested to improve the outcomes of HCV infections by enhancing ISG response [168]. This statement is in agreement with several other *in vitro* and *in vivo* studies [169 - 172]. Furthermore, Feld *et al.* analysed early HCV kinetics and markers of the ISG response in patients with chronic hepatitis C [173]. Ribavirin was demonstrated to improve the second phase of viral decline, solely in patients with an adequate initial response to peg-interferon. They also showed correlation between viral decline and enhancement of interferon-related cytokine induction [173]. This mechanism offers a very likely explanation on how ribavirin potentiates the effect of interferon alpha when applied together, and may also be at the root of its broad-spectrum antiviral activity [141]. In addition, it was suggested that the up-regulation of ISG by ribavirin is mediated by the inhibition of a transcriptional repressor, which is a novel mechanism that differs from those associated with interferon signalling [174].

The Future of Ribavirin as an Antivirus drug

What is the future of the utilisation of ribavirin against its approved targets: RSV, LASV and HCV?

De Clercq stated that between his two reviews about perspectives for the chemotherapy of RSV infections, one in 1996, and the other in 2015, modest progress has been made in the prophylaxis or therapy of this disease [55, 175]. Despite the fact that the effectiveness of ribavirin in the treatment of RSV infection is ambiguous, so far it is the only antiviral drug approved for this important respiratory disease in infants and young children [55]. However, ALN-RSV01, which is small interfering RNA directed against the mRNA of the RSV nucleocapsid protein, and GS-5806 which inhibits RSV fusion with the host cell, showed promising results in experimental RSV infection in humans [176, 177].

Similarly to RSV infection, there is no other approved antiviral drug for the treatment of Lassa fever than ribavirin; therefore, clinical trials with new agents are hampered by the economic and political instability of West African countries affected by the outbreak of LASV infection.

An ideal solution for managing HCV infection would be the development of a prophylactic HCV vaccine. However, high variability of the virus seems to be an insurmountable obstacle, for now. Nonetheless, the treatment of chronic HCV infection has been revolutionised by the development of all oral agents specifically targeting steps in the HCV life cycle (already mentioned DAAs). The question is whether the treatment era of interferon and ribavirin would definitely go into the past? Nowadays, the preferred and recommended treatment regimen is definitely with DAAs. However, there are certain scenarios where the administration of ribavirin with some existing DAAs could represent added value to the success of the therapy. Thus, ribavirin might be a part of alternative treatment option in order to shorten the duration of therapy and/or to prevent viral relapse in a few subgroups of patients (see Table **1** in Manns *et al.* [41]). These subgroups of patients are determined by viral genotype and subtype, the previous treatment experience and the presence of cirrhosis [41]. For instance, according to the World Health Organisation guidelines, a regimen consisting of sofosbuvir/pegylated interferon/ribavirin is recommended as an alternative treatment option for patients with HCV genotype 3 infection with cirrhosis, and for patients with genotypes 5 and 6 infection with and without cirrhosis [178]. DAAs achieve remarkable cure rates. However, they may be selected for resistant strains; hence, pegylated interferon might be required. In this case, ribavirin is an indispensable companion. Economic influence, *i.e.* drug cost, should be included as a factor that maintains the use of interferon/ribavirin based therapies. Namely, regardless of the low-cost production, DAAs are very expensive, causing a preference to use old regimen (with relatively good response rate) in some countries. For example, DAA regimens are preferred in Europe, North America and Japan, while China, which has the highest number of hepatitis C patients in the world, commonly uses interferon/ribavirin-based regimens [41, 179, 180].

In addition, ITPA polymorphisms were shown to be useful in predicting anaemia in individuals with chronic hepatitis C treated with DAA regimens that contained ribavirin or interferon/ribavirin [113, 114]. Therefore, in subgroups of patients that are candidates for alternative treatment options which include ribavirin, testing for the presence of ITPA polymorphisms might be of clinical significance, in terms of personal monitoring and dose management.

In summary, although ribavirin performance against RSV and LASV infections has many drawbacks, it still remains the only approved antiviral drug, without a worthy competitor in the case of Lassa fever. Hopefully, the more efficient agents will appear in the case of RSV-related disease. The final goal in the treatment of HCV infection is to avoid interferon-based regimens, which became possible with the introduction of DAAs. Thus, it seems that the role of ribavirin will fade away over time, or in the best case, its use will be marginalised, limited to specific cases. Nevertheless, it would be a good strategy to give ribavirin a backup role. Given its multiple mechanisms of action, ribavirin could serve to bypass viral resistance, or be given guidance in the search for new molecules with a broad-spectrum of antiviral activities, helping in never-ending battle against viruses.

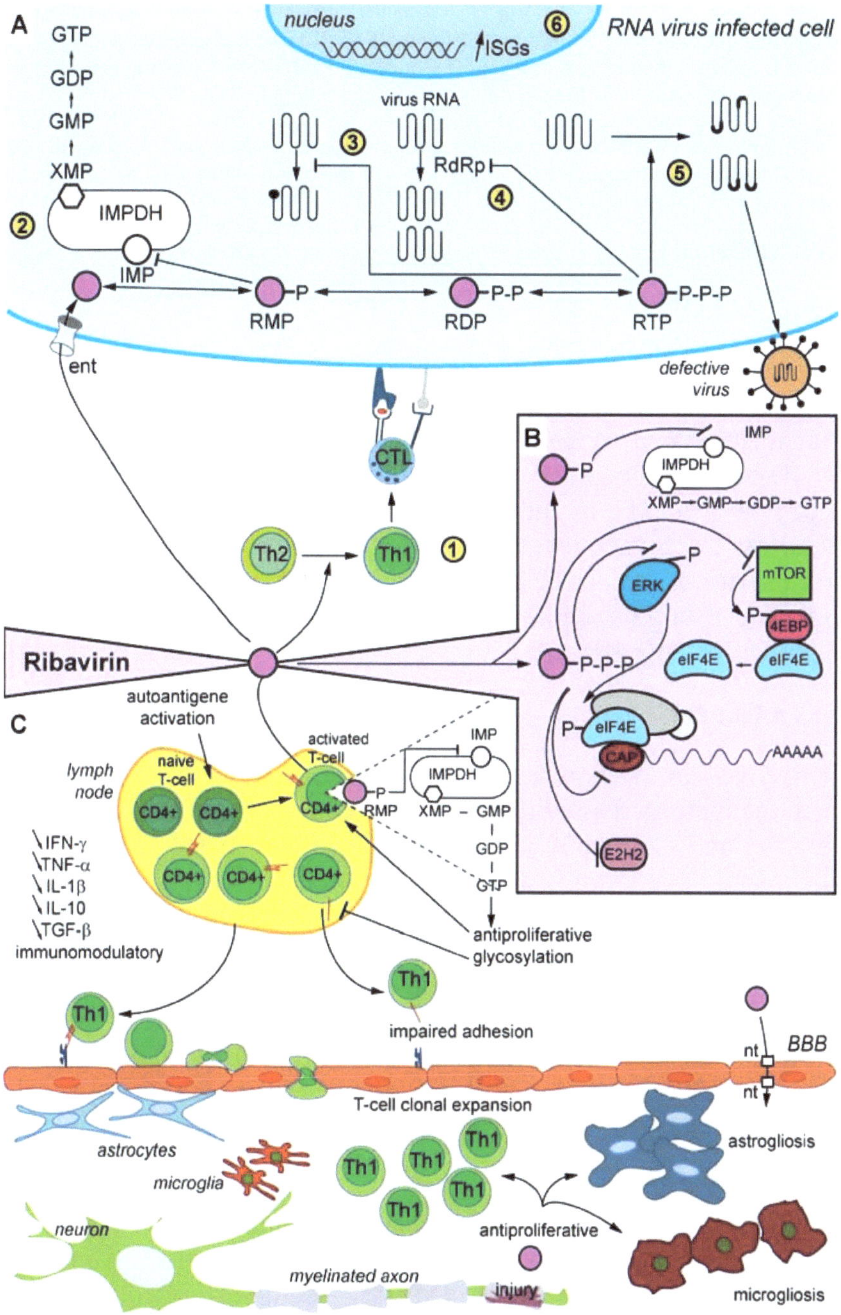

Fig. (4). Mechanisms of action of ribavirin. (**A**) Antiviral effect is achieved through several possible actions of ribavirin: 1 – immunomodulation by inducing Th1 response that will facilitate elimination of infected cells by cytotoxic T cells (CTL); 2 – inhibition of host IMPDH; 3 – inhibition of capping machinery; 4 – inhibition of virus replication by interaction with RNA-dependent RNA polymerase (RdRp); 5 – induction of mutagenesis; and 6 – activation of interferon-stimulated genes (scheme adapted from Feld and Hoofnagle

[311]). (**B**) The anticancer effect of ribavirin is achieved by affecting multiple cancer-targets. Ribavirin, through active metabolites, directly inhibits IMPDH and eIF4E. Ribavirin also prevents the phosphorylation of eIF4E by the down-regulation of Erk, favours the sequestration eIF4E by inhibiting the activity of mTOR, and decreases EZH2 expression. (**C**) Immunosuppressive and immunomodulatory actions of ribavirin during EAE on periphery (lymph node and outside BBB) and inside the CNS. Activated CD4+ T-cells are sensitive to IMPDH inhibition, and as a consequence cellularity of lymph node is decreased. Additionally, reduction in guanine nucleotides disturbs process of glycosylation of adhesion molecules, which are needed for transmigration of encephalitogenic T-cells across BBB. Ribavirin also alters cytokine milieu on the periphery by lowering levels of IFN-γ, TNF-α, IL-1β, IL-10 and TGF-β, and thus attenuating inflammation. Inside the CNS, ribavirin decreases the number of encephalitogenic T-cells and activated astrocytes and microglia.

RIBAVIRIN AGAINST CANCER CELLS

In the treatment of HCV infection, ribavirin is probably at the beginning of the end, to cite Feld [181]; however, some of the actions of ribavirin seem to be efficient against several cancer targets. Therefore, future investigation into this drug may be repositioned from the antiviral to the anticancer field. Here, the potential of ribavirin to be included in cancer therapy will be addressed. The first part summarises the consequences of IMPDH inhibition and guanine nucleotides depression in various cancer cells, while the second part deals with some novel promising cancer targets with particular focus on the eukaryotic translation initiation factor 4E (eIF4E).

IMPDH as a Cancer Target

Once again, ribavirin mechanism as competitive inhibitor of IMPDH comes into focus, since the increased activity of one of the isoforms of this enzyme is linked to proliferating and malignant cells [182, 183]. Back in 1983, Weber suggested that the neoplastic phenotype could be associated with the increased activity of the guanylate synthetic pathway, and more specifically with the increased activity of IMPDH, a crucial component of GTP biosynthesis [184]. When there is a high demand for guanine nucleotides, such as in the case of microbial infection or rapidly proliferating cells, this pathway becomes indispensable, since salvage pathways cannot respond to the challenge. Indeed, the elevated activity of IMPDH in tumour cells was confirmed in several studies [182, 183, 185 - 188]. Therefore, IMPDH is marked as a valid cancer target and the inhibition of this enzyme could represent one of the major approaches in cancer chemotherapy [189].

The IMPDH-dependent *de novo* biosynthesis of guanine nucleotides is ubiquitous in eukaryotes, bacteria and protozoa, with only a few exceptions. Some of the best characterised crystal structures of IMPDH include the following species: *Tritrichomonas foetus, Homo sapiens, Streptococcus pyogenes, Borrelia burgdorferi, Cricetulus griseus,* and *Pyrococcus horikoshii* (reviewed in [13]). Human and prokaryotic IMPDH share 30-40% sequence identity [190]. X-ray crystal structures of IMPDH showed that these enzymes exist as homotetramers regardless of the species they come from, and consist of monomers with molecular masses in the range from 56-58 kDa [13, 191]. Each monomer contains one $(b/a)_8$ barrel (the catalytic domain) and two CBS domains (named after cystathionine b-synthase because of the shared homology), which are not required for catalysis [13]. In humans, IMPDH exists as two isozymes, type I and type II, encoded by genes on chromosomes 7 and 3, respectively [192]. Both human isoforms contain 514 residues and are 84% identical in sequence structure, shearing similar kinetics, however with different roles [192, 193]. Type I is constitutively expressed in normal cells by most tissues; however, it is predominantly found in the retina and lymphocytes. It was shown that human lymphocytes critically depend on a *de novo* pathway of purine synthesis when they respond to mitogenic or antigenic stimulation [194]. It is considered that the expression of type II is significantly increased in highly proliferating cells [183, 186]. Pertinent to this, it was shown that type II mRNA predominates over type I in human ovarian tumours, human leukaemic cell lines, and in various types of leukaemic cells from patients [183, 185, 192]. In addition, it was shown that the down-regulation of type II mRNA is associated with the differentiation of leukaemic cells [186]. Differences between levels of the expression of type II IMPDH between normal lymphocytes and leukaemic cells suggested that selective inhibition of this isoform could represent an anti-leukaemic strategy. As mentioned in a previous section, through the inhibition of IMPDH, the biosynthesis of GTP is significantly affected and a pool of guanine nucleotides is reduced to the level at which cell functioning is endangered, because numerous biological processes depend on GTP (Fig. **4B**).

Here, a few concepts on how the depression of guanine nucleotides could affect cancer cells will be presented, based on studies under the guidance of Dr George Weber. Most obviously, GTP is of vital importance for RNA and protein synthesis. However, GTP is also necessary for DNA biosynthesis, since it is a positive regulator of ribonucleotide reductase and thus promotes the catalysis of GDP into dGTP. In these cases, the anticancer concept is clear. Namely, the reduction of GTP pools will cause an imbalance in nucleoside/deoxynucleoside triphosphates, which in turn will limit RNA, DNA and protein biosynthesis that will cause the slowdown or cessation of cell proliferation [14]. Next, GTP is involved in signal transduction as a co-factor of phospholipase C (PLC) [195]. The production of second messengers starts from phosphatidylinositol (PI) which is converted to 1-phosphatidylinositol-4-phosphate (PIP) and then to PI 4,5-bisphosphate (PIP2) through the actions of two synthetic enzymes, PI 4-kinase and PI 4-phosphate-5-kinase (PIP kinase), respectively. PIP2 is further metabolised into end products inositol-1,4,5-trisphosphate (IP3) and diacylglycerol (DAG) through the involvement of PLC. The synthetic activity of PI 4-kinase and PIP kinase is regulated by two phosphatases that degrade PIP and

PIP2 to PI. Elevated signal transduction activity is evidenced in animal (rat hepatomas) and human cancers (colon, breast and ovarian carcinoma cells and tissues) through the amplified production of IP3, which is a result of the increased activity of PI 4-kinase, PIP kinase and PLC and the decreased activity of catabolic enzymes [196, 197]. In this case, the anticancer concept is based on lowering the IP3 concentration *via* the reduced activity of PLC as a consequence of depressed levels of its coenzyme – GTP. Furthermore, it was shown that PI and PIP kinases have short half-lives; therefore, the activity of these enzymes could decrease if there is a restriction in macromolecular biosynthesis, as in the case of the inhibition of IMPDH and the limitation of purine biosynthesis [197, 198].

Guideposts for signalling pathways to be examined, with regard to the antitumor effects of ribavirin, arose from the investigation of tiazofurin. This is another nucleoside analogue which targets IMPDH, and was primarily designed as an anticancer drug, proving to have wide antitumor activity against murine and human neoplasms. Tiazofurin (2-b-D-ribofuranosylthiazole- 4-carboxamide), like ribavirin, is a prodrug that is converted to its active form thiazole-4-carboxamide adenine dinucleotide (TAD) inside the cell. As opposed to RMP, which inhibits IMPDH by binding to the IMP-XMP site, TAD is the NAD analogue that potently blocks IMPDH activity at the NAD–NADH ligand site. In the early eighties, tiazofurin was demonstrated to be efficient against Lewis lung carcinoma, as well as against murine P388 and human lymphoid tumour cell lines in culture [199 - 201]. The antitumor effects achieved by reduction of IMPDH activity and dGTP pool levels have been confirmed for tiazofurin in rat hepatomas as well [202]. In addition, treatment of rat hepatoma 3924A cells with tiazofurin resulted in rapid drop of IMPDH activity for 50% and subsequent decrease of GTP concentration to 55%, which was followed by a significant reduction of the already over-expressed c-myc and c-Ha-ras genes [203]. This result was also attained in K-562 human leukaemia cells, as well in MCF-7 breast and OVCAR-5 ovarian carcinoma cell cultures, in which tiazofurin induced differentiation and/or apoptosis accompanied with the down-regulation of the c-myc and K-ras oncogenes, all of which is critically dependent on the GTP supply [204]. Moreover, tiazofurin was shown to down-regulate the expression of the c-Ki-ras oncogene in a leukaemic patient [205], and was included in several clinical trials as a treatment for various end stage leukaemias, where it provided complete remission, a haematological improvement or an anti-leukaemic effect in a portion of patients [206 - 210]. However, the remissions were transient (from 1 to 10 months), and IMPDH activity after initial decrease returned to the high levels in the relapse phase [206 - 212]. Tiazofurin reached phase II/III clinical studies; however, it was never approved for treatment of leukaemia, due to severe toxicity [208].

The first reports about the antitumor effect of ribavirin dated from 1976 and 1977, and were performed on virus induced leukaemia in murine [213, 214]. One study claims that the treatment of spontaneous leukaemia in AKR mice with ribavirin had no obvious antitumor effect [214]. However, in combination with cytoreductive therapy, survival time for the leukaemic AKR mice was prolonged compared to the cytoreductive therapy alone, an effect that was attributed to the antiviral action of ribavirin [214]. In the other model of virus-induced murine leukaemia, some antitumor potential of ribavirin was recorded [213]. Further indications that ribavirin may be used as an anticancer drug came from Natsumeda *et al.*, who combined ribavirin with tiazofurin against rat hepatoma 3924A cells, since both drugs binds to IMPDH but at different ligand sites [215]. They demonstrated that ribavirin and tiazofurin induce the synergistic inhibition of *de novo* guanylate synthesis pathway and synergistic toxicity in hepatoma cells [215]. In addition, Prajda *et al.* showed that the inhibition of IMPDH activity in rat bone marrow generated by tiazofurin can be prolonged by the subsequent injection of ribavirin, while in only-the tiazofurin treated group did IMPDH activity rapidly return to the level of control (untreated) animals [216]. Dose-response studies showed that tiazofurin (IC50 = 29 mg/kg) was approximately twice as effective as ribavirin (IC50 = 65 mg/kg) in inhibiting IMPDH activity [216]. Moreover, tiazofurin depressed activities of other non-targeted enzymes in purine and pyrimidine biosynthesis, such as guanine phosphoribosyltransferase, thymidine kinase and deoxycytidine kinase, because of their short half-lives. Ribavirin again maintained the activities of these enzymes at lower levels when compared to only-tiazofurin treated rats [216]. The anticancer effects of ribavirin reflected in the induction of differentiation, growth suppression, reduction of signal transduction (IP3 concentration) and cell cycle arrest in G1 phase, were also demonstrated in combination therapies with IL3 in megakaryocytic leukaemia cell lines or with quercetin in human myeloma 8226 and OVCAR-5 cells [217, 218]. Kokeny *et al.* reported that ribavirin alone can induce anticancer effects in several human cancer cell lines [219]. Thus, ribavirin induced the significant growth inhibition of K562 (IC50=15 μM), OVCAR-5 (IC50=10 μM), MCF-7 (IC50=35 μM) and hepatocellular carcinoma cells (HepG2, IC50=45 μM) in a dose- and time-dependent manner [219]. Similarly to tiazofurin, ribavirin has the ability to induce erythroid differentiation in K562 human leukaemia cells, however only at high concentrations (145 μM), while lower concentrations (15 or 50 μM) were ineffective [219]. Nevertheless, ribavirin appears to be a potent inducer of apoptosis in K562 cells, valued through the portion of cells displaying morphological features typical of programmed cell death [219]. It was also shown that ribavirin changes the expression of many genes in K562 leukaemia cells, thus affecting the signalling pathways involved in guanylate metabolism, translation initiation, apoptosis, differentiation and cancer-related processes (key results are summarised in Table **2**, which represent tabular version of Figure five in [219]).

Table 2. Influence of ribavirin on gene expression in K562 leukaemia cells.

Anticancer effect of ribavirin	Down-regulated genes	Up-regulated genes
Inhibition of purine metabolism	IMPDH2, PARP1	
Inhibition of cell cycle and proliferation	CCND1, CCNE1, CHEAK1/2, ATR	TSC1/2, RB1, TP53
Induction of differentiation	MYC, TERT, AKT1	HBB, CD79A, MPO
Induction of apoptosis	Survivin, NFKB1, p21CIP, MYC	BAD, BAX, Caspases, PUMA, NOXA, TRADD
Inhibition of oncogenic signal, induction of tumour suppressors	MDM2, HRAS, MYC, TERT, ABL	TSC1/2, RB1, TP53
inhibition of translation initiation	EIF2A, EIF2B, EIF4G	

ABL - v-abl Abelson murine leukaemia viral oncogene homolog 1; AKT1 - v-akt murine thymoma viral oncogene homolog 1; BAD - Bcl-2-associated death promoter; BAX - BCL2 associated X protein; CCND - cyclin D1; CCNE1 - cyclin E1; CD79A - CD79a molecule, immunoglobulin-associated alpha; CHEK1/2 - CHK1 checkpoint homolog (S. pombe); EIF2A - eukaryotic translation initiation factor 2A; EIF2B - eukaryotic translation initiation factor 2B; EIF4G - eukaryotic translation initiation factor 4 gamma. HBB - haemoglobin, beta; HRAS - v-Ha-ras Harvey rat sarcoma viral oncogene homolog; MDM2 - Mdm2, transformed 3T3 cell double minute 2, p53 binding protein; MPO myeloperoxidase; MYC - v-myc myelocytomatosis viral oncogene homolog; NFKB1 - nuclear factor of kappa light polypeptide gene enhancer in B-cells 1; p21CIP - cyclin-dependent kinase inhibitor 1A; NOXA - phorbol-12-myristate-13-acetate-induced protein 1; PARP - poly (ADP-ribose) polymerase family, member 1; PUMA - BCL2 binding component 3; RB1- retinoblastoma 1; TP53 - tumour protein p53; TERT telomerase reverse transcriptase; TRADD - TNFRSF1A-associated *via* death domain; TSC1/2 - tuberous sclerosis ½. Tabular version of Figure five in [219].

eIF4E as the Cancer Target

Studies done after 2000, dealing with the ribavirin anti-tumour effect, are manly concentrated on the onco-protein eIF4E as the cancer target. It is estimated that the over-expression of eIF4E, which leads to the development of the neoplastic phenotype, is a hallmark of approximately 30% of the human malignancies and is associated with poor prognosis [220 - 224]. Thus, eIF4E is proved to be up-regulated in nearly all head and neck cancers, in at least half of breast cancers, as well as in the malignancies of bladder, colon, lung, prostate and skin [225 - 229]. In addition, protein and mRNA levels of eIF4E were highly elevated in all of the analysed samples from patients suffering from acute myeloid leukaemia (AML) subtypes M4 and M5 and chronic myeloid leukaemia (CML) [8, 230].

As far as the role of the eIF4E is concerned, it is important to point out that it is located in both the cytoplasm and the nucleus of most cells. In accordance to that, it affects the process of translation in two ways. First, in the cytoplasm at the stage of initiation, eIF4E is the part of heterotrimeric protein complex - eukaryotic initiation factor 4F (eIF4F), which is involved in the recruitment of transcripts to the ribosome. It acts through interaction with the cap structure on the 5′ terminus of mRNA, and thus regulates the rates of protein synthesis. Second, in the nucleus, eIF4E exports some specific mRNAs to the cytoplasm, thereby increasing translation material [231 - 233]. Apart from eIF4E, two other subunits of eIF4F are: eIF4A, which is a helicase that unwinds the secondary structure of mRNA; and eIF4G, which is a "scaffold" protein that serves as a docking site for eIF4A and eIF4E in the assembly of the eIF4F complex [234, 235]. Although nearly all eukaryotic mRNAs have a 7-methyl-GTP (m^7-GTP) type of cap structure at the 5′ end, and require eIF4E for the start of translation, they are not equally sensitive to this cap binding protein. Cellular mRNAs with lengthy and highly structured 5′ untranslated regions (UTRs) that are rich in G+C, critically rely on eIF4E, while mRNAs with short and unstructured 5′ UTRs (poor in C+G) are insensitive to it [236]. For instance, mRNAs with unstructured 5′ UTRs encode housekeeping proteins such as GAPDH or actin, whereas mRNAs with structured UTRs code for proteins involved in cell proliferation and growth, angiogenesis, cell survival, migration, *i.e.* processes that need to be strictly regulated for the normal development of the cell. Under normal conditions, eIF4E is expressed at a low level, hence the translation rate of mRNAs with structured 5′ UTRs is low. It is clear that a loss of the regulation of eIF4E in the sense that it is hyperactive or its expression is increased leads to phenomenon of malignant phenotype [236 - 238]. As previously stated, nuclear eIF4E regulates the export of certain mRNAs to the cytoplasm, wherein some of them carry information for the synthesis of proteins that are crucial for cell growth and survival [232, 239]. Therefore, the up-regulation of nuclear eIF4E contributes to tumorigenesis *via* an increased nuclear export of those specific mRNAs. As in the case of the initiation of translation in the cytoplasm, the nuclear export of housekeeping mRNAs is not affected by eIF4E dysregulation. Thus, gene products sensitive to eIF4E post-transcriptionally can be regulated at two points, translation and/or export from the nucleus; for both of them, the necessary step is binding to the m^7-GTP cap on mRNA [225, 239]. A detailed list of products of mRNAs whose increase lead to malignancy due to the up-regulation of eIF4E can be found in Benedetti and Graff [240]. This group includes: growth factors such as platelet-derived growth factor (PDGF), vascular endothelial growth factor (VEGF) and fibroblast growth factor-2 (FGF-2); c-myc, known onco-protein; cyclin D1, regulator of cell cycle progression; the enzyme ornithine decarboxylase (ODC) that catalyses the first and committed step in polyamine synthesis, which is important for growth, differentiation and survival; and matrix metalloproteinase 9 (MMP-9), a factor that enables tumour invasion through degradation of the extracellular matrix [240].

The up-regulation of eIF4E can occur due to multiple causes such as gene amplification, increased activation at the transcriptional level and the improved stability of eIF4E mRNA [241 - 244]. The activity of eIF4E is also regulated by sequestration and phosphorylation. Well known inhibitors of cap-dependent translation are eIF4E-binding proteins (4EBPs), which sequester free eIF4E and thus rest the translation of specific mRNAs [234, 245]. The bond between eIF4E and 4EBPs depends on the phosphorylation of the latter, which is induced upon stimuli like growth factors, mitogens, and hormones. Namely, the activation of PI3K/AKT/mTOR (phosphatidylinositide 3-kinases/protein kinase B/mammalian target of rapamycin) signalling pathways occurs, and mTOR Complex 1 (mTORC1) phosphorylates 4EBPs that in turn release eIF4E, which than becomes available for the initiation of translation (Fig. **4B**) [246, 247]. Post-translationally, eIF4E is also regulated by phosphorylation, which appears to be critical for its activity in cancer cells, while this type of control of the activity of eIF4E is not required in normal cells [248]. eIF4E is a substrate for mitogen-activated protein kinase (MAPK)-interacting kinase-1/2 (Mnk1/2), which are regulated by the ERK (extracellular regulated protein kinases) and p38 MAPK pathways (Fig. **4B**) [249, 250].

Because of its role in tumorigenesis, eIF4E is marked as a cancer target, and some of the approaches aimed towards this molecule are based on: affecting eIF4E synthesis; inhibition of the interactions between eIF4E and cap structure or eIF4G; inhibition of the phosphorylation of 4EBPs by targeting mTORC1 or PI3K and mTOR activity; and the inhibition of phosphorylation of eIF4E by targeting Mnk activity (reviewed in detail in [251]). Most of these agents are in the experimental/laboratory stage, but a few of them have entered clinical trials and are in phase I/II of investigations [251].

Ribavirin belongs to a group of agents that interfere with eIF4E-cap interactions (Fig. **4B**). The idea of ribavirin as an inhibitor of the capping machinery is not strictly related to anticancer studies. It came early in 1979, as one of the proposed mechanisms of ribavirin action against some RNA viruses that contain an m^7-GTP cap structure on the 5'-end [149]. Later, in an attempt to identify pharmacological agents able to restrain eIF4E-induced tumour growth, Kentsis *et al.* used biophysical studies *in vitro* to show that ribavirin and its active metabolite RTP act as physical mimics of m^7-GTP cap and directly bind to eIF4E with micromolar affinity [252]. Namely, m^7-GTP cap and ribavirin are positively charged and compete for cap binding sites of eIF4E, which is negatively charged, and placed between two tryptophan residues [239, 252, 253]. Furthermore, RTP induced the disruption of eIF4E nuclear bodies and was efficient in the inhibition of both translation and export of eIF4E-sensitive mRNAs, all at low micromolar concentrations [252]. Additionally, ribavirin suppressed the eIF4E-mediated transformation of murine cells *in vitro*; the eIF4E-dependent tumour growth in an animal model *in vivo*; and the growth of acute myelogenous leukaemic cells high in eIF4E derived from human patients [252]. These findings were challenged by two studies that were unable to confirm ribavirin function as a cap mimic [152, 153]. However, Kentsis and colleagues thoroughly explained that opposing outcomes are due to the differences in experimental methods and conditions [254]. In addition, the following studies from other groups presented results in support of ribavirin as inhibitor of eIF4E [229, 255]. Further, Tan *et al.* demonstrated that ribavirin was able to prevent apoptotic rescue by the inhibition of eIF4E-dependent Akt survival pathways [253]. Moreover, study of Shi and associates showed that the anti-proliferative and apoptotic properties of ribavirin in human leukaemic cell lines are due to the major inhibition of eIF4F complex assembly and the reduced translation of anti-apoptotic factor Mcl-1, which occurs as a consequence of ribavirin being a negative regulator of the mTOR/eIF4E and ERK/Mnk1/eIF4E signalling pathways [256].

In view of the fact that the over-expression of eIF4E was detected in M4/M5 subtypes of AML, ribavirin was the first agent whose effectiveness as an eIF4E inhibitor was tested in M4/M5 AML patients with poor prognoses [8]. Ribavirin induced encouraging, yet transient haematological responses (duration 2 to 4 months) in 11 evaluable patients. These responses were associated with the redistribution of eIF4E from the nucleus to the cytoplasm. Non-responders to ribavirin therapy or entry into relapse were characterised by the sustained or renewed nuclear localisation of eIF4E, respectively [8, 230]. In an attempt to improve outcomes, ribavirin was combined with cytarabine, a known chemotherapy drug for AML; however, it turns out that the search for a better partner is needed, since cytarabine impairs the absorption of ribavirin [257]. In addition, the problem of clinical and molecular resistance to ribavirin that occurred in all treated patients was addressed by developing ribavirin-resistant cell lines [258]. Two forms of resistance were identified: (i) one that arises because of impaired ribavirin uptake at the level of nucleoside transporter or because of the insufficient phosphorylation of ribavirin into active metabolite due to substantial reduction of ADK; and (ii) the other, which is more prevalent, occurs as a consequence of elevated levels of sonic hedgehog transcription factor glioma-associated protein 1 (GLI1) [258]. The over-expression of GLI1 induces an increase in the UDP glucuronosyltransferase family of enzymes, which then adds glucuronic acid to ribavirin and thus enables interactions between this drug and eIF4E [258].

Given that 50% of breast cancers have been shown to over-express eIF4E, which correlates with the disease severity, it was rational to test ribavirin effectiveness against this type of malignancy [229, 259]. First, ribavirin was shown to suppress proliferation and clonogenic potential of several examined breast cancer cell lines, as a result of targeting at eIF4E, as verified by the reduced export of mRNAs sensitive to eIF4E (cyclin D1, VEGF, Nijmegen breakage syndrome 1 protein

(NBS1) and phospho-AKT) and by reduced levels of these proteins [259]. Furthermore, *in vivo* studies on murine models demonstrated that ribavirin was successful in inhibiting the growth of primary breast tumours and in the reduction of pulmonary metastases [259]. In addition, ribavirin reduces the migration and invasion of breast cancer cells *in vitro*, concurrently with the decreased activity of MMP-3 and MMP-9 [259]. Of note, ribavirin was shown to restore the oestrogen receptor alpha (ESR1) expression and tamoxifen sensitivity in ESR1 negative breast cancer cell lines [260].

Targeting eIF4E expanded the possible application of ribavirin on various neoplasms associated with the increase of this initiation factor. Thus, Gu *et al.* presented data supportive of the up-regulation of eIF4E in high-grade astrocytic tumours [261]. Accordingly, two recent studies evaluated the anti-tumour potential of ribavirin in malignant glioma cell lines [7, 262]. Ogino and colleagues demonstrated that ribavirin suppressed the growth of seven investigated malignant glioma cell lines in a dose-dependent manner with varying levels of sensitivity to the drug between cell lines [262]. The half maximal inhibitory concentration of ribavirin for five of the malignant glioma cell lines was <100 μM, while for two lines it exceeded 250 μM. It was also suggested that PDGF receptor alpha (PDGFR-A) might be implicated in an anti-glioma effect of ribavirin based on a positive correlation between the level of PDGFR-A mRNA and ribavirin sensitivity [262]. Moreover, it was shown that the inhibition of malignant glioma by ribavirin *in vitro* may be achieved by its ability to influence the cell cycle, precisely by the induction of arrest at G0/G1 [262]. Of note, studies by other groups also reported ribavirin to be an inducer of cell cycle arrest (some of which were mentioned in previous sections [217, 218]). Further, Valle *et al.* registered halting of the cell cycle in G0/G1 in a melanoma cell line after ribavirin administration, while Teng *et al.* noted that ribavirin induced apoptosis and cell cycle arrest at the G2/M phase in mouse and human renal carcinoma cell lines [263, 264]. The second study confirmed the anti-glioma property of ribavirin, as cited by Volpin and associates [7]. Ribavirin, at a dosage of 30 or 50 μM, was demonstrated to decrease the proliferation of several human and murine glioma cell lines, as well as human glioblastoma stem-like cells through cell cycle arrest in G0 and apoptosis [7]. In addition, ribavirin also impaired the invasive potential of glioma cells, as well as their adhesion properties. It was determined that the mechanism underlying these effects is achieved by decreasing the phosphorylation of eIF4E, which correlated with reduced ERK phosphorylation [7]. The authors also addressed a new cancer target of ribavirin that emerged in the study of De la Cruz-Hernandez *et al.*, histone methyltransferase zeste homolog 2 (EZH2), which catalyses the addition of methyl groups to histone H3 at lysine 27 and has a role in regulation of gene expression [265]. Over-expression of EZH2 is associated with progression of various human cancers and is potential target for cancer therapy [266]. De la Cruz-Hernandez *et al.* used computational analysis that revealed a high structural similarity between ribavirin and 3-deazaneplanocin A, which is a known inhibitor of EZH2 [265]. Ribavirin indeed decreased EZH2 expression, inhibited histone methyltransferase activity and decreased H3K27 trimethylation in breast cancer cells [265]. Volpin *et al.* demonstrated that ribavirin induced the loss of EZH2 expression in glioma cells, and since it was shown that silencing EZH2 could be beneficial in the treatment of glioblastoma, this mechanism may also contribute to the anti-neoplastic effect of ribavirin [7, 267]. More importantly, this *in vitro* effect of ribavirin is translated to *in vivo* settings. Hence, ribavirin improved the survival of rats and mice with implanted gliosarcoma tumours or glioma stem-like cells, respectively [7]. In addition, another significant role of ribavirin was noted, as an enhancer of temozolomide and radiotherapy, which represent the usual therapeutic approach in the treatment of glioblastoma [7].

The most recent and first study to demonstrate the efficacy of ribavirin as a monotherapy or in combination with paclitaxel in oral tongue squamous cell carcinoma (OTSCC) *in vitro* and *in vivo* (OTSCC xenograft mouse models) reported that the gained beneficial effects are due to the inhibition of the mTOR/eIF4E signalling pathway. Namely, the inhibition of eIF4E by ribavirin was crucial for improvement of the anti-proliferative and pro-apoptotic effects of paclitaxel in OTSCC cells [268].

The Prospect of Ribavirin in Cancer

There are several properties of ribavirin that provide this drug with a future when it comes to the fight against cancer, whether it is investigated in the laboratory or trialled under clinical conditions.

First, from an economic point of view, the use of an old drug that is FDA approved and has been on the market for a long time, for a new indication, seems to be cost-effective, as it avoids the high financial expenses of developing a new drug. Therefore, the initial advantage of ribavirin would be the repurposing from an antivirus to anticancer drug.

Second, the IMPDH, eIF4E, EZH2, Erk/eIF4E and mTOR/eIF4E signalling pathways are known to be deregulated in various cancers, and ribavirin, due to its multivalent nature, has the ability to target all of them, individually or in combination, depending on the type of tumour (Fig. **4B**) [7, 184, 215, 225, 265, 266]. Its potential to act on multiple levels is certainly a desirable trait. It must be emphasised that the ability of ribavirin to inhibit eIF4E is the most deserving for its expanded use and testing against various cancers, which have shown elevated levels of this initiation factor. However, whether these different neoplasms that are high in eIF4E are equally susceptible to ribavirin treatment, and if not, which are the most sensitive to it, remains to be resolved in order to use ribavirin in the best possible way. As in the case of other anticancer chemotherapeutics, an obstacle in the form of resistance has appeared. However, it is encouraging that there is a strategy to overcome this issue [258].

Third, in addition to being a monotherapy, ribavirin has been shown to be useful as a partner in combination therapy with known anticancer drugs. This partnership is realised by inducing a synergistic effect or by restoring/enchasing the sensitivity to some standardised therapies [7, 215, 217, 218, 260, 269]. This property of the adjuvant, which avoids resistance to other drugs, further strengthens the position of ribavirin as an anti-tumour agent.

To summarise, several clinical studies evaluating ribavirin in the treatment of cancers are currently underway. Thus, two trials in phase I/ II test the response of AML patients to therapy with ribavirin and decitabine in combination with vismodegib, which is a hedgehog inhibitor (NCT02073838) or rapamycin, an inhibitor of mTOR regulatory kinase (NCT02109744). The other two are dealing with the effect of ribavirin in patients with head and neck cancer, precisely with oropharynx squamous cell cancer (NCT01721525) and tonsil and/or base of the tongue squamous cell carcinoma (NCT01268579). In addition, there are encouraging results which recommend the translation of ribavirin from *in vitro* experiments and *in vivo* animal models to clinical trials for patients with glioblastoma [7, 262].

RIBAVIRIN IN NEUROINFLAMMATION

There is another possibility of redirecting the use of ribavirin from an antiviral agent to an agent, which would be beneficial in combating neuroinflammation that accompanies a chronic autoimmune disease - multiple sclerosis (MS). However, all of the studies concerning the effects of ribavirin that could be useful in MS therapy were carried out on the *in vitro* model of neuroinflammation or *in vivo* on an animal model of this disease. Before summarising the results of these studies, basic facts about MS and its animal model, experimental autoimmune encephalomyelitis (EAE), will be briefly presented.

MS and EAE

MS is an immune-mediated chronic disease of the CNS, characterised by inflammation that causes neurological damage, which is primarily represented by demyelination and axonal injury. It affects about 2.5 million people worldwide. The average onset of disease is at the age of 30, with a women/men ratio of 2 to 1. The course of the disease is rarely predictable, characterised by periods of remissions and exacerbations (also called flare-ups, relapses or attacks). Most of the affected people (about 85%) usually possess the capacity to completely overcome some of these attacks, so the recovery and relapse stages alternately occur. This clinical course is referred to as relapse remitting MS (RRMS). Nevertheless, the neurological deterioration is constantly accumulating, leading to secondary progressive MS (SPMS), which is depicted by a partial disability with a significantly worsened quality of life. Unfortunately, nearly all RRMS cases develop SPMS. Approximately 15% of patients are affected by MS with a progressive type of clinical course from the start [270]. However, MS is very heterogeneous in its nature, so there is a wide range of intermediate forms ranging from benign types, allowing normal life and detected only upon autopsy, to the fulminant type, usually with a lethal outcome in a short period of time [270]. Despite all efforts, the cause of MS is still enigmatic. It is assumed that multiple overlapping factors such as immunological, infectious, environmental, genetic and epigenetic factors contribute to disease development. Accordingly, there is no cure for MS (for an extensive review of the current and prospective therapies see [271]). Also, many uncertainties concerning the pathophysiology of MS exist. The generally accepted view is that the unknown cause induces the activation of auto-reactive T-cells on the periphery, which then release a series of signals that enable them to proliferate, pass through the blood brain barrier (BBB) and enter the CNS [272]. Once in the CNS, they recognise their target autoantigen(s) presented by antigen-presenting cells (APCs) and become reactivated, *i.e.* able to produce pro-inflammatory chemokines and cytokines by which they attract and activate other immune cells (such as infiltrated macrophages or resident microglia), all in order to attack and degrade myelin [273]. It is believed that CD4+ T cells are the drivers of MS, while CD8+ T cells are responsible for maintaining and the progression of the disease [272]. B cells that have the ability to prime and regulate T-cells, which indicates their important role in the initiation and development of MS, also came into focus as potential therapeutic targets [274, 275]. The direct or indirect (through signals such as cytokines for example) pathological presence of immune system cells in the CNS substantially influences three major glia cell types: microglia, astrocytes and oligodendrocytes. Although the primary role of microglia and astrocytes is to ensure the normal functioning of the CNS and to protect it from any challenges, persistent inflammation overcomes their ability to resolve newly established conditions in favour of the CNS, and turns them from friends into foes [276, 277]. Coordinated interaction between glial cells is impaired, so they become key players in the pathological events associated with MS. However, during the course of the disease, the protective and pathological roles of microglia and astrocytes will alternate [276, 277]. The imperative is to discover precisely when and what exactly switches them from one state to another, in order to improve the therapeutic approach.

The rationale for using antiviral therapies in MS patients lies in the link between acute infections and relapse rate [278, 279]. It was hypothesised that upper respiratory infection, probably through mechanisms of immune modulation or molecular mimicry, induces the exacerbation of MS [280, 281]. Furthermore, it was speculated that the majority of MS relapses show seasonal preferences, mostly with the occurrence of the common cold [278, 282]. In fact, it was suggested that MS relapse may occur as a consequence of upper-respiratory infections with members of the *Picornaviridae* family (rhinoviruses and enteroviruses) [283]. Additionally, influenza virus, corona virus and RSV may induce upper-respiratory infections that can account for the MS relapse rate [284, 285]. A recent study shed light on the mechanism by which influenza infection provokes the development of EAE [286].

Therefore, animal models of MS are very helpful in revealing unknowns related to this disease. There are several ways to induce demyelination and an MS-like condition in animals (reviewed in [271]). Since all ribavirin studies concerning MS are performed on EAE, here we will introduce this model of inflammatory demyelination. EAE can be induced in primates or rodents in two ways: (i) actively through immunisation, by peripheral subcutaneous injection of the self-antigen (one of the myelin proteins) or CNS tissue homogenate; and (ii) passively, by the adoptive transfer of myelin-specific T-cells, derived from donor animals with actively induced EAE, into naïve animals [287, 288]. It is possible to induce EAE with different courses of disease [289 - 291]. Ribavirin is studied in acute monophasic and self-resolving EAE in rats, characterised by gradation of symptoms from flaccid tail to complete paralysis that is followed by a full recovery [10]. It is important to note that although EAE is a very convenient model that replicates some of the immunological and histological features of MS, it has limitations and has failed to explain all of the aspects and complexities of MS.

Targets of Ribavirin in EAE

Initial studies regarding ribavirin suggested that this drug has a poor ability to cross the BBB when intraperitoneal or intravenous routes of administration are used. However, it was later demonstrated that ribavirin could be useful against virus-induced encephalitis, if an efficient brain concentration is achieved, which is possible by intracerebral or intraventricular ribavirin therapy or using drug carriers such as cyclodextrin [2, 3, 26, 27, 292]. Furthermore, Gilbert and colleagues showed that ribavirin crosses the BBB when non-invasive methods of delivery are used, such as aerosols or intraperitoneal administration [82]. Of note, levels of ribavirin in the brain were not significantly different following aerosol compared to intraperitoneal administration, and were sufficiently higher to protect mice from a lethal intracerebral inoculation of the influenza A/WSN virus [82].

Furthermore, Colombo *et al.* demonstrated that the intranasal administration of ribavirin may be one way to increase the bioavailability of the drug in the CNS [293]. Most recently, Volpin *et al.* showed that the intraperitoneal application of ribavirin improves survival rates in animal models of glioblastoma [7]. The bottom-line is that ribavirin is apparently able to cross the BBB when applied as an intraperitoneal injection, yet it seems that its effectiveness is dependent on the applied dosage and type of condition being treated.

Besides the potential to cross the BBB, the other factors that contributed to the rationale of testing ribavirin in EAE were the known immunosuppressive actions of other IMPDH inhibitors, such as mycophenolate mofetil (MMF) or VX-148 [294, 295]. The results of a recently published clinical trial suggested that MMF has a beneficial effect in reducing relapses and can be used as an alternative therapy for MS [296]. Namely, due to the sensitivity of activated T- and B-cells on the *de novo* synthesis of guanosine nucleotides, the inhibition of IMPDH is a principal mechanism by which these agents suppress the proliferation of cells involved in adaptive immunity or prevent their recruitment into the inflammatory zone and graft rejection [297, 298]. The interaction between adhesion molecules on the surface of leukocytes and endothelial cells is very important for promotion of the immune response, *i.e.* for the capture, enrolment, arrest and infiltration of T-cells into targeted tissues [299]. For example, the interaction between alpha--integrin and vascular cell adhesion molecule 1 (VCAM-1) was shown to be very important for the transmigration of encephalitogenic T cells into the CNS during EAE [300]. Given that many adhesion molecules are glycoproteins (such alpha--integrin and VCAM-1), the process of glycosylation is crucial for their proper functioning. Since glycosylation is an enzymatic process that is highly dependent on guanosine nucleosides, lowering their level by IMPDH inhibition could be the strategy to diminish the interaction with adhesion molecules and thus to treat a specific immune disorder [299]. Accordingly, it was expected that IMPDH inhibitors which could pass through the BBB would be useful in the suppression of encephalitogenic T-cell functions, both on the periphery and in the CNS itself.

Indeed, the first nucleoside analogue that was shown to be beneficial in the treatment of actively induced EAE was tiazofurin [301]. Since tiazofurin did not influence the proliferation of autoreactive T-cells in EAE, it was suggested that the main mechanism by which tiazofurin postponed EAE onset, reduced disease severity and prevented lethal outcomes was by impairing adhesion interactions between encephalitogenic T-cells and endothelial cells of the brain vasculature due to IMPDH inhibition and the depression of GTP levels. In support of this was the *in vitro* results showing that mononuclear cells isolated from tiazofurin-treated animals have an impaired capacity of adhesion interactions and *in vivo* histological findings of the reduced infiltration of mononuclear cells in the spinal cord of tiazofurin-treated animals [301].

A beneficial effect of ribavirin in the treatment of actively induced EAE in rats was first recorded by Milicevic *et al.* and later verified in several other studies by our group [10, 302 - 305]. The key results of ribavirin treatment of EAE can be summarised as follows: (i) the suppression of all clinical parameters of disease, (ii) anti-proliferative and immunomodulatory effects on the cells of draining lymph nodes (DLN), as well as the modulation of cytokine production in targeted tissue, and (iii) favourable histological findings in the spinal cord specimens which are reflected by the reduced infiltration of mononuclear immune cells, absent or lower demyelination and axonal loss, the decreased expression of GFAP and vimentin mRNAs, as well as a reduced number of activated astrocytes and microglia in the white and grey matter [10, 303, 304, 306, 307]. Importantly, no haematological side effects of ribavirin application were recorded [10]. Furthermore, the treatment of EAE with a combination of ribavirin and tiazofurin was shown to be superior compared to the effects of each drug applied as a monotherapy [305 - 308].

The main mechanism of ribavirin anti-proliferative action in EAE is most likely the inhibition of IMPDH in activated T-cells, which deprives them of the essential guanosine nucleotides, since salvage pathways for the production of guanine nucleosides do not operate in T and B cells [10, 304]. In addition, the reduced migration of mononuclear immune cells into the CNS is also a consequence of the lowered GTP level, which is noted above as a prerequisite for adhesion interactions between T-cells and endothelial cells of the CNS vasculature. These early effects of ribavirin on the periphery, outside the CNS, might be the most responsible for further disease course (Fig. **4C**) [10, 304].

The majority of previous studies demonstrated that ribavirin acts as an immunomodulator, which changes the Th1/Th2 balance in favour of Th1 (see section: Additional indirect mechanisms of ribavirin). This is certainly not a desirable trait in the treatment of EAE/MS. However, we demonstrated that ribavirin substantially suppressed the release of Th1 pro-inflammatory cytokines (IFN-γ, IL1-β and TNF-α) and slightly reduced the production of anti-inflammatory cytokines (IL-10 and TGF-β) in the cells of the DLN in the initial phase of EAE (Fig. **4C**) [304]. Accordingly, the number of IFN-γ-, IL1-β- and TNF-α-positive cells in the spinal cord was significantly reduced at the peak and at the end of disease [304]. Of note, when an *in vitro* model of neuroinflammation, which is LPS activated microglia, was treated with ribavirin, a pleiotropic effect on cytokine production was recorded: the release of IL 1-β was significantly augmented; the release of IL-6 was slightly decreased; and the production of IL-10 was additionally stimulated [11]. A purely hypothetical explanation for the discrepancy between our *in vivo* studies and the studies of other groups may be due to different activator of T-cells. Namely, ribavirin was recorded as an inducer of Th1 phenotype in the case when the immune system was challenged by the virus, while in the case of EAE, autoantigens are inducers of the immune response.

In addition, upon treatment with ribavirin, the number of astrocytes and microglia was reduced and their morphology changed to a state that is an indicator of the silencing of inflammation (Fig. **4C**) [303, 306, 307]. Activated microglia cells *in vitro* were demonstrated to be sensitive to ribavirin administration as well. These *in vitro* studies confirmed that ribavirin is a regulator of the morphological profile of microglia and pointed to a potential role in attenuating nitrosative stress during neuroinflammation [11, 309]. However, a clinically achievable concentration of ribavirin (10 μM) reduced the viability of activated cells by 30% and induced the significant release of the pro-inflammatory cytokine IL1-β [11]. We have also reported that ribavirin has beneficial effects in the rat model of traumatic brain injury, which pointed to another possible mechanism by which ribavirin could influence astrocytes [9, 310]. Namely, ribavirin was shown to down-regulate ERK1/2 expression around the injury site, which correlated with the inhibition of reactive astrogliosis [9, 310]. Therefore, it is hypothesised that regulation of the ERK signalling pathway might be responsible for the effect of ribavirin on astrocytes during EAE [303, 306]. The fact that ribavirin is a regulator of the ERK signalling pathway is consistent with recently published reports in cancer studies, one of which is on glioma cells [7].

Results of *in vivo* studies are encouraging and recommend ribavirin as a potential new therapy for MS, which certainly requires new clinical trials. The beneficial effect of MMF, which is another IMPDH inhibitor, in MS patients, reinforces the rationale of ribavirin as an MS therapy [296]. However, the results from the *in vitro* model of neuroinflammation pointed to some questions that first need to be resolved in order to avoid potential adverse actions [11]. Namely, the cytokine milieu inside the CNS must be determined more precisely, with methods other than immunocytochemistry, in order to obtain a complete picture of the immunomodulatory effect of ribavirin. Second, the cytotoxic effect of ribavirin in clinically achievable dosages (5-20 μM) recorded on activated microglia *in vitro* suggests that additional toxicological screening of the *in vivo* effects of ribavirin on the CNS is necessary [11]. However, it must be emphasised that the concentration of ribavirin in the brain is several times lower than that in the serum following intraperitoneal or aerosol administration [82]. To summarise, it is most likely that the main location of action of ribavirin in EAE is outside the CNS, with encephalitogenic T-cells as the primary targets. Therefore, ribavirin could be a potential add-on therapy to existing monotherapies.

CONCLUDING REMARKS

Despite the several decades of extensive research, the complete spectrum of action of ribavirin is still not completely understood, although many mechanisms have been proposed. Therefore, the data presented in this chapter are meant to shed more light on the cellular and molecular mechanisms of actions underlying the effects of ribavirin against viral, neoplastic and inflammatory diseases. Antiviral mechanisms of ribavirin may rely on the interaction: (i) with the virus itself, or (ii) with the host cells. The first group of activities include: (a) interactions with viral RNA polymerase resulting in the lethal mutagenesis of RNA virus genomes or the inhibition of virus replication, and (b) inhibition of the capping of RNA virus genomes by acting as a functional mimic of the 7-methylguanosine cap structure. The second group of actions imply: (a) the inhibition of host IMPDH, (b) immunomodulatory effects on the host immune response, and (c) the activation of host ISGs. Available evidence indicates that lethal mutagenesis is a highly possible antiviral mechanism of action of ribavirin against RNA viruses, and that the lethal mutagenesis is probably enhanced by the inhibition of IMPDH, the rate-limiting enzyme in the *de novo* synthesis of guanine nucleotides. The activation of ISGs also emerged as a very likely explanation of the cooperative action of ribavirin with interferon alpha during treatment of hepatitis C. However, if the ribavirin action is directed toward cancer cells, its ability to act as a cap analogue, which results in the inhibition of the known onco-protein eIF4E, comes into the limelight. In neoplastic cells, as well as in viruses, ribavirin has the power to influence several other targets such as: IMPDH, EZH2, Erk and mTOR pathways, which may assist in the overall anticancer effect. The capacity of ribavirin to attenuate inflammation depends on the inhibition of IMPDH and immunomodulatory effects. The inhibition of IMPDH stands out as the potential mechanism of action of ribavirin against all three targets: viruses, and neoplastic and inflammatory diseases. This leads to a depletion of the GTP pool and may therefore affect many cellular processes, including adhesion interactions, the cell cycle, differentiation and proliferation. Since this pathway is essential for mitogenic functions, the inhibition of IMPDH probably represents the main mechanism of attenuating neuroinflammation by affecting highly proliferating cells such as activated lymphocytes, microglia and astrocytes. On the other hand, one of the consequences of the cytotoxic actions of ribavirin is the release of a myriad of toxic mediators that may be responsible for some of the side effects, such as leukopenia and anaemia, observed under therapeutic procedures containing ribavirin therapy.

Ribavirin might be fading away as an old antiviral drug, losing its central role in the therapy of HCV infection, and staying limited to the treatment of RSV and LASV infections in the expectation of a better solution. However, the multipotent nature of ribavirin reflected in various mechanisms of action provided possible new therapeutic applications, for indications such as cancer or MS. Considering the results from several clinical trials, redirecting ribavirin from an anti-viral to an anti-cancer drug seems most likely for the time being. Importantly, another advantage of ribavirin is that it can be applied in the treatment of various cancers, since its anti-proliferative action was recorded both in haematological malignancies and solid tumours.

Finally, the pharmacogenetic prediction of the outcome of ribavirin-based therapy may aid in the pre-treatment choice of therapy, thus contributing to the personalisation of therapy. Accordingly, genotyping could represent a valuable tool in future clinical application of this drug. Predicting the major side effect of ribavirin, haemolytic anaemia, would in that way facilitate decisions regarding the monitoring of patients and dose management.

CONSENT FOR PUBLICATION

Not applicable.

CONFLICT OF INTEREST

The author declares no conflict of interest, financial or otherwise.

ACKNOWLEDGEMENTS

The authors of this chapter are supported by Grant III 41014 obtained from Ministry of Education, Science and Technological Development of the Republic of Serbia. Danijela Savic conceived and wrote the chapter. Irena Lavrnja, Sanja Pekovic and Ivana Bjelobaba provided substantial content expertise and reviewed drafts of the chapter. The authors would like to acknowledge Dr. Ivan Spasojevic for preparing a drawing of Fig. (**1** and **4**). The authors would also like to thank their Professors Mirjana Stojiljkovic and Ljubisav Rakic on valuable advices while preparing this chapter.

REFERENCES

[1] Jordheim LP, Durantel D, Zoulim F, Dumontet C. Advances in the development of nucleoside and nucleotide analogues for cancer and viral diseases. Nat Rev Drug Discov 2013; 12(6): 447-64.

[2] Witkowski JT, Robins RK, Sidwell RW, Simon LN. Design, synthesis, and broad spectrum antiviral activity of 1- -D-ribofuranosyl-1,2,4-triazole-3-carboxamide and related nucleosides. J Med Chem 1972; 15(11): 1150-4.

[3] Sidwell RW, Huffman JH, Khare GP, Allen LB, Witkowski JT, Robins RK. Broad-spectrum antiviral activity of Virazole: 1-beta-D-ribofuranosyl-1,2,4-triazole-3-carboxamide. Science 1972; 177(4050): 705-6.

[4] Foster GR. Past, present, and future hepatitis C treatments. Semin Liver Dis 2004; 24 (Suppl. 2): 97-

104.

[5] Ventre K, Randolph AG. Ribavirin for respiratory syncytial virus infection of the lower respiratory tract in infants and young children. Cochrane Database Syst Rev 2007; (1): Cd000181.

[6] McCormick JB, King IJ, Webb PA, Scribner CL, Craven RB, Johnson KM, *et al.* Lassa fever. Effective therapy with ribavirin. N Engl J Med 1986; 314(1): 20-6.

[7] Volpin F, Casaos J, Sesen J, Mangraviti A, Choi J, Gorelick N, *et al.* Use of an anti-viral drug, Ribavirin, as an anti-glioblastoma therapeutic. Oncogene 2017; 36(21): 3037-47.

[8] Assouline S, Culjkovic B, Cocolakis E, Rousseau C, Beslu N, Amri A, *et al.* Molecular targeting of the oncogene eIF4E in acute myeloid leukemia (AML): a proof-of-principle clinical trial with ribavirin. Blood 2009; 114(2): 257-60.

[9] Pekovic S, Filipovic R, Subasic S, Lavrnja I, Stojkov D, Nedeljkovic N, *et al.* Downregulation of glial scarring after brain injury: the effect of purine nucleoside analogue ribavirin. Ann N Y Acad Sci 2005; 1048: 296-310.

[10] Milicevic I, Pekovic S, Subasic S, Mostarica-Stojkovic M, Stosic-Grujicic S, Medic-Mijacevic L, *et al.* Ribavirin reduces clinical signs and pathological changes of experimental autoimmune encephalomyelitis in Dark Agouti rats. J Neurosci Res 2003; 72(2): 268-78.

[11] Savic D, Stojiljkovic M, Lavrnja I, Parabucki A, Bjelobaba I, Nedeljkovic N, *et al.* Ribavirin shows immunomodulatory effects on activated microglia. Immunopharmacol Immunotoxicol 2014; 36(6): 433-41.

[12] Streeter DG, Witkowski JT, Khare GP, Sidwell RW, Bauer RJ, Robins RK, *et al.* Mechanism of action of 1- -D-ribofuranosyl-1,2,4-triazole-3-carboxamide (Virazole), a new broad-spectrum antiviral agent. Proc Natl Acad Sci USA 1973; 70(4): 1174-8.

[13] Hedstrom L. IMP dehydrogenase: structure, mechanism, and inhibition. Chem Rev 2009; 109(7): 2903-28.

[14] Weber G, Nakamura H, Natsumeda Y, Szekeres T, Nagai M. Regulation of GTP biosynthesis. Adv Enzyme Regul 1992; 32: 57-69.

[15] Wray SK, Gilbert BE, Noall MW, Knight V. Mode of action of ribavirin: effect of nucleotide pool alterations on influenza virus ribonucleoprotein synthesis. Antiviral Res 1985; 5(1): 29-37.

[16] De Clercq E, Li G. Approved Antiviral Drugs over the Past 50 Years. Clin Microbiol Rev 2016; 29(3): 695-747.

[17] Maugh TH II. Chemotherapy: antiviral agents come of age. Science 1976; 192(4235): 128-32.

[18] Parker WB. Metabolism and antiviral activity of ribavirin. Virus Res 2005; 107(2): 165-71.

[19] Canonico PG. Ribavirin.Modes and Mechanisms of Microbial Growth Inhibitors. Heidelberg: Springer-Verlag Berlin 1983.

[20] Smith RA. Mechanisms of action of ribavirin.Ribavirin A broad spectrum antiviral agent. London, New York: Academic Press 1980; pp. 99-118.

[21] Crotty S, Maag D, Arnold JJ, Zhong W, Lau JY, Hong Z, *et al.* The broad-spectrum antiviral ribonucleoside ribavirin is an RNA virus mutagen. Nat Med 2000; 6(12): 1375-9.

[22] Sakharov V, Baykov S, Konstantinova I, Esipov R, Dorogov M. An Efficient Chemoenzymatic Process for Preparation of Ribavirin 2015.

[23] Ma Y, Yang S, Liu L, Xu Q, Chenglin Zhang NC, Xie X. Fermentative production of ribavirin by overexpressing purine nucleoside phosphorylase in Bacillus. J Chem Pharm Res 2014; 6(5): 1377-84.

[24] Kumar R, Tripathi P, Baranwal M, Singh S, Tripathi S, Banerjee G. Randomized, controlled trial of oral ribavirin for Japanese encephalitis in children in Uttar Pradesh, India. Clin Infect Dis 2009; 48(4): 400-6.

[25] McJunkin JE, Nahata MC, De Los Reyes EC, Hunt WG, Caceres M, Khan RR, *et al.* Safety and pharmacokinetics of ribavirin for the treatment of la crosse encephalitis. Pediatr Infect Dis J 2011; 30(10): 860-5.

[26] Solbrig MV, Schlaberg R, Briese T, Horscroft N, Lipkin WI. Neuroprotection and Reduced Proliferation of Microglia in Ribavirin-Treated Bornavirus-Infected Rats. Antimicrob Agents Chemother 2002; 46(7): 2287-91.

[27] Hosoya M, Mori S, Tomoda A, Mori K, Sawaishi Y, Kimura H, *et al.* Pharmacokinetics and Effects of Ribavirin following Intraventricular Administration for Treatment of Subacute Sclerosing Panencephalitis. Antimicrob Agents Chemother 2004; 48(12): 4631-5.

[28] Eriksson B, Helgstrand E, Johansson NG, Larsson A, Misiorny A, Noren JO, *et al.* Inhibition of influenza virus ribonucleic acid polymerase by ribavirin triphosphate. Antimicrob Agents Chemother 1977; 11(6): 946-51.

[29] Stein DS, Creticos CM, Jackson GG, Bernstein JM, Hayden FG, Schiff GM, *et al.* Oral ribavirin treatment of influenza A and B. Antimicrob Agents Chemother 1987; 31(8): 1285-7.

[30] Seo S. Combination therapy with amantadine, oseltamivir and ribavirin for influenza A infection: safety and pharmacokinetics. 2013; 18(3): 377-86.

[31] Hayden FG, Sable CA, Connor JD, Lane J. Intravenous ribavirin by constant infusion for serious influenza and parainfluenzavirus infection. Antivir Ther 1996; 1(1): 51-6.

[32] Rodriguez WJ, Hall CB, Welliver R, Simoes EA, Ryan ME, Stutman H, *et al.* Efficacy and safety of aerosolized ribavirin in young children hospitalized with influenza: a double-blind, multicenter, placebo-controlled trial. J Pediatr 1994; 125(1): 129-35.

[33] Wilson SZ, Gilbert BE, Quarles JM, Knight V, McClung HW, Moore RV, *et al.* Treatment of influenza A (H1N1) virus infection with ribavirin aerosol. Antimicrob Agents Chemother 1984; 26(2): 200-3.

[34] Smith CB, Charette RP, Fox JP, Cooney MK, Hall CE. Lack of effect of oral ribavirin in naturally occurring influenza A virus (H1N1) infection. J Infect Dis 1980; 141(5): 548-54.

[35] Fernandez-Larsson R, Patterson JL. Ribavirin is an inhibitor of human immunodeficiency virus reverse transcriptase. Mol Pharmacol 1990; 38(6): 766-70.

[36] Baba M, Pauwels R, Balzarini J, Herdewijn P, De Clercq E, Desmyter J. Ribavirin antagonizes inhibitory effects of pyrimidine 2′,3′-dideoxynucleosides but enhances inhibitory effects of purine 2′,3′-dideoxynucleosides on replication of human immunodeficiency virus *in vitro.* Antimicrob Agents Chemother 1987; 31(10): 1613-7.

[37] Comparison of ribavirin and placebo in CDC group III human immunodeficiency virus infection. Spanish Ribavirin Trial Group. Lancet 1991; 338(8758): 6-9.

[38] Roberts RB, Hollinger FB, Parks WP, Rasheed S, Laurence J, Heseltine PN, *et al.* A multicenter clinical trial of oral ribavirin in HIV-infected people with lymphadenopathy: virologic observations. Ribavirin-LAS Collaborative Group. AIDS 1990; 4(1): 67-72.

[39] Roberts RB, Jurica K, Meyer WA III, Paxton H, Makuch RW. A phase 1 study of ribavirin in human immunodeficiency virus-infected patients. J Infect Dis 1990; 162(3): 638-42.

[40] Schulof RS, Parenti DM, Simon GL, Paxton H, Meyer WA III, Schlesselman SB, *et al.* Clinical, virologic, and immunologic effects of combination therapy with ribavirin and isoprinosine in HIV-infected homosexual men. J Acquir Immune Defic Syndr 1990; 3(5): 485-92.

[41] Manns MP, Buti M, Gane E, Pawlotsky JM, Razavi H, Terrault N, *et al.* Hepatitis C virus infection. Nat Rev Dis Primers 2017; 3: 17006.

[42] Simmonds P. The origin of hepatitis C virus. Curr Top Microbiol Immunol 2013; 369: 1-15.

[43] Grebely J, Matthews GV, Dore GJ. Treatment of acute HCV infection. Nat Rev Gastroenterol Hepatol

2011; 8(5): 265-74.

[44] Kalafateli M, Buzzetti E, Thorburn D, Davidson BR, Tsochatzis E, Gurusamy KS. Pharmacological interventions for acute hepatitis C infection: an attempted network meta-analysis. Cochrane Database Syst Rev 2017; 3: Cd011644.

[45] Gower E, Estes C, Blach S, Razavi-Shearer K, Razavi H. Global epidemiology and genotype distribution of the hepatitis C virus infection. J Hepatol 2014; 61(1) (Suppl.): S45-57.

[46] Smith DB, Bukh J, Kuiken C, Muerhoff AS, Rice CM, Stapleton JT, *et al.* Expanded Classification of Hepatitis C Virus Into 7 Genotypes and 67 Subtypes: Updated Criteria and Genotype Assignment Web Resource. Hepatology 2014; 59(1): 318-27.

[47] Messina JP, Humphreys I, Flaxman A, Brown A, Cooke GS, Pybus OG, *et al.* Global Distribution and Prevalence of Hepatitis C Virus Genotypes. Hepatology 2015; 61(1): 77-87.

[48] Catanese MT, Uryu K, Kopp M, Edwards TJ, Andrus L, Rice WJ, *et al.* Ultrastructural analysis of hepatitis C virus particles. Proc Natl Acad Sci USA 2013; 110(23): 9505-10.

[49] Chevaliez S, Pawlotsky JM. HCV Genome and Life Cycle.Hepatitis C Viruses: Genomes and Molecular Biology. Norfolk, UK: Horizon Bioscience 2006.

[50] McHutchison JG, Gordon SC, Schiff ER, Shiffman ML, Lee WM, Rustgi VK, *et al.* Interferon Alfa-2b Alone or in Combination with Ribavirin as Initial Treatment for Chronic Hepatitis C. N Engl J Med 1998; 339(21): 1485-92.

[51] Poynard T, Marcellin P, Lee SS, Niederau C, Minuk GS, Ideo G, *et al.* Randomised trial of interferon alpha2b plus ribavirin for 48 weeks or for 24 weeks *versus* interferon alpha2b plus placebo for 48 weeks for treatment of chronic infection with hepatitis C virus. International Hepatitis Interventional Therapy Group (IHIT). Lancet 1998; 352(9138): 1426-32.

[52] Manns MP, McHutchison JG, Gordon SC, Rustgi VK, Shiffman M, Reindollar R, *et al.* Peginterferon alfa-2b plus ribavirin compared with interferon alfa-2b plus ribavirin for initial treatment of chronic hepatitis C: a randomised trial. Lancet 2001; 358(9286): 958-65.

[53] Gaeta GB, Precone DF, Felaco FM, Bruno R, Spadaro A, Stornaiuolo G, *et al.* Premature discontinuation of interferon plus ribavirin for adverse effects: a multicentre survey in 'real world' patients with chronic hepatitis C. Aliment Pharmacol Ther 2002; 16(9): 1633-9.

[54] Loustaud-Ratti V, Debette-Gratien M, Jacques J, Alain S, Marquet P, Sautereau D, *et al.* Ribavirin: Past, present and future. World J Hepatol 2016; 8(2): 123-30.

[55] De Clercq E. Chemotherapy of respiratory syncytial virus infections: the final breakthrough. Int J Antimicrob Agents 2015; 45(3): 234-7.

[56] Bitko V, Shulyayeva O, Mazumder B, Musiyenko A, Ramaswamy M, Look DC, *et al.* Nonstructural Proteins of Respiratory Syncytial Virus Suppress Premature Apoptosis by an NF-κB-Dependent, Interferon-Independent Mechanism and Facilitate Virus Growth. J Virol 2007; 81(4): 1786-95.

[57] Turner TL, Kopp BT, Paul G, Landgrave LC, Hayes D, Thompson R. Respiratory syncytial virus: current and emerging treatment options. Clinicoecon Outcomes Res 2014; 6: 217-25.

[58] Beaird OE, Freifeld A, Ison MG. Current practices for treatment of respiratory syncytial virus and other non-influenza respiratory viruses in high-risk patient populations: a survey of institutions in the Midwestern Respiratory Virus Collaborative 2016.

[59] Nair H, Nokes DJ, Gessner BD, Dherani M, Madhi SA, Singleton RJ, *et al.* Global burden of acute lower respiratory infections due to respiratory syncytial virus in young children: a systematic review and meta-analysis. Lancet •••; 375(9725): 1545-55.

[60] Ng MC, How CH. Recurrent wheeze and cough in young children: is it asthma? Singapore Med J 2014; 55(5): 236-41.

[61] Andabaka T, Nickerson JW, Rojas-Reyes MX, Rueda JD, Bacic Vrca V, Barsic B. Monoclonal

antibody for reducing the risk of respiratory syncytial virus infection in children. Cochrane Database Syst Rev 2013; (4): Cd006602.

[62] Shook BC, Lin K. Recent Advances in Developing Antiviral Therapies for Respiratory Syncytial Virus. Top Curr Chem (J) 2017; 375(2): 40.

[63] Chemaly RF, Aitken SL, Wolfe CR, Jain R, Boeckh MJ. Aerosolized ribavirin: the most expensive drug for pneumonia. Transpl Infect Dis 2016; 18(4): 634-6.

[64] Waghmare A, Campbell AP, Xie H, Seo S, Kuypers J, Leisenring W, *et al.* Respiratory syncytial virus lower respiratory disease in hematopoietic cell transplant recipients: viral RNA detection in blood, antiviral treatment, and clinical outcomes. Clin Infect Dis 2013; 57(12): 1731-41.

[65] Bausch DG, Hadi CM, Khan SH, Lertora JJ. Review of the literature and proposed guidelines for the use of oral ribavirin as postexposure prophylaxis for Lassa fever. Clin Infect Dis 2010; 51(12): 1435-41.

[66] Hastie KM, Liu T, Li S, King LB, Ngo N, Zandonatti MA, *et al.* Crystal structure of the Lassa virus nucleoprotein-RNA complex reveals a gating mechanism for RNA binding. Proc Natl Acad Sci USA 2011; 108(48): 19365-70.

[67] Yun NE, Walker DH. Pathogenesis of Lassa Fever. Viruses 2012; 4(10): 2031-48.

[68] Lassa Fever 2015.https://www.cdc.gov/vhf/lassa/index.html

[69] Monath TP, Newhouse VF, Kemp GE, Setzer HW, Cacciapuoti A. Lassa virus isolation from Mastomys natalensis rodents during an epidemic in Sierra Leone. Science 1974; 185(4147): 263-5.

[70] Fisher-Hoch SP, Tomori O, Nasidi A, Perez-Oronoz GI, Fakile Y, Hutwagner L, *et al.* Review of cases of nosocomial Lassa fever in Nigeria: the high price of poor medical practice. BMJ 1995; 311(7009): 857-9.

[71] Monath TP, Mertens PE, Patton R, Moser CR, Baum JJ, Pinneo L, *et al.* A hospital epidemic of Lassa fever in Zorzor, Liberia, March-April 1972. Am J Trop Med Hyg 1973; 22(6): 773-9.

[72] Lassa fever fact sheet 2016.http://www.who.int/mediacentre/factsheets/fs179/en/

[73] 2017.http://www.who.int/csr/don/28-june-2017-lassa-fever-nigeria/en/

[74] Khakoo S, Glue P, Grellier L, Wells B, Bell A, Dash C, *et al.* Ribavirin and interferon alfa-2b in chronic hepatitis C: assessment of possible pharmacokinetic and pharmacodynamic interactions. Br J Clin Pharmacol 1998; 46(6): 563-70.

[75] Preston SL, Drusano GL, Glue P, Nash J, Gupta SK, McNamara P. Pharmacokinetics and absolute bioavailability of ribavirin in healthy volunteers as determined by stable-isotope methodology. Antimicrob Agents Chemother 1999; 43(10): 2451-6.

[76] Wade JR, Snoeck E, Duff F, Lamb M, Jorga K. Pharmacokinetics of ribavirin in patients with hepatitis C virus. Br J Clin Pharmacol 2006; 62(6): 710-4.

[77] Strader DB, Wright T, Thomas DL, Seeff LB. Diagnosis, management, and treatment of hepatitis C. Hepatology 2004; 39(4): 1147-71.

[78] Paroni R, Del Puppo M, Borghi C, Sirtori CR, Galli Kienle M. Pharmacokinetics of ribavirin and urinary excretion of the major metabolite 1,2,4-triazole-3-carboxamide in normal volunteers. Int J Clin Pharmacol Ther Toxicol 1989; 27(6): 302-7.

[79] Glue P, Schenker S, Gupta S, Clement RP, Zambas D, Salfi M. The single dose pharmacokinetics of ribavirin in subjects with chronic liver disease. Br J Clin Pharmacol 2000; 49(5): 417-21.

[80] Connor J. Ribavirin pharmacokinetics. Pediatr Infect Dis J 1990; 9(9) (Suppl.): S91-2.

[81] Gilbert BE, Wyde PR. Pharmacokinetics of ribavirin aerosol in mice. Antimicrob Agents Chemother 1988; 32(1): 117-21.

[82] Gilbert BE, Wyde PR, Wilson SZ, Robins RK. Aerosol and intraperitoneal administration of ribavirin

and ribavirin triacetate: pharmacokinetics and protection of mice against intracerebral infection with influenza A/WSN virus. Antimicrob Agents Chemother 1991; 35(7): 1448-53.

[83] Englund JA, Piedra PA, Jefferson LS, Wilson SZ, Taber LH, Gilbert BE. High-dose, short-duration ribavirin aerosol therapy in children with suspected respiratory syncytial virus infection. J Pediatr 1990; 117(2 Pt 1): 313-20.

[84] Kong W, Engel K, Wang J. Mammalian nucleoside transporters. Curr Drug Metab 2004; 5(1): 63-84.

[85] Young JD, Yao SY, Baldwin JM, Cass CE, Baldwin SA. The human concentrative and equilibrative nucleoside transporter families, SLC28 and SLC29. Mol Aspects Med 2013; 34(2-3): 529-47.

[86] Pastor-Anglada M, Cano-Soldado P, Errasti-Murugarren E, Casado FJ. SLC28 genes and concentrative nucleoside transporter (CNT) proteins. Xenobiotica 2008; 38(7-8): 972-94.

[87] Young JD. The SLC28 (CNT) and SLC29 (ENT) nucleoside transporter families: a 30-year collaborative odyssey. Biochem Soc Trans 2016; 44(3): 869-76.

[88] Yamamoto T, Kuniki K, Takekuma Y, Hirano T, Iseki K, Sugawara M. Ribavirin uptake by cultured human choriocarcinoma (BeWo) cells and Xenopus laevis oocytes expressing recombinant plasma membrane human nucleoside transporters. Eur J Pharmacol 2007; 557(1): 1-8.

[89] Moss AM, Endres CJ, Ruiz-Garcia A, Choi DS, Unadkat JD. Role of the equilibrative and concentrative nucleoside transporters in the intestinal absorption of the nucleoside drug, ribavirin, in wild-type and Ent1(-/-) mice. Mol Pharm 2012; 9(9): 2442-9.

[90] Choi MK, Kim MH, Maeng HJ, Song IS. Contribution of CNT1 and ENT1 to ribavirin uptake in human hepatocytes. Arch Pharm Res 2015; 38(5): 904-13.

[91] Fukuchi Y, Furihata T, Hashizume M, Iikura M, Chiba K. Characterization of ribavirin uptake systems in human hepatocytes. J Hepatol 2010; 52(4): 486-92.

[92] Endres CJ, Moss AM, Govindarajan R, Choi DS, Unadkat JD. The role of nucleoside transporters in the erythrocyte disposition and oral absorption of ribavirin in the wild-type and equilibrative nucleoside transporter 1-/- mice. J Pharmacol Exp Ther 2009; 331(1): 287-96.

[93] Jarvis SM, Thorn JA, Glue P. Ribavirin uptake by human erythrocytes and the involvement of nitrobenzylthioinosine-sensitive (es)-nucleoside transporters. Br J Pharmacol 1998; 123(8): 1587-92.

[94] Gallois-Montbrun S, Chen Y, Dutartre H, Sophys M, Morera S, Guerreiro C, *et al.* Structural analysis of the activation of ribavirin analogs by NDP kinase: comparison with other ribavirin targets. Mol Pharmacol 2003; 63(3): 538-46.

[95] Willis RC, Carson DA, Seegmiller JE. Adenosine kinase initiates the major route of ribavirin activation in a cultured human cell line. Proc Natl Acad Sci USA 1978; 75(7): 3042-4.

[96] Zimmerman TP, Deeprose RD. Metabolism of 5-amino-1-β-d-ribofuranosyl- imidazole---carboxamide and related five-membered heterocycles to 5′-triphosphates in human blood and L5178Y cells. Biochem Pharmacol 1978; 27(5): 709-16.

[97] Miller RL, Adamczyk DL, Miller WH, Koszalka GW, Rideout JL, Beacham LM III, *et al.* Adenosine kinase from rabbit liver. II. Substrate and inhibitor specificity. J Biol Chem 1979; 254(7): 2346-52.

[98] Streeter DG, Simon LN, Robins RK, Miller JP. The phosphorylation of ribavirin by deoxyadenosine kinase from rat liver. Differentiation between adenosine and deoxyadenosine kinase. Biochemistry 1974; 13(22): 4543-9.

[99] Wu JZ, Larson G, Walker H, Shim JH, Hong Z. Phosphorylation of Ribavirin and Viramidine by Adenosine Kinase and Cytosolic 5′-Nucleotidase II: Implications for Ribavirin Metabolism in Erythrocytes. Antimicrob Agents Chemother 2005; 49(6): 2164-71.

[100] Page T, Connor JD. The metabolism of ribavirin in erythrocytes and nucleated cells. Int J Biochem 1990; 22(4): 379-83.

[101] Dixit NM, Perelson AS. The metabolism, pharmacokinetics and mechanisms of antiviral activity of

ribavirin against hepatitis C virus. Cell Mol Life Sci 2006; 63(7-8): 832-42.

[102] Karasawa T, Saito T, Ueno Y, Sugimoto M, Soga T. Metabolome Analysis of Erythrocytes from Patients with Chronic Hepatitis C Reveals the Etiology of Ribavirin-Induced Hemolysis. Int J Med Sci 2013; 10(11): 1575-7.

[103] De Franceschi L, Fattovich G, Turrini F, Ayi K, Brugnara C, Manzato F, *et al.* Hemolytic anemia induced by ribavirin therapy in patients with chronic hepatitis C virus infection: role of membrane oxidative damage. Hepatology 2000; 31(4): 997-1004.

[104] Naik GS, Tyagi MG. A pharmacological profile of ribavirin and monitoring of its plasma concentration in chronic hepatitis C infection. J Clin Exp Hepatol 2012; 2(1): 42-54.

[105] Stanke-Labesque F, Loustaud-Ratti V, Babany G, Gagnieu MC, Marquet P. Ribavirin therapeutic drug monitoring: why, when and how? Fundam Clin Pharmacol 2010; 24(4): 401-6.

[106] Arase Y, Ikeda K, Tsubota A, Suzuki F, Suzuki Y, Saitoh S, *et al.* Significance of serum ribavirin concentration in combination therapy of interferon and ribavirin for chronic hepatitis C. Intervirology 2005; 48(2-3): 138-44.

[107] Maeda Y, Kiribayashi Y, Moriya T, Maruhashi A, Omoda K, Funakoshi S, *et al.* Dosage adjustment of ribavirin based on renal function in Japanese patients with chronic hepatitis C. Ther Drug Monit 2004; 26(1): 9-15.

[108] Fellay J, Thompson AJ, Ge D, Gumbs CE, Urban TJ, Shianna KV, *et al.* ITPA gene variants protect against anaemia in patients treated for chronic hepatitis C. Nature 2010; 464(7287): 405-8.

[109] Hitomi Y, Cirulli ET, Fellay J, McHutchison JG, Thompson AJ, Gumbs CE, *et al.* Inosine triphosphate protects against ribavirin-induced adenosine triphosphate loss by adenylosuccinate synthase function. Gastroenterology 2011; 140(4): 1314-21.

[110] Thompson AJ, Fellay J, Patel K, Tillmann HL, Naggie S, Ge D, *et al.* Variants in the ITPA gene protect against ribavirin-induced hemolytic anemia and decrease the need for ribavirin dose reduction. Gastroenterology 2010; 139(4): 1181-9.

[111] Thompson AJ, Santoro R, Piazzolla V, Clark PJ, Naggie S, Tillmann HL, *et al.* Inosine triphosphatase genetic variants are protective against anemia during antiviral therapy for HCV2/3 but do not decrease dose reductions of RBV or increase SVR. Hepatology 2011; 53(2): 389-95.

[112] Ochi H, Maekawa T, Abe H, Hayashida Y, Nakano R, Kubo M, *et al.* ITPA polymorphism affects ribavirin-induced anemia and outcomes of therapy--a genome-wide study of Japanese HCV virus patients. Gastroenterology 2010; 139(4): 1190-7.

[113] Ogawa E, Furusyo N, Nakamuta M, Kajiwara E, Nomura H, Dohmen K, *et al.* Clinical milestones for the prediction of severe anemia by chronic hepatitis C patients receiving telaprevir-based triple therapy. J Hepatol 2013; 59(4): 667-74.

[114] Asselah T, Zeuzem S, Soriano V, Bronowicki JP, Lohse AW, Mullhaupt B, *et al.* ITPA Genotypes Predict Anemia but Do Not Affect Virological Response with Interferon-Free Faldaprevir, Deleobuvir, and Ribavirin for HCV Infection. PLoS One 2015; 10(12): e0144004.

[115] Akamatsu S, Hayes CN, Tsuge M, Murakami E, Hiraga N, Abe H, *et al.* Ribavirin dose reduction during telaprevir/ribavirin/peg-interferon therapy overcomes the effect of the ITPA gene polymorphism. J Viral Hepat 2015; 22(2): 166-74.

[116] Doehring A, Hofmann WP, Schlecker C, Zeuzem S, Sarrazin C, Berg T, *et al.* Role of nucleoside transporters SLC28A2/3 and SLC29A1/2 genetics in ribavirin therapy: protection against anemia in patients with chronic hepatitis C. Pharmacogenet Genomics 2011; 21(5): 289-96.

[117] Rau M, Stickel F, Russmann S, Manser CN, Becker PP, Weisskopf M, *et al.* Impact of genetic SLC28 transporter and ITPA variants on ribavirin serum level, hemoglobin drop and therapeutic response in patients with HCV infection. J Hepatol 2013; 58(4): 669-75.

[118] Ampuero J, Del Campo JA, Rojas L, Calleja JL, Cabezas J, Lens S, *et al.* Role of ITPA and SLC28A2

genes in the prediction of anaemia associated with protease inhibitor plus ribavirin and peginterferon in hepatitis C treatment. J Clin Virol 2015; 68: 56-60.

[119] Lindahl K, Stahle L, Bruchfeld A, Schvarcz R. High-dose ribavirin in combination with standard dose peginterferon for treatment of patients with chronic hepatitis C. Hepatology 2005; 41(2): 275-9.

[120] Tsubota A, Hirose Y, Izumi N, Kumada H. Pharmacokinetics of ribavirin in combined interferon-alpha 2b and ribavirin therapy for chronic hepatitis C virus infection. Br J Clin Pharmacol 2003; 55(4): 360-7.

[121] Maynard M, Pradat P, Gagnieu MC, Souvignet C, Trepo C. Prediction of sustained virological response by ribavirin plasma concentration at week 4 of therapy in hepatitis C virus genotype 1 patients. Antivir Ther 2008; 13(4): 607-11.

[122] Pedersen C, Alsio A, Lagging M, Langeland N, Farkkila M, Buhl MR, *et al.* Ribavirin plasma concentration is a predictor of sustained virological response in patients treated for chronic hepatitis C virus genotype 2/3 infection. J Viral Hepat 2011; 18(4): 245-51.

[123] Brok J, Gluud LL, Gluud C. Ribavirin monotherapy for chronic hepatitis C. Cochrane Database Syst Rev 2009; (4): Cd005527.

[124] Dusheiko G, Main J, Thomas H, Reichard O, Lee C, Dhillon A, *et al.* Ribavirin treatment for patients with chronic hepatitis C: results of a placebo-controlled study. J Hepatol 1996; 25(5): 591-8.

[125] Pawlotsky JM, Dahari H, Neumann AU, Hezode C, Germanidis G, Lonjon I, *et al.* Antiviral action of ribavirin in chronic hepatitis C. Gastroenterology 2004; 126(3): 703-14.

[126] Dixit NM, Layden-Almer JE, Layden TJ, Perelson AS. Modelling how ribavirin improves interferon response rates in hepatitis C virus infection. Nature 2004; 432(7019): 922-4.

[127] Lowe JK, Brox L, Henderson JF. Consequences of inhibition of guanine nucleotide synthesis by mycophenolic acid and virazole. Cancer Res 1977; 37(3): 736-43.

[128] Lanford RE, Chavez D, Guerra B, Lau JY, Hong Z, Brasky KM, *et al.* Ribavirin induces error-prone replication of GB virus B in primary tamarin hepatocytes. J Virol 2001; 75(17): 8074-81.

[129] Leyssen P, Balzarini J, De Clercq E, Neyts J. The predominant mechanism by which ribavirin exerts its antiviral activity *in vitro* against flaviviruses and paramyxoviruses is mediated by inhibition of IMP dehydrogenase. J Virol 2005; 79(3): 1943-7.

[130] Olschlager S, Neyts J, Gunther S. Depletion of GTP pool is not the predominant mechanism by which ribavirin exerts its antiviral effect on Lassa virus. Antiviral Res 2011; 91(2): 89-93.

[131] Smee DF, Bray M, Huggins JW. Antiviral activity and mode of action studies of ribavirin and mycophenolic acid against orthopoxviruses *in vitro*. Antivir Chem Chemother 2001; 12(6): 327-35.

[132] Smee DF, Matthews TR. Metabolism of ribavirin in respiratory syncytial virus-infected and uninfected cells. Antimicrob Agents Chemother 1986; 30(1): 117-21.

[133] Zhou S, Liu R, Baroudy BM, Malcolm BA, Reyes GR. The effect of ribavirin and IMPDH inhibitors on hepatitis C virus subgenomic replicon RNA. Virology 2003; 310(2): 333-42.

[134] McHutchison JG, Shiffman ML, Cheung RC, Gordon SC, Wright TL, Pottage JC Jr, *et al.* A randomized, double-blind, placebo-controlled dose-escalation trial of merimepodib (VX-497) and interferon-alpha in previously untreated patients with chronic hepatitis C. Antivir Ther 2005; 10(5): 635-43.

[135] Bisaillon M, Lemay G. Viral and cellular enzymes involved in synthesis of mRNA cap structure. Virology 1997; 236(1): 1-7.

[136] Maag D, Castro C, Hong Z, Cameron CE. Hepatitis C virus RNA-dependent RNA polymerase (NS5B) as a mediator of the antiviral activity of ribavirin. J Biol Chem 2001; 276(49): 46094-8.

[137] Vo NV, Young KC, Lai MM. Mutagenic and inhibitory effects of ribavirin on hepatitis C virus RNA polymerase. Biochemistry 2003; 42(35): 10462-71.

[138] Jones ME. Pyrimidine nucleotide biosynthesis in animals: genes, enzymes, and regulation of UMP biosynthesis. Annu Rev Biochem 1980; 49: 253-79.

[139] Toltzis P, O'Connell K, Patterson JL. Effect of phosphorylated ribavirin on vesicular stomatitis virus transcription. Antimicrob Agents Chemother 1988; 32(4): 492-7.

[140] Rankin JT Jr, Eppes SB, Antczak JB, Joklik WK. Studies on the mechanism of the antiviral activity of ribavirin against reovirus. Virology 1989; 168(1): 147-58.

[141] Thomas E, Ghany MG, Liang TJ. The application and mechanism of action of ribavirin in therapy of hepatitis C. Antivir Chem Chemother 2012; 23(1): 1-12.

[142] Domingo E, Holland JJ. RNA virus mutations and fitness for survival. Annu Rev Microbiol 1997; 51: 151-78.

[143] Lauring AS, Andino R. Quasispecies Theory and the Behavior of RNA Viruses. PLoS Pathog 2010; 6(7)

[144] Asahina Y, Izumi N, Enomoto N, Uchihara M, Kurosaki M, Onuki Y, *et al.* Mutagenic effects of ribavirin and response to interferon/ribavirin combination therapy in chronic hepatitis C. J Hepatol 2005; 43(4): 623-9.

[145] Chevaliez S, Brillet R, Lázaro E, Hézode C, Pawlotsky JM. Analysis of Ribavirin Mutagenicity in Human Hepatitis C Virus Infection. J Virol 2007; 81(14): 7732-41.

[146] Lutchman G, Danehower S, Song BC, Liang TJ, Hoofnagle JH, Thomson M, *et al.* Mutation rate of the hepatitis C virus NS5B in patients undergoing treatment with ribavirin monotherapy. Gastroenterology 2007; 132(5): 1757-66.

[147] Young KC, Lindsay KL, Lee KJ, Liu WC, He JW, Milstein SL, *et al.* Identification of a ribavirin-resistant NS5B mutation of hepatitis C virus during ribavirin monotherapy. Hepatology 2003; 38(4): 869-78.

[148] Graci JD, Cameron CE. Mechanisms of action of ribavirin against distinct viruses. Rev Med Virol 2006; 16(1): 37-48.

[149] Goswami BB, Borek E, Sharma OK, Fujitaki J, Smith RA. The broad spectrum antiviral agent ribavirin inhibits capping of mRNA. Biochem Biophys Res Commun 1979; 89(3): 830-6.

[150] Bougie I, Bisaillon M. The broad spectrum antiviral nucleoside ribavirin as a substrate for a viral RNA capping enzyme. J Biol Chem 2004; 279(21): 22124-30.

[151] Benarroch D, Egloff MP, Mulard L, Guerreiro C, Romette JL, Canard B. A structural basis for the inhibition of the NS5 dengue virus mRNA 2′-O-methyltransferase domain by ribavirin 5′-triphosphate. J Biol Chem 2004; 279(34): 35638-43.

[152] Westman B, Beeren L, Grudzien E, Stepinski J, Worch R, Zuberek J, *et al.* The antiviral drug ribavirin does not mimic the 7-methylguanosine moiety of the mRNA cap structure *in vitro*. RNA 2005; 11(10): 1505-13.

[153] Yan Y, Svitkin Y, Lee JM, Bisaillon M, Pelletier J. Ribavirin is not a functional mimic of the 7-methyl guanosine mRNA cap. RNA 2005; 11(8): 1238-44.

[154] Neumann-Haefelin C, Thimme R. Adaptive immune responses in hepatitis C virus infection. Curr Top Microbiol Immunol 2013; 369: 243-62.

[155] Jacobson Brown PM, Neuman MG. Immunopathogenesis of hepatitis C viral infection: Th1/Th2 responses and the role of cytokines. Clin Biochem 2001; 34(3): 167-71.

[156] Antonaci S, Piazzolla G, Napoli N, Vella FS, Fiore G, Schiraldi O. Relationship between T lymphocyte responsiveness and T-helper1/T-helper2 type cytokine release in chronic hepatitis C: a critical reappraisal. Microbios 2001; 106(415): 203-12.

[157] Tsai SL, Liaw YF, Chen MH, Huang CY, Kuo GC. Detection of type 2-like T-helper cells in hepatitis

C virus infection: implications for hepatitis C virus chronicity. Hepatology 1997; 25(2): 449-58.

[158] Reiser M, Marousis CG, Nelson DR, Lauer G, Gonzalez-Peralta RP, Davis GL, *et al.* Serum interleukin 4 and interleukin 10 levels in patients with chronic hepatitis C virus infection. J Hepatol 1997; 26(3): 471-8.

[159] Cacciarelli TV, Martinez OM, Gish RG, Villanueva JC, Krams SM. Immunoregulatory cytokines in chronic hepatitis C virus infection: pre- and posttreatment with interferon alfa. Hepatology 1996; 24(1): 6-9.

[160] Napoli J, Bishop GA, McGuinness PH, Painter DM, McCaughan GW. Progressive liver injury in chronic hepatitis C infection correlates with increased intrahepatic expression of Th1-associated cytokines. Hepatology 1996; 24(4): 759-65.

[161] Tam RC, Pai B, Bard J, Lim C, Averett DR, Phan UT, *et al.* Ribavirin polarizes human T cell responses towards a Type 1 cytokine profile. J Hepatol 1999; 30(3): 376-82.

[162] Hultgren C, Milich DR, Weiland O, Sallberg M. The antiviral compound ribavirin modulates the T helper (Th) 1/Th2 subset balance in hepatitis B and C virus-specific immune responses. J Gen Virol 1998; 79(Pt 10): 2381-91.

[163] Rigopoulou EI, Abbott WG, Williams R, Naoumov NV. Direct evidence for immunomodulatory properties of ribavirin on T-cell reactivity to hepatitis C virus. Antiviral Res 2007; 75(1): 36-42.

[164] Martin J, Navas S, Quiroga JA, Pardo M, Carreno V. Effects of the ribavirin-interferon alpha combination on cultured peripheral blood mononuclear cells from chronic hepatitis C patients. Cytokine 1998; 10(8): 635-44.

[165] Cramp ME, Rossol S, Chokshi S, Carucci P, Williams R, Naoumov NV. Hepatitis C virus-specific T-cell reactivity during interferon and ribavirin treatment in chronic hepatitis C. Gastroenterology 2000; 118(2): 346-55.

[166] Schoggins JW, Rice CM. Interferon-stimulated genes and their antiviral effector functions. Curr Opin Virol 2011; 1(6): 519-25.

[167] Taniguchi T, Takaoka A. The interferon-alpha/beta system in antiviral responses: a multimodal machinery of gene regulation by the IRF family of transcription factors. Curr Opin Immunol 2002; 14(1): 111-6.

[168] Feld JJ, Nanda S, Huang Y, Chen W, Cam M, Pusek SN, *et al.* Hepatic Gene Expression During Treatment with Peginterferon and Ribavirin: Identifying Molecular Pathways for Treatment Response. Hepatology 2007; 46(5): 1548-63.

[169] Liu WL, Su WC, Cheng CW, Hwang LH, Wang CC, Chen HL, *et al.* Ribavirin up-regulates the activity of double-stranded RNA-activated protein kinase and enhances the action of interferon-alpha against hepatitis C virus. J Infect Dis 2007; 196(3): 425-34.

[170] Su WC, Liu WL, Cheng CW, Chou YB, Hung KH, Huang WH, *et al.* Ribavirin enhances interferon signaling *via* stimulation of mTOR and p53 activities. FEBS Lett 2009; 583(17): 2793-8.

[171] Tokumoto Y, Hiasa Y, Uesugi K, Watanabe T, Mashiba T, Abe M, *et al.* Ribavirin regulates hepatitis C virus replication through enhancing interferon-stimulated genes and interleukin 8. J Infect Dis 2012; 205(7): 1121-30.

[172] Zhang Y, Jamaluddin M, Wang S, Tian B, Garofalo RP, Casola A, *et al.* Ribavirin treatment up-regulates antiviral gene expression *via* the interferon-stimulated response element in respiratory syncytial virus-infected epithelial cells. J Virol 2003; 77(10): 5933-47.

[173] Feld JJ, Lutchman GA, Heller T, Hara K, Pfeiffer JK, Leff RD, *et al.* Ribavirin improves early responses to peginterferon through enhanced interferon signaling. Gastroenterology 2010; 139(1): 154-62.e4.

[174] Thomas E, Feld JJ, Li Q, Hu Z, Fried MW, Liang TJ. Ribavirin Potentiates Interferon Action by Augmenting Interferon-Stimulated Gene Induction in Hepatitis C Virus Cell Culture Models.

Hepatology 2011; 53(1): 32-41.

[175] De Clercq E. Perspectives for the chemotherapy of respiratory syncytial virus (RSV) infections. Int J Antimicrob Agents 1996; 7(3): 193-202.

[176] DeVincenzo J, Lambkin-Williams R, Wilkinson T, Cehelsky J, Nochur S, Walsh E, *et al.* A randomized, double-blind, placebo-controlled study of an RNAi-based therapy directed against respiratory syncytial virus. Proc Natl Acad Sci USA 2010; 107(19): 8800-5.

[177] DeVincenzo JP, Whitley RJ, Mackman RL, Scaglioni-Weinlich C, Harrison L, Farrell E, *et al.* Oral GS-5806 activity in a respiratory syncytial virus challenge study. N Engl J Med 2014; 371(8): 711-22.

[178] 2016.http://www.who.int/hepatitis/publications/hepatitis-c-guidelines-2016/en/

[179] Bian DD, Zhou HY, Liu S, Liu M, Duan C, Zhang JY, *et al.* Current treatment status and barriers for patients with chronic HCV infection in mainland China: A national multicenter cross-sectional survey in 56 hospitals. Medicine (Baltimore) 2017; 96(34)

[180] Han QY, Liu ZW. Current treatment of chronic hepatitis C in China: Dilemma and potential problems. World J Gastroenterol 2016; 22(19): 4615-8.

[181] Feld JJ. The beginning of the end: what is the future of interferon therapy for chronic hepatitis C? Antiviral Res 2014; 105: 32-8.

[182] Zimmermann A, Gu JJ, Spychala J, Mitchell BS. Inosine monophosphate dehydrogenase expression: transcriptional regulation of the type I and type II genes. Adv Enzyme Regul 1996; 36: 75-84.

[183] Nagai M, Natsumeda Y, Konno Y, Hoffman R, Irino S, Weber G. Selective up-regulation of type II inosine 5′-monophosphate dehydrogenase messenger RNA expression in human leukemias. Cancer Res 1991; 51(15): 3886-90.

[184] Weber G. Biochemical strategy of cancer cells and the design of chemotherapy: G. H. A. Clowes Memorial Lecture. Cancer Res 1983; 43(8): 3466-92.

[185] Konno Y, Natsumeda Y, Nagai M, Yamaji Y, Ohno S, Suzuki K, *et al.* Expression of human IMP dehydrogenase types I and II in Escherichia coli and distribution in human normal lymphocytes and leukemic cell lines. J Biol Chem 1991; 266(1): 506-9.

[186] Nagai M, Natsumeda Y, Weber G. Proliferation-linked regulation of type II IMP dehydrogenase gene in human normal lymphocytes and HL-60 leukemic cells. Cancer Res 1992; 52(2): 258-61.

[187] Natsumeda Y, Ikegami T, Murayama K, Weber G. De novo guanylate synthesis in the commitment to replication in hepatoma 3924A cells. Cancer Res 1988; 48(3): 507-11.

[188] Senda M, Natsumeda Y. Tissue-differential expression of two distinct genes for human IMP dehydrogenase (E.C.1.1.1.205). Life Sci 1994; 54(24): 1917-26.

[189] Jackson RC, Weber G, Morris HP. IMP dehydrogenase, an enzyme linked with proliferation and malignancy. Nature 1975; 256(5515): 331-3.

[190] Natsumeda Y, Carr SF. Human type I and II IMP dehydrogenases as drug targets. Ann N Y Acad Sci 1993; 696: 88-93.

[191] Carr SF, Papp E, Wu JC, Natsumeda Y. Characterization of human type I and type II IMP dehydrogenases. J Biol Chem 1993; 268(36): 27286-90.

[192] Natsumeda Y, Ohno S, Kawasaki H, Konno Y, Weber G, Suzuki K. Two distinct cDNAs for human IMP dehydrogenase. J Biol Chem 1990; 265(9): 5292-5.

[193] Shu Q, Nair V. Inosine monophosphate dehydrogenase (IMPDH) as a target in drug discovery. Med Res Rev 2008; 28(2): 219-32.

[194] Allison AC, Hovi T, Watts RW, Webster AD. Immunological observations on patients with Lesch-Nyhan syndrome, and on the role of de-novo purine synthesis in lymphocyte transformation. Lancet 1975; 2(7946): 1179-83.

[195] Weber G, Shen F, Orban TI, Kokeny S, Olah E. Targeting signal transduction. Adv Enzyme Regul 2003; 43: 47-56.

[196] Weber G, Shen F, Yang H, Prajda N, Li W. Amplification of signal transduction capacity and down-regulation by drugs. Adv Enzyme Regul 1999; 39: 51-66.

[197] Weber G. Down-regulation of increased signal transduction capacity in human cancer cells. Adv Enzyme Regul 2005; 45: 37-51.

[198] Weber G, Singhal RL, Prajda N, Yeh YA, Look KY, Sledge GW Jr. Regulation of signal transduction. Adv Enzyme Regul 1995; 35: 1-21.

[199] Earle MF, Glazer RI. Activity and metabolism of 2-beta-D-ribofuranosylthiazole-4-carboxamide in human lymphoid tumor cells in culture. Cancer Res 1983; 43(1): 133-7.

[200] Jayaram HN, Smith AL, Glazer RI, Johns DG, Cooney DA. Studies on the mechanism of action of 2-beta-D-ribofuranosylthiazole-4-carboxamide (NSC 286193)--II. Relationship between dose level and biochemical effects in P388 leukemia *in vivo*. Biochem Pharmacol 1982; 31(23): 3839-45.

[201] Robins RK, Srivastava PC, Narayanan VL, Plowman J, Paull KD. 2-beta-D-Ribofuranosylthiazo-e-4-carboxamide, a novel potential antitumor agent for lung tumors and metastases. J Med Chem 1982; 25(2): 107-8.

[202] Lui MS, Faderan MA, Liepnieks JJ, Natsumeda Y, Olah E, Jayaram HN, *et al.* Modulation of IMP dehydrogenase activity and guanylate metabolism by tiazofurin (2-beta-D-ribofuranosylthiazo-e-4-carboxamide). J Biol Chem 1984; 259(8): 5078-82.

[203] Olah E, Kote Z, Natsumeda Y, Yamaji Y, Jarai G, Lapis E, *et al.* Down-regulation of c-myc and c-H--ras gene expression by tiazofurin in rat hepatoma cells. Cancer Biochem Biophys 1990; 11(2): 107-17.

[204] Olah E, Natsumeda Y, Ikegami T, Kote Z, Horanyi M, Szelenyi J, *et al.* Induction of erythroid differentiation and modulation of gene expression by tiazofurin in K-562 leukemia cells. Proc Natl Acad Sci USA 1988; 85(17): 6533-7.

[205] Weber G, Nagai M, Natsumeda Y, Eble JN, Jayaram HN, Paulik E, *et al.* Tiazofurin down-regulates expression of c-Ki-ras oncogene in a leukemic patient. Cancer Commun 1991; 3(3): 61-6.

[206] Malek K, Boosalis MS, Waraska K, Mitchell BS, Wright DG. Effects of the IMP-dehydrogenase inhibitor, Tiazofurin, in bcr-abl positive acute myelogenous leukemia. Part I. *In vivo* studies. Leuk Res 2004; 28(11): 1125-36.

[207] Tricot G, Weber G. Biochemically targeted therapy of refractory leukemia and myeloid blast crisis of chronic granulocytic leukemia with Tiazofurin, a selective blocker of inosine 5′-phosphate dehydrogenase activity. Anticancer Res 1996; 16(6a): 3341-7.

[208] Tricot GJ, Jayaram HN, Lapis E, Natsumeda Y, Nichols CR, Kneebone P, *et al.* Biochemically directed therapy of leukemia with tiazofurin, a selective blocker of inosine 5′-phosphate dehydrogenase activity. Cancer Res 1989; 49(13): 3696-701.

[209] Tricot GJ, Jayaram HN, Nichols CR, Pennington K, Lapis E, Weber G, *et al.* Hematological and biochemical action of tiazofurin (NSC 286193) in a case of refractory acute myeloid leukemia. Cancer Res 1987; 47(18): 4988-91.

[210] Wright DG, Boosalis MS, Waraska K, Oshry LJ, Weintraub LR, Vosburgh E. Tiazofurin effects on IMP-dehydrogenase activity and expression in the leukemia cells of patients with CML blast crisis. Anticancer Res 1996; 16(6a): 3349-51.

[211] Tricot G, Jayaram HN, Weber G, Hoffman R. Tiazofurin: biological effects and clinical uses. Int J Cell Cloning 1990; 8(3): 161-70.

[212] Weber G. Critical issues in chemotherapy with tiazofurin. Adv Enzyme Regul 1989; 29: 75-95.

[213] Jolley WB, Chu WT, Salter JM, Hinshaw DB, Randhawa AS. Antitumor effects of ribavirin on L1210

leukemia in the DBA/2 mouse. Ann N Y Acad Sci 1977; 284: 585-90.

[214] Bekesi JG, Roboz JP, Zimmerman E, Holland JF. Treatment of spontaneous leukemia in AKR mice with chemotherapy, immunotherapy, or interferon. Cancer Res 1976; 36(2 pt 2): 631-9.

[215] Natsumeda Y, Yamada Y, Yamaji Y, Weber G. Synergistic cytotoxic effect of tiazofurin and ribavirin in hepatoma cells. Biochem Biophys Res Commun 1988; 153(1): 321-7.

[216] Prajda N, Hata Y, Abonyi M, Singhal RL, Weber G. Sequential impact of tiazofurin and ribavirin on the enzymic program of the bone marrow. Cancer Res 1993; 53(24): 5982-6.

[217] Li W, Shen F, Weber G. Ribavirin and quercetin synergistically downregulate signal transduction and are cytotoxic in human ovarian carcinoma cells. Oncol Res 1999; 11(5): 243-7.

[218] Majumdar A, Kerby S, Mullikin B, Beckstead JH, Stenberg PE, Seidman MM. IL-3 and ribavirin induce differentiation and growth suppression during long-term treatment of a megakaryocytic leukemia cell line. J Cell Physiol 1995; 165(3): 530-7.

[219] Kokeny S, Papp J, Weber G, Vaszko T, Carmona-Saez P, Olah E. Ribavirin acts *via* multiple pathways in inhibition of leukemic cell proliferation. Anticancer Res 2009; 29(6): 1971-80.

[220] Byrnes K, White S, Chu Q, Meschonat C, Yu H, Johnson LW, *et al.* High eIF4E, VEGF, and microvessel density in stage I to III breast cancer. Ann Surg 2006; 243(5): 684-90.

[221] Coleman LJ, Peter MB, Teall TJ, Brannan RA, Hanby AM, Honarpisheh H, *et al.* Combined analysis of eIF4E and 4E-binding protein expression predicts breast cancer survival and estimates eIF4E activity. Br J Cancer 2009; 100(9): 1393-9.

[222] Culjkovic B, Borden KL. 2009.

[223] Holm N, Byrnes K, Johnson L, Abreo F, Sehon K, Alley J, *et al.* A prospective trial on initiation factor 4E (eIF4E) overexpression and cancer recurrence in node-negative breast cancer. Ann Surg Oncol 2008; 15(11): 3207-15.

[224] Siddiqui N, Borden KL. mRNA export and cancer. Wiley Interdiscip Rev RNA 2012; 3(1): 13-25.

[225] De Benedetti A, Harris AL. eIF4E expression in tumors: its possible role in progression of malignancies. Int J Biochem Cell Biol 1999; 31(1): 59-72.

[226] Mamane Y, Petroulakis E, Rong L, Yoshida K, Ler LW, Sonenberg N. eIF4E--from translation to transformation. Oncogene 2004; 23(18): 3172-9.

[227] Nathan CO, Franklin S, Abreo FW, Nassar R, de Benedetti A, Williams J, *et al.* Expression of eIF4E during head and neck tumorigenesis: possible role in angiogenesis. Laryngoscope 1999; 109(8): 1253-8.

[228] Nathan CO, Liu L, Li BD, Abreo FW, Nandy I, De Benedetti A. Detection of the proto-oncogene eIF4E in surgical margins may predict recurrence in head and neck cancer. Oncogene 1997; 15(5): 579-84.

[229] Pettersson F, Yau C, Dobocan MC, Culjkovic-Kraljacic B, Retrouvey H, Puckett R, *et al.* Ribavirin treatment effects on breast cancers overexpressing eIF4E, a biomarker with prognostic specificity for luminal B-type breast cancer. Clin Cancer Res 2011; 17(9): 2874-84.

[230] Borden KL, Culjkovic-Kraljacic B. Ribavirin as an anti-cancer therapy: acute myeloid leukemia and beyond? Leuk Lymphoma 2010; 51(10): 1805-15.

[231] Pestova TV, Hellen CU. The structure and function of initiation factors in eukaryotic protein synthesis. Cell Mol Life Sci 2000; 57(4): 651-74.

[232] Topisirovic I, Siddiqui N, Lapointe VL, Trost M, Thibault P, Bangeranye C, *et al.* Molecular dissection of the eukaryotic initiation factor 4E (eIF4E) export-competent RNP. EMBO J 2009; 28(8): 1087-98.

[233] von der Haar T, Gross JD, Wagner G, McCarthy JE. The mRNA cap-binding protein eIF4E in post-

transcriptional gene expression. Nat Struct Mol Biol 2004; 11(6): 503-11.

[234] Fischer PM. Cap in hand: targeting eIF4E. Cell Cycle 2009; 8(16): 2535-41.

[235] Hsieh AC, Ruggero D. Targeting eukaryotic translation initiation factor 4E (eIF4E) in cancer. Clin Cancer Res 2010; 16(20): 4914-20.

[236] Davuluri RV, Suzuki Y, Sugano S, Zhang MQ. CART classification of human 5′ UTR sequences. Genome Res 2000; 10(11): 1807-16.

[237] Bitterman PB. Attacking a Nexus of the Oncogenic Circuitry by Reversing Aberrant eIF4F-Mediated Translation 2012.

[238] Jia Y, Polunovsky V, Bitterman PB, Wagner CR. Cap-dependent translation initiation factor eIF4E: an emerging anticancer drug target. Med Res Rev 2012; 32(4): 786-814.

[239] Osborne MJ, Borden KL. The eukaryotic translation initiation factor eIF4E in the nucleus: taking the road less traveled. Immunol Rev 2015; 263(1): 210-23.

[240] De Benedetti A, Graff JR. eIF-4E expression and its role in malignancies and metastases. Oncogene 2004; 23(18): 3189-99.

[241] Lin CJ, Cencic R, Mills JR, Robert F, Pelletier J. c-Myc and eIF4F are components of a feedforward loop that links transcription and translation. Cancer Res 2008; 68(13): 5326-34.

[242] Rosenwald IB, Chen JJ, Wang S, Savas L, London IM, Pullman J. Upregulation of protein synthesis initiation factor eIF-4E is an early event during colon carcinogenesis. Oncogene 1999; 18(15): 2507-17.

[243] Schmidt EV. The role of c-myc in regulation of translation initiation. Oncogene 2004; 23(18): 3217-21.

[244] Topisirovic I, Siddiqui N, Orolicki S, Skrabanek LA, Tremblay M, Hoang T, *et al.* Stability of eukaryotic translation initiation factor 4E mRNA is regulated by HuR, and this activity is dysregulated in cancer. Mol Cell Biol 2009; 29(5): 1152-62.

[245] Richter JD, Sonenberg N. Regulation of cap-dependent translation by eIF4E inhibitory proteins. Nature 2005; 433(7025): 477-80.

[246] Tee AR, Blenis J. mTOR, translational control and human disease. Semin Cell Dev Biol 2005; 16(1): 29-37.

[247] Wullschleger S, Loewith R, Hall MN. TOR signaling in growth and metabolism. Cell 2006; 124(3): 471-84.

[248] Hay N. Mnk earmarks eIF4E for cancer therapy. Proc Natl Acad Sci USA 2010; 107(32): 13975-6.

[249] Scheper GC, Proud CG. Does phosphorylation of the cap-binding protein eIF4E play a role in translation initiation? Eur J Biochem 2002; 269(22): 5350-9.

[250] Ueda T, Watanabe-Fukunaga R, Fukuyama H, Nagata S, Fukunaga R. Mnk2 and Mnk1 are essential for constitutive and inducible phosphorylation of eukaryotic initiation factor 4E but not for cell growth or development. Mol Cell Biol 2004; 24(15): 6539-49.

[251] Lu C, Makala L, Wu D, Cai Y. Targeting translation: eIF4E as an emerging anticancer drug target. Expert Rev Mol Med 2016; 18: e2.

[252] Kentsis A, Topisirovic I, Culjkovic B, Shao L, Borden KL. Ribavirin suppresses eIF4E-mediated oncogenic transformation by physical mimicry of the 7-methyl guanosine mRNA cap. Proc Natl Acad Sci USA 2004; 101(52): 18105-10.

[253] Tan K, Culjkovic B, Amri A, Borden KL. Ribavirin targets eIF4E dependent Akt survival signaling. Biochem Biophys Res Commun 2008; 375(3): 341-5.

[254] Kentsis A, Volpon L, Topisirovic I, Soll CE, Culjkovic B, Shao L, *et al.* Further evidence that ribavirin interacts with eIF4E. RNA 2005; 11(12): 1762-6.

[255] Bollmann F, Fechir K, Nowag S, Koch K, Art J, Kleinert H, *et al.* Human inducible nitric oxide synthase (iNOS) expression depends on chromosome region maintenance 1 (CRM1)- and eukaryotic translation initiation factor 4E (elF4E)-mediated nucleocytoplasmic mRNA transport. Nitric Oxide 2013; 30: 49-59.

[256] Shi F, Len Y, Gong Y, Shi R, Yang X, Naren D, *et al.* Ribavirin Inhibits the Activity of mTOR/eIF4E, ERK/Mnk1/eIF4E Signaling Pathway and Synergizes with Tyrosine Kinase Inhibitor Imatinib to Impair Bcr-Abl Mediated Proliferation and Apoptosis in Ph+ Leukemia. PLoS One 2015; 10(8): e0136746.

[257] Assouline S, Culjkovic-Kraljacic B, Bergeron J, Caplan S, Cocolakis E, Lambert C, *et al.* A phase I trial of ribavirin and low-dose cytarabine for the treatment of relapsed and refractory acute myeloid leukemia with elevated eIF4E. Haematologica 2015; 100(1): e7-9.

[258] Zahreddine HA, Culjkovic-Kraljacic B, Assouline S, Gendron P, Romeo AA, Morris SJ, *et al.* The sonic hedgehog factor GLI1 imparts drug resistance through inducible glucuronidation. Nature 2014; 511(7507): 90-3.

[259] Pettersson F, Del Rincon SV, Emond A, Huor B, Ngan E, Ng J, *et al.* Genetic and pharmacologic inhibition of eIF4E reduces breast cancer cell migration, invasion, and metastasis. Cancer Res 2015; 75(6): 1102-12.

[260] Sappok A, Mahlknecht U. Ribavirin restores ESR1 gene expression and tamoxifen sensitivity in ESR1 negative breast cancer cell lines. Clin Epigenetics 2011; 3(1): 8.

[261] Gu X, Jones L, Lowery-Norberg M, Fowler M. Expression of eukaryotic initiation factor 4E in astrocytic tumors. Appl Immunohistochem Mol Morphol 2005; 13(2): 178-83.

[262] Ogino A, Sano E, Ochiai Y, Yamamuro S, Tashiro S, Yachi K, *et al.* Efficacy of ribavirin against malignant glioma cell lines. Oncol Lett 2014; 8(6): 2469-74.

[263] Teng L, Ding D, Chen Y, Dai H, Liu G, Qiao Z, *et al.* Anti-tumor effect of ribavirin in combination with interferon-alpha on renal cell carcinoma cell lines *in vitro.* Cancer Cell Int 2014; 14: 63.

[264] Vallee S, Fouchier F, Braguer D, Marvaldi J, Champion S. Ribavirin-induced resistance to heat shock, inhibition of the Ras-Raf-1 pathway and arrest in G(1). Eur J Pharmacol 2000; 404(1-2): 49-62.

[265] De la Cruz-Hernandez E, Medina-Franco JL, Trujillo J, Chavez-Blanco A, Dominguez-Gomez G, Perez-Cardenas E, *et al.* Ribavirin as a tri-targeted antitumor repositioned drug. Oncol Rep 2015; 33(5): 2384-92.

[266] Yan KS, Lin CY, Liao TW, Peng CM, Lee SC, Liu YJ, *et al.* EZH2 in Cancer Progression and Potential Application in Cancer Therapy: A Friend or Foe? Int J Mol Sci 2017; 18(6)

[267] Fan TY, Wang H, Xiang P, Liu YW, Li HZ, Lei BX, *et al.* Inhibition of EZH2 reverses chemotherapeutic drug TMZ chemosensitivity in glioblastoma. Int J Clin Exp Pathol 2014; 7(10): 6662-70.

[268] Dai D, Chen H, Tang J, Tang Y. Inhibition of mTOR/eIF4E by anti-viral drug ribavirin effectively enhances the effects of paclitaxel in oral tongue squamous cell carcinoma. Biochem Biophys Res Commun 2017; 482(4): 1259-64.

[269] Hayman TJ, Williams ES, Jamal M, Shankavaram UT, Camphausen K, Tofilon PJ. Translation initiation factor eIF4E is a target for tumor cell radiosensitization. Cancer Res 2012; 72(9): 2362-72.

[270] Compston A, Coles A. Multiple sclerosis. Lancet 2008; 372(9648): 1502-17.

[271] Bjelobaba I, Savic D, Lavrnja I. Multiple Sclerosis and Neuroinflammation: The Overview of Current and Prospective Therapies. Curr Pharm Des 2017; 23(5): 693-730.

[272] McFarland HF, Martin R. Multiple sclerosis: a complicated picture of autoimmunity. Nat Immunol 2007; 8(9): 913-9.

[273] Lassmann H. Mechanisms of white matter damage in multiple sclerosis. Glia 2014; 62(11): 1816-30.

[274] von Budingen HC, Palanichamy A, Lehmann-Horn K, Michel BA, Zamvil SS. Update on the autoimmune pathology of multiple sclerosis: B-cells as disease-drivers and therapeutic targets. Eur Neurol 2015; 73(3-4): 238-46.

[275] Hauser SL, Waubant E, Arnold DL, Vollmer T, Antel J, Fox RJ, *et al.* B-cell depletion with rituximab in relapsing-remitting multiple sclerosis. N Engl J Med 2008; 358(7): 676-88.

[276] Bogie JF, Stinissen P, Hendriks JJ. Macrophage subsets and microglia in multiple sclerosis. Acta Neuropathol 2014; 128(2): 191-213.

[277] Correale J, Farez MF. The Role of Astrocytes in Multiple Sclerosis Progression. Front Neurol 2015; 6: 180.

[278] Andersen O, Lygner PE, Bergstrom T, Andersson M, Vahlne A. Viral infections trigger multiple sclerosis relapses: a prospective seroepidemiological study. J Neurol 1993; 240(7): 417-22.

[279] Sibley WA, Bamford CR, Clark K. Clinical viral infections and multiple sclerosis. Lancet 1985; 1(8441): 1313-5.

[280] Buljevac D, Flach HZ, Hop WC, Hijdra D, Laman JD, Savelkoul HF, *et al.* Prospective study on the relationship between infections and multiple sclerosis exacerbations. Brain 2002; 125(Pt 5): 952-60.

[281] Tremlett H, van der Mei IA, Pittas F, Blizzard L, Paley G, Mesaros D, *et al.* Monthly ambient sunlight, infections and relapse rates in multiple sclerosis. Neuroepidemiology 2008; 31(4): 271-9.

[282] Jin Y, de Pedro-Cuesta J, Soderstrom M, Stawiarz L, Link H. Seasonal patterns in optic neuritis and multiple sclerosis: a meta-analysis. J Neurol Sci 2000; 181(1-2): 56-64.

[283] Kriesel JD, White A, Hayden FG, Spruance SL, Petajan J. Multiple sclerosis attacks are associated with picornavirus infections. Mult Scler 2004; 10(2): 145-8.

[284] Edwards S, Zvartau M, Clarke H, Irving W, Blumhardt LD. Clinical relapses and disease activity on magnetic resonance imaging associated with viral upper respiratory tract infections in multiple sclerosis. J Neurol Neurosurg Psychiatry 1998; 64(6): 736-41.

[285] Panitch HS. Influence of infection on exacerbations of multiple sclerosis. Ann Neurol 1994; 36 (Suppl.): S25-8.

[286] Blackmore S, Hernandez J, Juda M, Ryder E, Freund GG, Johnson RW, *et al.* Influenza infection triggers disease in a genetic model of experimental autoimmune encephalomyelitis. Proc Natl Acad Sci USA 2017; 114(30): E6107-16.

[287] Stromnes IM, Goverman JM. Passive induction of experimental allergic encephalomyelitis. Nat Protoc 2006; 1(4): 1952-60.

[288] Stromnes IM, Goverman JM. Active induction of experimental allergic encephalomyelitis. Nat Protoc 2006; 1(4): 1810-9.

[289] Berard JL, Wolak K, Fournier S, David S. Characterization of relapsing-remitting and chronic forms of experimental autoimmune encephalomyelitis in C57BL/6 mice. Glia 2010; 58(4): 434-45.

[290] Mohajeri M, Sadeghizadeh M, Javan M. Pertussis toxin promotes relapsing-remitting experimental autoimmune encephalomyelitis in Lewis rats. J Neuroimmunol 2015; 289: 105-10.

[291] Stosic-Grujicic S, Ramic Z, Bumbasirevic V, Harhaji L, Mostarica-Stojkovic M. Induction of experimental autoimmune encephalomyelitis in Dark Agouti rats without adjuvant. Clin Exp Immunol 2004; 136(1): 49-55.

[292] Jeulin H, Venard V, Carapito D, Finance C, Kedzierewicz F. Effective ribavirin concentration in mice brain using cyclodextrin as a drug carrier: evaluation in a measles encephalitis model. Antiviral Res 2009; 81(3): 261-6.

[293] Colombo G, Lorenzini L, Zironi E, Galligioni V, Sonvico F, Balducci AG, *et al.* Brain distribution of ribavirin after intranasal administration. Antiviral Res 2011; 92(3): 408-14.

[294] Allison AC, Eugui EM. Mycophenolate mofetil and its mechanisms of action. Immunopharmacology 2000; 47(2-3): 85-118.

[295] Jain J, Almquist SJ, Heiser AD, Shlyakhter D, Leon E, Memmott C, *et al.* Characterization of pharmacological efficacy of VX-148, a new, potent immunosuppressive inosine 5′-monophosphate dehydrogenase inhibitor. J Pharmacol Exp Ther 2002; 302(3): 1272-7.

[296] Michel L, Vukusic S, De Seze J, Ducray F, Ongagna JC, Lefrere F, *et al.* Mycophenolate mofetil in multiple sclerosis: a multicentre retrospective study on 344 patients. J Neurol Neurosurg Psychiatry 2014; 85(3): 279-83.

[297] Allison AC, Eugui EM. Mechanisms of action of mycophenolate mofetil in preventing acute and chronic allograft rejection. Transplantation 2005; 80(2) (Suppl.): S181-90.

[298] Allison AC, Eugui EM. Purine metabolism and immunosuppressive effects of mycophenolate mofetil (MMF). Clin Transplant 1996; 10(1 Pt 2): 77-84.

[299] Ren G, Roberts AI, Shi Y. Adhesion molecules: Key players in Mesenchymal stem cell-mediated immunosuppression. Cell Adhes Migr 2011; 5(1): 20-2.

[300] Vajkoczy P, Laschinger M, Engelhardt B. Alpha4-integrin-VCAM-1 binding mediates G protein-independent capture of encephalitogenic T cell blasts to CNS white matter microvessels. J Clin Invest 2001; 108(4): 557-65.

[301] Stosic-Grujicic S, Savic-Radojevic A, Maksimovic-Ivanic D, Markovic M, Bumbasirevic V, Ramic Z, *et al.* Down-regulation of experimental allergic encephalomyelitis in DA rats by tiazofurin. J Neuroimmunol 2002; 130(1-2): 66-77.

[302] Lavrnja I, Nedeljkovic N, Bjelobaba I, Stojkov D, Dacic S, Pekovic S, *et al.* Ribavirin administration alters ectonucleotidase activities in experimental autoimmune encephalomyelitis. Gen Physiol Biophys 2009; 28(Spec No): 69-76.

[303] Lavrnja I, Savic D, Bjelobaba I, Dacic S, Bozic I, Parabucki A, *et al.* The effect of ribavirin on reactive astrogliosis in experimental autoimmune encephalomyelitis. J Pharmacol Sci 2012; 119(3): 221-32.

[304] Lavrnja I, Stojkov D, Bjelobaba I, Pekovic S, Dacic S, Nedeljkovic N, *et al.* Ribavirin ameliorates experimental autoimmune encephalomyelitis in rats and modulates cytokine production. Int Immunopharmacol 2008; 8(9): 1282-90.

[305] Stojkov D, Lavrnja I, Šubašić S, Bjelobaba I, Peković S, Gadjanski I, *et al.* Therapeutic effect of nucleoside analogs on experimental autoimmune encephalomyelitis in dark agouti rats. Arch Biol Sci 2006; 58(1): 13-20.

[306] Savić D, Lavrnja I, Dacić S, Bjelobaba I, Nedeljković N, Peković S, *et al.* Combined treatment with ribavirin and tiazofurin attenuates response of glial cells in experimental autoimmune encephalomyelitis. Arch Biol Sci 2012; 64(3): 843-50.

[307] Stojkov D, Lavrnja I, Pekovic S, Dacic S, Bjelobaba I, Mostarica-Stojkovic M, *et al.* Therapeutic effects of combined treatment with ribavirin and tiazofurin on experimental autoimmune encephalomyelitis development: clinical and histopathological evaluation. J Neurol Sci 2008; 267(1-2): 76-85.

[308] Lavrnja I, Stojkov D, Pekovic S, Subasic S, Mostarica-Stojkovic M, Stosic-Grujicic S, *et al.* Combination of nucleoside analogues tiazofurin and ribavirin downregulates experimental autoimmune encephalomyelitis. Ann N Y Acad Sci 2005; 1048: 392-5.

[309] Bozic I, Savic D, Jovanovic M, Bjelobaba I, Laketa D, Nedeljkovic N, *et al.* 2015.

[310] Pekovic S, Subasic S, Nedeljkovic N, Bjelobaba I, Filipovic R, Milenkovic I, *et al.* Molecular basis of brain injury and repair.Neurobiological Studies - from Genes to Behaviour. Kerala, India: Research Signpost 2006; pp. 143-65.

[311] Feld JJ, Hoofnagle JH. Mechanism of action of interferon and ribavirin in treatment of hepatitis C. Nature 2005; 436(7053): 967-72.

CHAPTER 5

Carbon Nanotubes and Drug Targets

Malay K. Das* and **Anupam Sarma**

Department of Pharmaceutical Sciences, Dibrugarh University, Dibrugarh 786004 INDIA

Abstract: Carbon nanotubes (CNTs), a carbon allotrope with various physicochemical properties, have gained worldwide interest since last two decades. CNTs are made up of cylindrical shaped graphene foils with nanometric dimension and several micrometres in length. The application of CNTs in the field of pharmacy and medicine is based on their large surface area capable to adsorb or conjugate various classes of drugs, genes, vaccines, and antibodies. This review focuses on the applicability of CNTs as targeted drug delivery vehicle considering the biopharmaceutics and toxicokinetics including advantages, obstacles, and perspectives of various CNTs.

Keywords: Biodistribution, Carbon nanotubes, CNS targeting, Drug targeting, Gene therapy, Lymphatic targeting, Macrophage targeting, PEGylated CNTs, Pharmacokinetics, Toxicity, Tumor targeting, Vaccine delivery.

INTRODUCTION

Since the first scientific report on Carbon nanotubes (CNTs) published in the year 1991, the extensive research on the characteristics of CNTs has provided fundamental advancements in its application in material chemistry as well as in medicine [1]. During the last few decades, increasing attention has been paid to drug targeting concepts using nanotechnology to improve the overall therapeutic efficacy by reducing the side effects while minimizing degradation or elimination of drugs. In the current scenario, nanotechnology is a multidisciplinary, rapidly expanding scientific zone and has achieved breakthrough in molecular biology, diagnostics, and imaging, bioengineering, nanomedicines, and therapeutics. Nanotechnology has paved the way to explore its impact on various fields and it has a great contribution on the future of bio-nanomedicines [2 - 5]. Great progress has been made in the area of nanomaterial such as liposomes, dendrimers, nanostructured lipid carriers (NLCs), carbon nanotubes (CNTs) and polymer-

* **Corresponding author Malay K Das:** Department of Pharmaceutical Sciences, Dibrugarh University, Dibrugarh 786004 INDIA; Tel/Fax: +91-9954229317; E-mail: mkdps@dibru.ac.in

Atta-ur-Rahman & M. Iqbal Choudhary (Eds.)

mediated therapeutic delivery strategies to target drug at the specific sites for boosting its safety and therapeutic efficacy.

The one important property of the CNTs is that they can penetrate cell membranes with no cytotoxicity. The higher entrapment efficiency throughout the length of CNTs makes them superior from other varieties of round nanoparticles. These have boosted the *in vivo* utility of CNTs in the delivery of various categories of therapeutic agents including drugs, genes, and vaccines [6].

Types of Carbon Nanotubes

In Carbon nanotubes, the carbon atoms are organized in a sequence of fused benzene ring to form a circular hollow tube. The CNTs belongs to fullerenes family of carbon allotrope along with graphite (*sp2* planar) and diamond (*sp3* cubic) [9 - 11]. The CNTs can be categorized into two types considering the number of structural layers: single-walled carbon nanotubes (SWCNTs) and multi-walled carbon nanotubes (MWCNTs) (Fig. **1**).

The CNTs consisting of single graphene sheet are known as SWCNTs. The MWCNTs are concentric cylinders composed of multilayered graphene sheets bound together by Vander Waals forces with a typical distance of 0.34 nm between the layers of graphene. Thus, SWCNTs have a larger diameter (10-200 nm) than SWCNTs (1-3 nm) and can be made up of several hundred cylinders [12 - 13]. The methods of production with catalytic metal particles influence the diameter of CNTs, particularly for SWCNTs. Further, the manufacturing and purification process influence the length of CNTs in the order of micrometers. The longest CNTs of 18 cm have been reported in the literature [14 - 15].

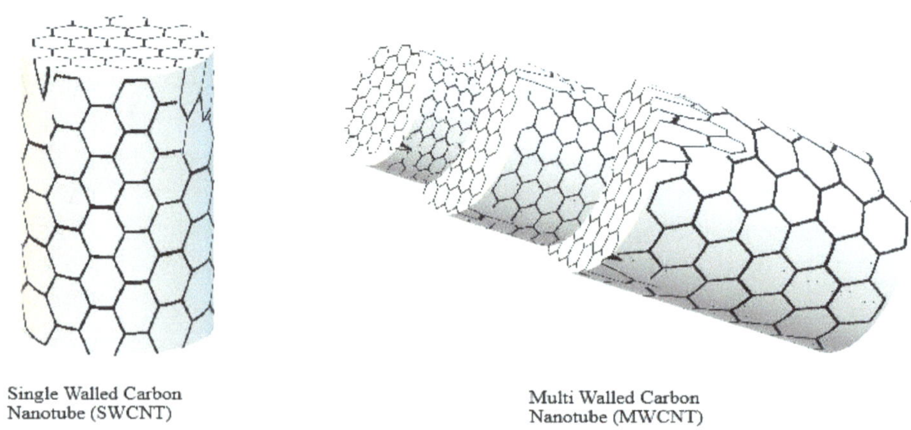

Fig. (1). Schematic diagram of SWCNT and MWCNT.

The atomic structure of carbon nanotubes depends upon its chiral angle. Multi-walled carbon nanotubes with concentric graphene layers usually have varying chiral angles leading to MWCNTs of different diameters. The chiral angle has effects on mechanical, optical and electrical properties of CNTs. Graphene itself is a semiconductor, so nanotubes can be either metallic or semiconductors depending on the chiral angle [16].

It is very difficult to disperse individual SWCNTs; they form the large bundles of 5-50 nm in diameter held together by van der Waals forces [17]. MWCNTs appear as individual tubes with fewer tendencies to form bundles [18]. However, the cohesive van der Waals forces are weaker in MWCNTs than SWCNT [19]. The degree of bundle formation and the diameter of CNTs determine the ratio of surface area to mass of the tubes. It is apparent that the CNTs have very high surface area in comparison to their mass. Individual SWCNTs have a surface area to mass ratio of about 1300 m^2/g, while MWCNTs typically have a surface area to mass ratio of around a few hundred m^2/g [20]. Since SWCNTs form themselves into bundles; the effective ratio of surface area to mass is often smaller or ~ 300 m^2/g [133] (Table **1**).

Table 1. Comparison between SWCNTs and MWCNTs.

SWCNTs	MWCNTs
Black colored granule or powder with metallic shine	Mostly Black colored granule or powder
Composed of one layer of graphene sheet	Composed of more than one layer of graphene sheet
No structural deformities	Have structural deformities
Conducts electricity properly	Deficient of electrical conductance
Synthesized by catalytic reaction	Need no catalyst for synthesis
Synthesis is difficult in bulk amount	Synthesis is easy in bulk amount
More defection occurs during functionalization	Less defection, but difficult to improve
High impurity	Less impurity
Less accumulation observed inside body	More accumulation observed inside body
Easy to characterize and evaluate	Difficult to characterize and evaluate
Easy to bend	Difficult to bend

Production of Carbon Nanotubes

The CNTs are the novel nanocarriers having the great potential of use in drug delivery/targeting and in biomedical application as well. For the use in pharmaceutical field, the CNTs should be of high standard, devoid of impurities and carbonaceous matter possessing the intact structures [7 - 8][122 - 125].

The simplest and the natural way of production of CNTs are the application of heat to carbon black and graphite under a flame controlled environment. But the CNTs manufactured in this way are generally different in shape, size, mechanical property, purity and quality due to the uncontrollable natural environment [126]. However, nowadays CNTs can be synthesized by artificial methods considering the interest of researcher, which includes electric arc discharge (EAD), laser ablation (LA) and chemical vapor deposition (CVD) methods. Besides these methods, some other methods like plasma enhanced chemical vapor deposition (PE-CVD) and high-pressure carbon monoxide disproportionation process (HiPCO) technique are also utilized for the manufacturing of CNTs. In HiPCO technique SWCNTs can be produced by creating catalytic reaction in a constant-flow of carbon monoxide (CO) as a carbon feedstock and $Fe(CO)_5$ (iron pentacarbonyl) as an iron-containing catalyst precursor. The size distribution of CNTs produced by this technique can be controlled by adjusting the pressure of the CO used. This technique is appropriate to manufacture CNTs in bulk amount [21]. In the case of electric arc discharge method of CNTs production, both anode and cathode are used. The CNTs are manufactured by the use of high voltage (around 100 amps) of electrons generated from an electric arc, which bombards the graphite surface (anode). CNTs are created at the foot of this anode [22 - 23]. In case of laser ablation technique, the CNTs are manufactured by striking a laser beam on the graphite using transition metal as catalyst. This method utilizes two different laser sources, one as primary laser and other one as secondary laser beam. The bombardment is done with the primary laser followed by a secondary laser beam. This method has the advantage of producing a high quality of SWCNTs and MWCNTs along with the disadvantage of time-consuming and costly [24 - 26].

The catalytic chemical vapor deposition (CVD) technique works in a bit different principle from the above techniques. In this method, the mixture of vaporized carbon along with an inert gas is passed through a hot furnace(approximately 700°C) where it decomposes to give CNTs deposited on the surface of a substrate which is generally made by embedding nanometer-sized nickel or cobalt particles, or a combination of both, as a catalyst on its surface [24][27, 28]. The variables, including nanotube diameter and tensile strength, depend on the size of the metal particles. These can be controlled by masked deposition of the metal, by annealing or by plasma etching of a metal layer. For commercial production, the nano-sized metal particles are mixed with MgO or Al_2O_3 to increase catalyst support and increase the surface area for higher yield [29]. Additionally, several newer techniques namely plasma-enhanced chemical vapor deposition (PECVD), CoMoCat process, thermal CVD, laser-assisted CVD and high-pressure CVD have been developed for high-quality CNT production [27][30 - 33].

PHARMACOLOGY OF CARBON NANOTUBES

Mechanism of Cellular Uptake of Carbon Nanotubes

The mechanisms involved in cellular uptake of the functionalized CNTs (f-CNTs) have been studied extensively from twenty years back [34 - 35]. The f-CNTs can penetrate into the cells through an endosomally-mediated, energy-dependent approach and by passive diffusion. The f-CNTs have the ability to pass the blood brain barrier (BBB) without the help of any external device. Scientists postulated the main mechanisms involved in the cellular uptakes of f-CNTs *viz* (i) phagocytosis, (ii) passive diffusion, (iii) caveolae-mediated endocytosis, (iv) clathrin-mediated endocytosis, and (v) caveolae-clathrin mediated endocytosis. The understanding of the mechanisms of intracellular uptake of CNTs is vital for the controlled uptake and devising new strategies in the field of novel drug delivery [36 - 37].

Toxicity of Carbon Nanotubes

The toxicological aspect of formulation is extremely crucial in case of drug delivery. The various reports from literatures on toxicological assays of CNTs are found contradictory. Some basic *in vitro* experiment has implied that CNTs are toxicologically inert to certain types of cells, whereas some other studies have showed that CNTs, especially the pristine forms are potentially hazardous to living systems [38 - 56]. There are very few numbers of reports available on pharmacological activities of CNTs conjugated with therapeutic moieties applied in human and consequently their clinical toxicity.

In vitro Toxicological Studies of Carbon Nanotubes

Many *in vitro* cytotoxicity studies of water-dispersible SWCNTs on A549 cells, showed no intracellular accumulation of SWCNTs in A549 cells but it was demonstrated that SWCNTs could induce an cytotoxicity indirectly by transformation of cell culture medium which created a false-positive toxic effect [41 - 43]. Dumortier *et al.* found that fluorescein labelled water soluble SWCNTs were nontoxic for B and T lymphocytes and macrophages of mouse [44]. Some other studies investigated the toxicity and the mechanism involved in water soluble MWCNTs induced cytotoxicity in C6 rat glioma cell line. It was observed that small sized MWCNTs are comparatively more toxic than the bigger one. Because of the high oxidative stress, MWCNTs induces cytotoxicity in C6 rat glioma cells [45]. Besides these, pristine CNTs, water insoluble CNTs showed high *in vitro* toxicity to various cells, *viz* human keratinocytes, rat brain neuronal cells, human embryonic kidney cells and human lung cancer cells [40][46 - 48].

In vivo Toxicological Studies of Carbon Nanotubes

The toxicological assessments of CNTs on animals (rat & mice) have been performed by IV, SC injections and gastrointestinal exposure with functionalized or dispersed SWCNTs and/ or MWCNTs [40]. The safety data available in the literature have collectively indicated the low toxicity of CNTs for biomedical applications by various ways. Polyethylene-Glycol-MWCNTs (PEG-MWCNTs) showed toxic effect after a dosage of 60 mg/kg was applied in mice [40][49]. The toxic effects of SWCNTs are because of the oxidative stress, not the routes of administration [58]. When CNTs were subcutaneously implanted for tissue repairing it showed very excellent biocompatibility without causing serious toxicity [59]. However, Fraczek *et al.* observed that SWCNTs and MWCNTs can cause inflammation after implantation [51]. Folkmann *et al.* observed the oxidative damage to DNA after oral administration of SWCNTs in mice [50]. Takanashi *et al.* reported no carcinogenic activity of MWCNTs when inserted subcutaneously in rasH2 mice [53]. However, the other scientists have found the asbestos-like pathogenicity when CNTs were inserted into the peritoneal space of mice [46]. There are many parameters due to which the CNTs shows toxicity *in vivo*. Metal impurities present in CNTs have partial contribution to the oxidative stress and therefore, purification of CNTs is very much essential. The chemically modified/functionalized CNTs exhibited higher biocompatibility in comparison to pristine CNTs. Yang *et al.* reported that f-CNTs are suitable for biomedical application due to its biocompatible nature along with low toxicity [40]. However, more toxicological studies are required to optimize the safety range of different CNTs and elucidate the toxicological mechanism.

Human Toxicity of Carbon Nanotubes

The clinical studies of f-CNTs linked with therapeutic moieties are not evaluated yet in man. Most of the literature says that pristine CNTs might be the cause of occupational lung diseases in employs of CNT industries [40][52][54]. Several studies on rodent has been performed to estimate the lung toxicity of CNTs, by administering the test dusts intratracheally or intrapharyngeally, which results in the formation of inflammation, epithelioid granulomas, fibrosis, and biochemical alteration in the lungs [46 - 48][52 - 55]. Ali-Boucetta *et al.* have concluded that the asbestos-like toxicity caused by long, pristine CNTs can be eradicate completely if the surface is functionalized and length is shortened by treatment with chemical, like tri (ethylene glycol) (TEG) [56]. However, discussions on the potential risks of pristine CNTs as well as their impurities are still going on. Moreover, the partial similarity between pristine MWCNTs and asbestos fibers has raised many doubts about their safety profile. Besides their huge potential benefits in therapeutics, the toxicity of CNTs is the major point to be taken in consideration for discussion.

PHARMACOKINETICS OF CARBON NANOTUBES

During the development process of nanomedicines, a prime emphasis must be given to the pharmacokinetics of nanoparticles due to some differences from pharmacokinetics of conventional drugs. The *in vivo* pharmacokinetics, as well as biodistribution of CNTs, is solely related to the surface property and the functionalization. Functionalization with PEG can modulate the pharmacokinetics, biodistribution, and biosafety of CNTs [57]. Size of CNTs is also an important factor as particle size < 20 nm may be excreted by the kidneys and particle size > 50 nm may be trapped by the Reticuloendothelial system (RES). By PEGylation of CNTs, the biocompatibility and drug delivery capability can be enhanced while the toxicity can be reduced, which implies the improvement of the pharmacodynamics of drugs.

Most of the nanoparticles have experienced very short biological half-lives in the bloodstream *in vivo*. This is due to extraction by the tissues and retention rather than excretion. Nanomaterials may circulate likely through the lymphatic system in the body and it may cause immunological implications [58]. However, it has been found that there is negligible accumulation of radiolabeled functionalized SWCNTs and MWCNTs in mice body and excreted through kidney in intact form. [59]. The attraction of indigenous proteins for the surface of SWCNTs modulates the uptake and toxicity of CNTs. However, the binding of CNTs to albumin could be blocked by functionalized with Pluronic F127 surfactant [60]. Despite of some arguments, most of the research on pharmacokinetics indicates that functionalized CNTs are excreted through urine or feces leaving a residual amount remaining in the tissues.

Metabolism of CNTs has the impact on its therapeutic effects as well as toxicity of CNT derived nanomedicines. To study the metabolism patern of carbon nanomaterials is very tough. The backbone of CNT is stable and their metabolism pathways generally involve defunctionalization of the surface modified groups. Metabolisms of CNTs are different from organ to organ. But liver is the most general metabolism organ as it contains abundance of metabolic enzymes [57].

Factors Affecting CNT Biodistribution

CNT Surface Modification

The pristine CNTs have some undesirable effect in the body, so to reduce the toxic level and to manipulate the biodistribution of CNTs the surface modification is very much essential. Modification of CNTs surfaces has helped for dispersion in physiological environments and revealed their interaction with the biological system [61 - 71]. The biodistribution pattern of different types of CNTs depends on the nature of surface functionalization. Depending upon the kind of surface modification of CNTs the biodistribution studies are categorized in two types.

Non-covalently Functionalized (or Coated) CNT

Cherukuri *et al* had performed an experiment to study the intracellular localization of non-covalently coated CNTs in rabbits. He had applied intrinsic near-infrared fluorescence for the tracking of Pluronic F108 coated SWCNTs. From the study, it was found that SWCNTs were mainly localized in liver and exhibited a very fast blood clearance with a half-life of one hour, which was due to the development of SWCNT-protein complexes [72]. In another study by Yang S. *et al* studied the biodistribution of 13C-enriched backbone SWCNTs coated with Tween 80 taking benefit of the isotopic abundance of 13C/12C to quantify the CNT in organs using mass spectroscopy. From the study, it was observed that the coated SWCNTs showed high affinity towards liver, lungs, and spleen with a residence time of 28 days in these organs [73]. The probable route of elimination of the surface modified SWCNTs was thought to be from the lungs (from 15 to 9.4%) towards the spleen.

Liu *et al.* have studied the pattern of biodistribution of PEGylated phospholipid coated SWCNTs with the help of positron emission tomography (PET) and Raman spectroscopy [74]. In this study, the impact of the chain length (2000 or 5400) of PEG to target cancer cells using the RGD peptide as a ligand was investigated. Here, a notable amount of surface modified SWCNTs with lipid was detected in liver as well as in spleen. While these lipid-coated SWCNTs were not found in the tumor of xenograft-bearing animals, but integrin targeted SWCNT-PEG5400-RGD demonstrated higher tumor uptake about 10-15% of ID per gram tissue. Further, Liu *et al.* studied the effect of the length and branching of PEG on the biodistribution pattern of PEG-lipid conjugated SWCNTs [68]. From the study, it was seen that PEG-lipid SWCNTs accumulated in liver and spleen, but there was an increase in blood circulatory time up to 15 h with reduced uptake by the liver was observed of branched (7 kDa) PEG conjugated SWCNTs. The biliary excretion is the main method of elimination for these non-covalently PEGylated SWCNTs over a period of 2-month, but the SWCNTs of smaller size (50 nm in length, 1-2 nm in diameter) are excreted out through urine. Prencipe *et al.* have synthesized different types of surfactant polymers from poly (y-glutamic acid; yPGA) and poly (maleic anhydride-alt-1- octadecene; PMHC18) and increased PEGylation. It was found that SWCNTs coated with the polymers shows a very long circulatory half-life of up to 22 h after intravenous administration. This is because of the highly dense coating of PEG [75].

Similarly, Liu *et al.* applied PMHC18 polymer anchored with different PEG chain lengths (2 kDa-5 kDa) at several densities (5%-100%). Following intravenous administration of PMHC18-PEG-coated SWCNTs in 4T1 murine cancer bearing Balb/C mice, the tumor and skin uptake of SWCNTs is increased with the increase in degree of PEGylation (length and density of PEG chains). The circulatory half-life about 12-13 h was observed with 100% 5 kDa (PEG-PMHC18)-SWCNT, while RES organ uptake was still observed. [76].

In these studies, the same type of CNTs and the same strategy for noncovalent functionalization was applied using different types of coating. However, preservation of the carbon backbone pristine is essential to investigate the biodistribution pattern of functionalized CNTs, since any conformational changes may damage the electronic properties of CNTs which would give false result in near-infrared or Raman spectroscopy. In general, the pristine, non-covalently functionalized SWCNTs are accumulated significantly in liver.

Covalently-functionalized CNT

Several investigations have been found in literature which describe the biodistribution of covalently functionalized CNT *in vivo*, most of which used SWCNTs [67][77 - 81]. Wang *et al.* investigated the biodistribution of iodine-125 labeled hydroxylated SWCNTs (125I-SWCNT-OH) administered intraperitoneally (i.p.) [79]. From the study, the distribution of 125I-SWCNT-OH was found in most of the organs with the strong affinity towards stomach, kidney, and bone. 125I-SWCNT-OH was excreted mainly *via* urine (94%). It has been found that there is no alteration in distribution in tissues with the change in route of administration (subcutaneous, stomach gavage and intravenous). Further, Wang *et al.* investigated the *in vivo* fate of 131I-SWCNT-OH following i. v. and i. p. administrations and found their fast distribution into organs within 2-60 min after injection [81]. Interestingly, 24% ID uptake was found after 2 min of injection in kidneys with a small reduction in the signal within 60 min of the experiment.

Guo *et al.* reported a biodistribution study in which MWCNTs were functionalized with glucosamine, labeled with Technetium-99 (99mTc-MWCNT-glu) and injected i. p. into mice [78]. 99mTc-MWCNT-glu has showed a fast biodistribution and a circulatory half-life of 5.5 h. Notable amounts of MWCNT was accumulated in kidney over 24 h with more than 75% of radioactivity found in the urine and faeces. In a later study of taurine-functionalized MWCNT (14C-taurine-MWCNT), showed a high affinity to the liver (80% of ID) after i. v. injection with retention up to 28 days and complete elimination before 3 months [80]. Deng *et al.* studied the effect of route of administration through stomach gavage and intratracheal instillation (i. t.) to study the *in vivo* fate of *f*-MWCNT.

It was observed the higher accumulation in the small and large intestines for the stomach gavage and higher affinity to the lungs after i. t. administration [24]. In a further study, they also found predominant liver accumulation of 125I-taurine-MWCNT where a novel radiotracer technique was adopted for analysis [82]. Yang *et al.* used covalently PEGylated SWCNT to observe 30% ID retained in the blood compartment one day post-injection, quantified by isotope-MS [83].

Singh *et al.* formulated SWCNT-NH3+ which was further functionalized with the chelate diethylenetriaminepentacetate (DTPA) and radiolabeled with Indium-111 ([111In] DTPA-SWCNT) followed by injection into mice through i.v. route. The study showed tissue distribution of *f*-SWCNT in the kidney, muscle, skin, bone, and blood after 30 min of injection. Interestingly, *f*-SWCNT was observed to be excreted *via* urine with a maximum circulatory half-life of 3.5 h [77].

In a further study, [111In] DTPA-MWCNTs were injected intravenously in rats and observed the accumulation in the kidneys within 30 min of injection and subsequently excreted *via* urine as detected by dynamic whole body single photon emission computed tomography (SPECT/CT) [84]. It has been observed that the degree of chemical surface functionalization is a key factor of nanotube individualization that leads to urinary excretion. All these observations were broadly in agreement with the data reported by others on hydroxylated (125I-SWCNT-OH) and glucosamine functionalized (99mTc-MWCNT-glu) despite the very different chemical processing, chemical treatment and starting material [64 - 65][85].

Following these initial investigations, McDevitt *et al.* repeated the same functionalization chemistry as Singh *et al.* [77] and Lacerda *et al.* [84 - 85] to prepare ammonium functionalized 111In labelled SWCNTs, using 2-(4-isothiocyanatobenzyl)-1,4,7,10- tetraazacyclodecane (DOTA) as the chelating molecule. For tissue targeting approach monoclonal antibodies (Rituximab) were conjugated with the SWCNTs. The differences in tissue biodistribution of [111In] DOTA-SWCNT-NH3+ and the antibody-targeted [111In] DOTA-SWCNT-Rituximab were investigated. The [111In] DOTA-SWCNT was found to be accumulated mostly in the kidney, liver, spleen, and bone showing a rapid blood clearance. While antibody conjugation altered the biodistribution by decreasing the kidney residence time with a 2-fold increase in affinity towards liver along with no change in spleen accumulation [67]. This study and a subsequent similar investigation by the same group are generally in good agreement with the original observations of Singh *et al.* and confirmed that rapid clearance and urinary excretion of CNTs occur, provided appropriate chemical functionalization has been performed [66].

Although pharmacokinetic studies are quantitative in nature, very few data have been in currently available biodistribution studies with CNTs. The blood circulatory half-life of CNTs is the only parameter reported in most CNTs pharmacokinetic studies.

Surface Density of Functionalization

It is reported in literature that by manipulating the degree of surface modification of MWCNTs, the organ distribution and clearance could be controlled [65]. It was reported that with increase in the functionalization density on the surface of CNTs, leads to lesser accumulation by RES and increase the urinary excretion of CNTs. This systematic study correlated with previous studies [64][85] as well as Ruggiero *et al.* [86], who showed that high ammonium functionalized SWCNTs (*via* the 1,3-dipolar cycloaddition) were excreted in intact form *via* the glomerular filtration system. Jain *et al.* studied the importance of the density of functionalization on carboxylated MWCNTs on the tissue distribution and toxicokinetic profile. It was observed that MWCNTs of short lengths (500 nm) and a high degree of oxidation (surface carboxyl density N 3 μmol/mg) were rapidly excreted from the body through renal excretion without causing nephrotoxicity. Pristine MWCNT and those with lower degrees of carboxylation exhibit higher accumulation in RES organs such as the liver, spleen, and lungs [87].

PEGylation of CNTs

To enhance the circulatory time of nanoparticles a polymeric coat of polyethylene glycol (PEG) is very commonly used on their surface. This PEG layer can act as a protective barrier against protein adsorption and detection by macrophages and RES, thus, their circulatory half life can be enhanced [88]. PEGylation on CNTs has been described for the development of CNTs in biomedical applications [71][74][89 - 90]. PEGylation can be done either by physisorption and coating of PEG around the CNTs [68][74] or by covalent binding of the polymer on the CNT backbone. It is reported that the PEG coat on the nanoparticles is stable in blood circulation, especially non-covalent PEGylation provides longer circulatory half-life when PEGylation density is increased [68][75 - 76].

Most of the covalently functionalized PEGylated CNTs have been conjugated due to the amine termination of PEG with the oxidized or hydroxy-terminated CNTs. Yang *et al.* studied the covalently functionalized PEGylated CNTs using 13C-enriched SWCNTs with diamine terminated PEG (PEG1500N). The author had reported high aqueous dispersibility (at ~20 mg/ml) with 15.3 h circulatory half-life. However, liver and spleen uptake up to 7 days' of post-injection was confirmed by isotope ratio mass spectroscopy [83]. Bhirde *et al.* further studied the *in vivo* fate of PEGylated SWCNTs and their potential as anticancer therapeutics. From the Raman spectroscopy of excretion samples of PEGylated SWCNT indicates the 7 days circulatory period in body, whereas their pristine forms were unable to detect in the excretion after 2 days post-injection which may be due to their accumulation in vital organs. From the histopathological examination, lung was found to be the preferred site of accumulation for the oxidized SWCNTs with signs of inflammatory responses [91].

Zhang *et al.* studied the hepatic toxicity of PEGylated MWCNTs and pristine MWCNTs dispersed in 1% Tween 20 after i. v. administration. It was found that the PEGylated MWCNTs caused lesser hepatic injury compared to pristine MWCNTs as indicated by no significant changes of hepatic enzymes level. These results indicated that PEGylation of CNTs could partially but not exclusively improve the biocompatibility of CNTs [92].

Yang *et al.* studied biodistribution of MWCNTs functionalized with hydroxyl terminated PEG in mice. It was observed that MWCNT-PEG4000-OH had a short blood circulatory half-life of about 5.2 min, due to the change in the type of CNT from SWCNT to MWCNT and hence the difference in length/diameter ratio. Another possible reason could be the difference in the terminal functional group of the PEG [93]. Interestingly, increases in PEG chain length to 20 kDa also led to a significant reduction in hepatic accumulation (for MWCNT-PEG20000-OH). Previously, amine terminated PEG [94] was used for long blood circulatory half-life, while in the latter study, hydroxyl-terminated PEG was used. It was also reported that the amino group terminated PEG functionalized quantum dots (QDs) which creates longer circulation half-lives in comparison to hydroxyl and carboxylic group terminated PEG-QDs [95].

The second prime factor which has crucial effect on biodistribution of CNT is the aqueous dispersibility of the administered CNT population. Highly dispersed CNTs exhibit better pharmacokinetic profiles as compared to agglomerated CNTs because it can bypass the RES system. To make the dispersion of pristine CNTs, they are generally sonicated or vortexed in serum, surfactants or block copolymers, therefore can be considered 'coated' CNTs. Lacerda *et al.* reported that the pristine MWCNTs coated with serum were accumulated primarily in the liver, spleen, and lungs compared to ammonium functionalized MWCNTs. MWCNT aggregates were also observed in the lung which was not the case for their functionalized counterpart [85]. Cherukuri *et al.* reported that serum proteins may compete with the noncovalently functionalized group of CNT surface which lead to desorption and aggregation in the blood which will favour recognition by macrophages. In general, a well-functionalized CNTs population should also achieve effective individualization of the nanotubes and stable dispersion in physiologically-relevant environments [72].

Other Factors Affecting the Biodistribution of CNTs

Biodistribution of CNTs depends upon the methods applied for the preparation of CNTs and various functionalization strategies. It was found that both ammonium functionalized SWCNTs and MWCNTs showed a similar biodistribution profile that at high degrees of functionalization can lead to predominant urinary excretion [64][71]. The length of CNTs also has impact on the biodistribution profile. Murphy *et al.* had reported that shorter CNTs have a great application in nanomedicine, as they are cleared from the pulmonary pleural space and into the lymph nodes after intrapleural administration [96]. Jain *et al.* had reported well-individualized and short MWCNTs (500 nm) were rapidly excreted through urinary route without causing nephrotoxicity. Whereas, pristine MWCNTs and longer MWCNTs showed a stronger entrapment in RES organs such as the liver, spleen and lungs and thus induced significant hepatotoxicity in mice [87].

TARGETED AND CONTROLLED DRUG DELIVERY *VIA* CNTS

Targeted drug delivery is a strategy adopted to deliver an active therapeutic agent to a part of the body for a long period. There are two strategies that are used to target nanoscale particles: passive targeting and active targeting [97]. Passive targeting is depended upon the size of the carrier. Because of the hydrophobic property, the CNTs can stay in the circulation for a prolonged period of time. This synergy leads to enhanced permeability and retention (EPR). The second mechanism involves using a CNT carrier to actively target and selectively deliver the drug to diseased part of the body. For proper targeting, CNTs are functionalized with targeting ligands so that they can easily recognize the target apart from surrounding healthy cells [98]. For the delivery of drug, controlled release occurs in response to certain external and/or internal stimuli. Changes in the temperature, pH, enzyme activity and redox potential are all internal stimuli. While changes in magnetic field, electric field, light irradiation, and temperature are categorized as external stimuli [99 - 101].

CNTs Drug Delivery to Lymphatic System

Most of the cancers metastasis occurs mainly through lymphatic canal. The targeted drug delivery systems towards the lymphatic system can effectively block the metastasis of cancers. Yang D *et al* prepared a lymphatic targeted CNT drug delivery system. He used polyacrylic acid (PAA) to conjugate with CNTs and made them highly hydrophilic. By employing the coprecipitation method, Fe_3O_4-based magnetic nanoparticles were adsorbed on the PAA-CNT surface. Gemcitabine was incorporated into PAA-CNT; Fe_3O_4 magnetic nanoparticles simply by stirring the solution of the nanosystem for 24 h, with a loading efficiency of 62%. It was observed that CNTs were detected only in the local lymph nodes and were found to be absent in the major organs, such as liver, kidney, heart, spleen, and lungs, after 3 h of subcutaneous administration [102].

CNTs Drug Delivery to Tumor

To show the anticancer effect of drugs, it must reach the cancer cells in its effective concentration. Whereas, some anticancer drugs cannot enter cancer tissues because of their short circulatory time in blood. For example, the use of paclitaxel (PTX), a widely used potent chemotherapeutic agent, is limited by its poor solubility in aqueous medium and nonspecific cytotoxicity. For this reason, Liu *et al.* coupled the PTX to PEGylated SWCNTs through a cleavable ester bond, which forms a water-soluble SWCNT-PTX complex. It was noticed that the efficacy of SWCNT-PTX conjugate was higher in comparison to the clinically used Taxol in the suppression of tumor growth in a murine 4T1 breast cancer model. The blood circulatory time of that complex was found almost six-fold higher than Taxol, due to the functionalization by PEG. The uptake of PTX in tumor cells for SWCNT-based PTX delivery system is possibly through an enhanced permeability and retention effect (EPR effects) [103]. Wu *et al.* developed a novel MWCNT as a drug delivery system by employing hydroxycamptothecin (HCPT) onto the surface of MWCNTs. The surface of MWCNT was modified by carboxylation and then by amidation with a hydrophilic diaminotriethylene glycol. Hydroxyl derivatized MWCNTs were then conjugated with succinylated HCPT through a cleavable ester bond. This formulation had exhibited higher antitumor activity and low toxicity in comparison with the clinical used HCPT formulation both *in vitro* as well as *in vivo*. From the *in vivo* single photon emission computed tomography imaging and *ex vivo* g-scintillation counting analysis it was found that the formulation had longer blood circulation time and higher tumor-specific drug accumulation [104]. Bhirde *et al.* illustrated the superiority of drug-SWCNT complex over non targeted conjugates. To target the squamous cell cancer, cisplatin and epidermal growth factor (EGF) were linked with SWCNTs. SWCNT-cisplatin without EGF was used as a nontargeted control. It was seen t cells as confirmed by confocal microscopy. Control cells showed limited uptake which reveals the importance of EGF in the system. It was noticed; SWCNT quantum dot was cleared from the tumor region within 20 min. The squamous cancer cells were destroyed selectively by SWCNT-cisplatin-EGF, whereas control systems showed no effect on the proliferation of these cells [105]. It is known that dendritic cells are important for the induction and regulation of immune responses. Yang *et al* prepared cationic 1, 6-diaminohexane functionalized CNTs to carry *si*-RNA into the antigen-presenting dendritic cells *in vivo*. Splenic CD11c+ dendritic cells,

CD11b+ cells and Gr-1+CD11b+ cells comprising dendritic cells, macrophages, and other myeloid cells showed active take up for the complexes. These complexes silenced the suppressor of cytokine signaling 1 (SOCS1) expression and reduced the growth of B16 tumor in mice [106]. It is known that the telomerase reverse transcriptase plays a crucial role in tumor growth and development through the maintenance of telomere structure. Zhang *et al* had conjugated-RNA with SWCNTs through a CONH-(CH2)6-NH3+ Cl- spacer. They had demonstrated that *si*-RNA delivered by SWCNT complexes silences the expression of telomerase reverse transcriptase and thus inhibits the proliferation and growth of tumor cells *in vitro* as well as *in vivo* in mouse models. It was seen that by 48 h, all the cells, such as LLC, TC-1, and 1H8 cells, treated with telomerase reverse transcriptase, SWCNT-*si*RNA showed an almost complete inhibition of proliferation in murine tumor models [107]. Unmodified *si*-RNA with pristine SWCNTs has also been complexed noncovalently to deliver *si*-RNA to cancer cells in a more recent study. The complex was prepared by simple sonication of pristine SWCNTs in a solution of *si*-RNA that plays the dual role of cargo and dispersant for CNTs. It was envisioned that there was strong specific inhibition of cellular HIF-1 activity when *si*-RNA complexes targeted to hypoxia-inducible factor 1 a (HIF-1a) were added in serum-containing culture media. The ability to respond biologically to SWCNT-*si*RNA complexes have been observed in a wide variety of cancer cell types. Moreover, intratumoral administration of SWCNT-HIF-1a *si*-RNA complexes in mice bearing MiaPaCa-2/HRE tumors significantly inhibit the activity of tumor HIF-1a, suggesting that such SWCNT-*si*RNA complexes promise considerably as therapeutic agents [108].

CNTs Drug Delivery to Central Nervous System (CNS)

Central nervous system disorders consist mainly of neurodegenerative diseases and brain tumors. Due to the unique and complex environment and the restricted anatomical access (blood–brain barrier) of the CNS, it is more difficult to treat CNS disorders than any other diseases. But nanotechnology is a promising field to revolutionize this problem. As a promising biomedical material, CNTs have been used in neuroscience [109 - 111]. Various studies have indicated that both pristine and chemically functionalized CNTs have positive effects on neuronal growth, although they showed cytotoxicity. One study has reported that the charge on the nanotubes could be manipulated to control neurite outgrowth [112], and it has been reported that CNTs linked with nerve growth factor or brain derived neurotrophic factor could stimulate the growth of neurons on the nanotube scaffold [127]. Shityakov *et al.* investigated the dynamic transport of FITC-labeled MWCNTs across immortalized murine microvascular cEND cells using the Transwell device. The accumulation of FITC-MWCNTs in cells was detected by fluorescence and phase-contrast microscopy. FITC-MWCNT aggregates were nonfluorescent and localized on the cell surface whereas the dispersed FITC-MWNTs were located intracellularly or associated with cell membrane [128]. Yang *et al.* reported that SWCNTs were internalized by absorptive cells and macrophages in the intestine but were also detected in other organs such as liver, brain, and heart when SWCNTs were administered orally in mice. SWCNTs were observed in lysosomes of neurons and neurites from TEM imaging. SWCNTs could deliver acetylcholine, which was linked with SWCNTs, leading to improvement in the learning and memory capability of an experimentally induced Alzheimer disease mouse model [129].Stereotactic delivery of f-CNTs into the brain is the most common technique employed in the early studies involving f-CNTs. Lee *et al.* reported that amine-modified SWCNT (a-SWCNT) has the ability to protect neurons from injury using a rat stroke model. In the study, rats were given a-SWCNTs *via* lateral ventricle injection before the induction of stroke. A-SWNTs-treated animals were found to have smaller infarct regions in the brain compared with the PBS treated group. A-SWCNTs-treated animals performed better in the behavioral functions, which was assessed by Rota-Rod treadmill test, indicating the enhancement of recovery from stroke injury [130]. While stereotactic injection offers direct access to the brain, its invasive nature creates necessity of a systemic administration route to facilitate clinical translation of f-CNTs. Researchers developed f-CNTs associated with a brain-targeting ligand to facilitate brain delivery after intravenous injection. Ren *et al.* developed dual-targeted f-CNTs that carry the brain targeting ligand, Angiopep-2 (ANG) as well as the anticancer drug doxorubicin (DOX) [131]. Oxidized MWCNTs (O-MWCNTs) were coated with polyethylene glycol (PEG) and ANG was conjugated to its terminal. DOX was adsorbed MWCNT sidewalls. This f-CNTs were administered intravenously in mice, and an enhanced brain uptake of DOX-O-MWNTs-PEG-ANG was observed as compared to non-ANG and DOX alone counterparts. In a recent study, researcher examined quantitatively the *in vivo* biodistribution and brain uptake of f-CNTs following intravenous injection using γ-scintigraphy. Depletion of capillaries confirmed the presence of MWNT-NH3+ in brain capillaries and parenchyma. The radiolabeled MWNT-NH3+ exhibited intrinsic properties to cross BBB both in vitro and *in vivo*. Thus, it has paved the way for future applications of f-CNTs as nanocarriers for brain delivery [132].

CNTs Drug Delivery for Gene Therapy

Gene therapy is a novel and sophisticated method for delivery of therapeutics. Gene therapy focuses on the use of genetic material for the treatment of diseased cells by acting on the cause of the disease. But genes and nucleic acids require a delivery vehicle to deliver *in vivo*. CNTs have very much potential as a delivery system for gene therapy. It is possible to conjugate DNA/*si*RNA with CNTs. Because of the huge surface area, unique optical and electrical properties, CNTs not only deliver DNA/*si*RNA into cells but also provide a controlled delivery to the disease sites [113]. Cai *et al.* had reported the use of nickel particles as gene delivery system which was enclosed on the ends of CNTs and immobilized the pDNA on the surface of CNTs. It was found that the CNT-pDNA complex get entered the mammalian cell and showed gene expression in 80-100% of cell population [114]. Kam *et al.* had reported the method of gene slicing by *si*RNA-CNT conjugates [115]. McDevitt *et al.* had successfully attached tumor specific monoclonal antibodies, radiometal ion chelates, and fluorescent probes to SWCNTs [116]. Meng *et al.* had found increased anti-tumor response when a tumor lysate protein was conjugated to MWCNTs [117]. Angiogenesis targeting antibodies were attached to SWCNTs *via* radiometal ion chelates which reduced the volume of tumor cells and prolonged the lifetime of animal models [118].

CNTs for Vaccine Delivery

Vaccination is the process of antibody generation against an antigen. Delivery of vaccine can be achieved by coupling an antigen with CNTs without losing its conformation and by inducing antibody response with right specificity. Pantarotto*et al.* inspected the use of CNTs in developing the immune response. It was reported that protein CNT complex could enhance the immune response when conjugated with the antigen [119]. Bianco *et al.* illustrated the use of cationic CNTs as a delivery vehicle for deliver synthetic oligodeoxynucleotides, which improve the immune stimulatory responses. Synthetic oligonucleotide shows nonspecific protection against cellular pathogens and enhances antigen specific immune response [123].

CNTs for Macrophages Targeting

Reports available in literature has told that CNTs might be a promising nanocarrier in macrophage targeted drug delivery system. Prajapati *et al* had reported that AmB linked CNTs had better antileishmanial activity in comparison to free AmB in J774A.1 macrophage cell line [121]. Further, mannose linked CNTs loaded with AmB exhibited effective targeting in J774 macrophage cell lines with lesser toxicity in RBCs, kidney, and liver [120].

REGULATORY CONSIDERATIONS OF CNTS

The toxic effect of CNTs in biological system is yet unclear and several investigations are still going on. Several research findings showed the toxicology of pristine CNTs, which could be minimized by surface modification which make them more biologically safe and non-immunogenic. The structural properties of CNTs *viz* needle-shape, long and thin and insoluble nature may create adverse effect in lungs but this hazard could be overcome by surface modification of

CNTs. Eco-toxicological aspect of CNTs are controversial due to the lack of reliable information as per Occupational Health and Safety regulations. Therefore, a significant research and development in the field of analytical and detection methods need to be performed. Additionally, strict laws are necessary in the development of new therapeutic products for being safe, effective and economic in generally regarded as safe (GRAS) prominence.

CONCLUSION AND FUTURE PERSPECTIVES

The carbon nanotechnology is one of the potential techniques to deliver the potent drugs for the treatment of acute as well as chronic diseases. Numerous amount of basic research has been performed on CNTs during the past decades, but a few has been transferred into human applications. Till date, no CNT-based medicine has reached to the clinical trials. The toxicity report of CNTs available till date is not considered sufficient to get approval for human. Carbon nanotube linked with other therapeutic agents has increased the scope for their potential therapeutic applications. The PEGylation of CNTs extend the survival of CNTs in blood and overcomes the problem common to nanotherapeutics, *ie* uptake by the RES. The CNT-protein complexes are beneficial for *in vivo* protein delivery applications. The CNT linked with DNA would be suitable for gene delivery application. Thus, the CNTs have the promising potential for effective targeted delivery of novel therapeutic compounds.

CONSENT FOR PUBLICATION

Not applicable.

CONFLICT OF INTEREST

The author (editor) declares no conflict of interest, financial or otherwise.

ACKNOWLEDGEMENT

Declared none.

REFERENCES

[1] Iijima S. Helical Microtubules of Graphitic Carbon. Nature 1991; 354(7): 56-8. [http://dx.doi.org/10.1038/354056a0]

[2] Mousa SA, Bharali DJ. Nanotechnology-based detection and targeted therapy in cancer: nano-bio paradigms and applications. Cancers (Basel) 2011; 3(3): 2888-903. [http://dx.doi.org/10.3390/cancers3032888] [PMID: 24212938]

[3] Jain NK, Mishra V, Mehra NK. Targeted drug delivery to macrophages. Expert Opin Drug Deliv 2013; 10(3): 353-67. [http://dx.doi.org/10.1517/17425247.2013.751370] [PMID: 23289618]

[4] Mehra NK, Jain AK, Lodhi N, *et al.* Challenges in the use of carbon nanotubes for biomedical applications. Crit Rev Ther Drug Carrier Syst 2008; 25(2): 169-206. [http://dx.doi.org/10.1615/CritRevTherDrugCarrierSyst.v25.i2.20] [PMID: 18540838]

[5] Jain AK, Mehra NK, Lodhi N, *et al.* Carbon nanotubes and their toxicity. Nanotoxicology 2007; 1(3): 167-97. [http://dx.doi.org/10.1080/17435390701639688]

[6] Kostarelos K, Lacerda L, Pastorin G, *et al.* Cellular uptake of functionalized carbon nanotubes is independent of functional group and cell type. Nat Nanotechnol 2007; 2(2): 108-13. [http://dx.doi.org/10.1038/nnano.2006.209] [PMID: 18654229]

[7] Wong BS, Yoong SL, Jagusiak A, *et al.* Carbon nanotubes for delivery of small molecule drugs. Adv Drug Deliv Rev 2013; 65(15): 1964-2015. [http://dx.doi.org/10.1016/j.addr.2013.08.005] [PMID: 23954402]

[8] Tran PA, Zhang L, Webster TJ. Carbon nanofibers and carbon nanotubes in regenerative medicine. Adv Drug Deliv Rev 2009; 61(12): 1097-114. [http://dx.doi.org/10.1016/j.addr.2009.07.010] [PMID: 19647768]

[9] Liu Z, Sun X, Nakayama-Ratchford N, Dai H. Supramolecular chemistry on water-soluble carbon nanotubes for drug loading and delivery. ACS Nano 2007; 1(1): 50-6. [http://dx.doi.org/10.1021/nn700040t] [PMID: 19203129]

[10] Hirlekar R, Yamagar M, Garse H, Vij M. Carbon Nanotubes And Its Applications: A Review. Asian J Pharm Clin Res 2009; 2(4): 17-27.

[11] Bhagat Singh GP, Baburao C, Pispati V, Muthy N, Prassana SR, Ganesh Rathode B. Carbon Nanotubes – A Novel Drug Delivery System. Int J Res Pharm Chem 2012; 2(2): 523-32.

[12] Popov VN. Carbon nanotubes: properties and application. Mater Sci Eng Rep 2004; 43(3): 61-102. [http://dx.doi.org/10.1016/j.mser.2003.10.001]

[13] Hou Px. Xu St, Ying Z, Yang Q, Liu C, Cheng Hm. Hydrogen adsorption/desorption behavior of multi-walled carbon nanotubes with different diameters. Carbon 2003; 41: 2471-6. [http://dx.doi.org/10.1016/S0008-6223(03)00271-9]

[14] Wang X, Li Q, Xie J, *et al.* Fabrication of ultralong and electrically uniform single-walled carbon nanotubes on clean substrates. Nano Lett 2009; 9(9): 3137-41. [http://dx.doi.org/10.1021/nl901260b] [PMID: 19650638]

[15] Liu J, Rinzler AG, Dai H, *et al.* Fullerene pipes. Science 1998; 280(5367): 1253-6. [http://dx.doi.org/10.1126/science.280.5367.1253] [PMID: 9596576]

[16] Thostenson ET, Ren ZF, Chou TW. Advances in the science and technology of carbon nanotubes and their composites: a review. Compos Sci Technol 2001; 61: 1899-912. [http://dx.doi.org/10.1016/S0266-3538(01)00094-X]

[17] Maynard AD, Ku BK, Emery M, Stolzenburg M, Mcmurry PH. Measuring particle size-dependent physicochemical structure in airborne single walled carbon nanotube agglomerates. . J Nanopart Res 2007; 9: 85-92. [http://dx.doi.org/10.1007/s11051-006-9178-2]

[18] Lam CW, James JT, McCluskey R, Arepalli S, Hunter RL. A review of carbon nanotube toxicity and assessment of potential occupational and environmental health risks. Crit Rev Toxicol 2006; 36(3): 189-217. [http://dx.doi.org/10.1080/10408440600570233] [PMID: 16686422]

[19] Yu MF, Files BS, Arepalli S, Ruoff RS. Tensile loading of ropes of single wall carbon nanotubes and their mechanical properties. Phys Rev Lett 2000; 84(24): 5552-5. [http://dx.doi.org/10.1103/PhysRevLett.84.5552] [PMID: 10990992]

[20] Peigney A, Laurent C, Flahaut E, Bacsa RR, Rousset A. Specific surface area of carbon nanotubes and

bundles of carbon nanotubes. Carbon 2001; 39: 507-14.
[http://dx.doi.org/10.1016/S0008-6223(00)00155-X]

[21] Bronikowski MJ, Willis PA, Colbert DT, Smith KA, Richard E. Gas-phase production of carbon single-walled nanotubes from carbon monoxide via the HiPco process : A parametric study. J Vac Sci Technol A 2001; 19(4): 1800-5.
[http://dx.doi.org/10.1116/1.1380721]

[22] Wang Y, Zhang Z, Liu H, *et al.* The effect of catalyst concentration on the synthesis of single-wall carbon nanotubes. Spectrochim Acta A Mol Biomol Spectrosc 2002; 58(10): 2089-95.
[http://dx.doi.org/10.1016/S1386-1425(01)00691-6] [PMID: 12212733]

[23] Ajayan PM, Terrones M, de la Guardia A, *et al.* Nanotubes in a flash--ignition and reconstruction. Science 2002; 296(5568): 705.
[http://dx.doi.org/10.1126/science.296.5568.705] [PMID: 11976446]

[24] Nagy JB, Bister G, Fonseca A, *et al.* On the growth mechanism of single-walled carbon nanotubes by catalytic carbon vapor deposition on supported metal catalysts. J Nanosci Nanotechnol 2004; 4(4): 326-45.
[http://dx.doi.org/10.1166/jnn.2004.069] [PMID: 15296223]

[25] Thess A, Lee R, Nikolaev P, *et al.* Crystalline Ropes of Metallic Carbon Nanotubes. Science 1996; 273(5274): 483-7.
[http://dx.doi.org/10.1126/science.273.5274.483] [PMID: 8662534]

[26] Conceicao J, Laaksonen RT, Wang L, Guo T, Nordlander P, Smalley RE. Photoelectron spectroscopy of transition-metal clusters: Correlation of valence electronic structure to reactivity. Phys Rev B Condens Matter 1995; 51(7): 4668-71.
[http://dx.doi.org/10.1103/PhysRevB.51.4668] [PMID: 9979322]

[27] Joséyacamán M, Mikiyoshida M, Rendón L, Santiesteban JG. Catalytic growth of carbon microtubules with fullerene structure. Appl Phys Lett 1993; 62(6): 657-9.
[http://dx.doi.org/10.1063/1.108857]

[28] Abdulkareem AS, Afolabi AS, Iyuke SE, Vz Pienaar HC. Synthesis of carbon nanotubes by swirled floating catalyst chemical vapour deposition method. J Nanosci Nanotechnol 2007; 7(9): 3233-8.
[http://dx.doi.org/10.1166/jnn.2007.685] [PMID: 18019155]

[29] Inami N, Mohamed MA, Shikoh E, *et al.* Synthesis-condition dependence of carbon nanotube growth by alcohol catalytic chemical vapor deposition method. Sci Technol Adv Mater 2007; 8: 292-5.
[http://dx.doi.org/10.1016/j.stam.2007.02.009]

[30] Choi EC, Park YS, Hong B. Synthesis of carbon nanotubes on diamond-like carbon by the hot filament plasma-enhanced chemical vapor deposition method. Micron 2009; 40(5-6): 612-6.
[http://dx.doi.org/10.1016/j.micron.2009.02.009] [PMID: 19318258]

[31] Kleinsorge B, Golovko VB, Hofmann S, *et al.* Growth of aligned carbon nanofibres over large areas using colloidal catalysts at low temperatures. Chem Commun (Camb) 2004; 1416–1417(12): 1416-7.
[http://dx.doi.org/10.1039/b401785d] [PMID: 15179492]

[32] Vajtai R, Wei BQ, Ajayan PM, Carbon N, Oct RM. Controlled growth of carbon nanotubes. Philos Trans A Math Phys Eng Sci 2004; 362(1823): 2143-60.
[http://dx.doi.org/10.1098/rsta.2004.1433] [PMID: 15370475]

[33] Baddour CE, Briens C. Carbon Nanotube Synthesis : A Review. Int J Chem React Eng 2005; 3: 1-20.
[http://dx.doi.org/10.2202/1542-6580.1279]

[34] Kam NW, O'Connell M, Wisdom JA, Dai H. Carbon nanotubes as multifunctional biological transporters and near-infrared agents for selective cancer cell destruction. Proc Natl Acad Sci USA 2005; 102(33): 11600-5.
[http://dx.doi.org/10.1073/pnas.0502680102] [PMID: 16087878]

[35] Lacerda L, Ali-Boucetta H, Kraszewski S, *et al.* How do functionalized carbon nanotubes land on,

bind to and pierce through model and plasma membranes. Nanoscale 2013; 5(21): 10242-50. [http://dx.doi.org/10.1039/c3nr03184e] [PMID: 24056765]

[36] Lacerda L, Bianco A, Prato M, Kostarelos K. Carbon nanotubes cell translocation and delivery of nucleic acids *in vitro* and *in vivo.* J Mater Chem 2008; 18: 17-22. [http://dx.doi.org/10.1039/B711554G]

[37] Lacerda L, Russier J, Pastorin G, *et al.* Translocation mechanisms of chemically functionalised carbon nanotubes across plasma membranes. Biomaterials 2012; 33(11): 3334-43. [http://dx.doi.org/10.1016/j.biomaterials.2012.01.024] [PMID: 22289266]

[38] Digge MS, Moon RS, Gattani SG. Applications of carbon nanotubes in drug delivery: a review. Int J Pharm Tech Res 2012; 4(2): 839-47.

[39] Yang W, Thordarson P, Gooding JJ, Ringer SP, Braet F. Carbon nanotubes for biological and biomedical applications. Nanotechnology 2007; 18: 1-12. [http://dx.doi.org/10.1088/0957-4484/18/41/412001]

[40] Yang S, Luo J, Zhou Q, Wang H. Pharmacokinetics, metabolism and toxicity of carbon nanotubes for bio-medical purposes. Theranostics 2012; 2(3): 271-82. [http://dx.doi.org/http://dx.doi.org/10.7150/thno.3618] [PMID: 22509195]

[41] Chang Y, Yang ST, Liu JH, *et al. In vitro* toxicity evaluation of graphene oxide on A549 cells. Toxicol Lett 2011; 200(3): 201-10. [http://dx.doi.org/10.1016/j.toxlet.2010.11.016] [PMID: 21130147]

[42] Firme CP III, Bandaru PR. Toxicity issues in the application of carbon nanotubes to biological systems. Nanomedicine (Lond) 2010; 6(2): 245-56. [http://dx.doi.org/10.1016/j.nano.2009.07.003] [PMID: 19699321]

[43] Casey A, Herzog E, Lyng FM, Byrne HJ, Chambers G, Davoren M. Single walled carbon nanotubes induce indirect cytotoxicity by medium depletion in A549 lung cells. Toxicol Lett 2008; 179(2): 78-84. [http://dx.doi.org/10.1016/j.toxlet.2008.04.006] [PMID: 18502058]

[44] Dumortier H, Lacotte S, Pastorin G, *et al.* Functionalized carbon nanotubes are non-cytotoxic and preserve the functionality of primary immune cells. Nano Lett 2006; 6(7): 1522-8. [http://dx.doi.org/10.1021/nl061160x] [PMID: 16834443]

[45] Han YG, Xu J, Li ZG, Ren GG, Yang Z. *In vitro* toxicity of multi-walled carbon nanotubes in C6 rat glioma cells. Neurotoxicology 2012; 33(5): 1128-34. [http://dx.doi.org/10.1016/j.neuro.2012.06.004] [PMID: 22728153]

[46] Shvedova AA, Kisin ER, Porter D, *et al.* Mechanisms of pulmonary toxicity and medical applications of carbon nanotubes: Two faces of Janus? Pharmacol Ther 2009; 121(2): 192-204. [http://dx.doi.org/10.1016/j.pharmthera.2008.10.009] [PMID: 19103221]

[47] Fisher C, Rider AE, Han ZJ, Kumar S, Levchenko I, Ostrikov KK. Applications and nanotoxicity of carbon nanotubes and graphene in biomedicine. J Nanomater 2012; 1-19. [http://dx.doi.org/http://dx.doi.org/10.1155/2012/315185]

[48] Ursini CL, Cavallo D, Fresegna AM, *et al.* Study of cytotoxic and genotoxic effects of hydroxyl-functionalized multiwalled carbon nanotubes on human pulmonary cells. J Nanomater 2012; 1-9. [http://dx.doi.org/http://dx.doi.org/10.1155/2012/815979]

[49] Zhang D, Deng X, Ji Z, *et al.* Long-term hepatotoxicity of polyethylene-glycol functionalized multi-walled carbon nanotubes in mice. Nanotechnology 2010; 21(17): 175101. [http://dx.doi.org/10.1088/0957-4484/21/17/175101] [PMID: 20357413]

[50] Folkmann JK, Risom L, Jacobsen NR, Wallin H, Loft S, Møller P. Oxidatively damaged DNA in rats exposed by oral gavage to C60 fullerenes and single-walled carbon nanotubes. Environ Health Perspect 2009; 117(5): 703-8. [http://dx.doi.org/10.1289/ehp.11922] [PMID: 19479010]

[51] Fraczek A, Menaszek E, Paluszkiewicz C, Blazewicz M. Comparative *in vivo* biocompatibility study of single- and multi-wall carbon nanotubes. Acta Biomater 2008; 4(6): 1593-602. [http://dx.doi.org/10.1016/j.actbio.2008.05.018] [PMID: 18585111]

[52] Lacerda L, Bianco A, Prato M, Kostarelos K. Carbon nanotubes as nanomedicines: from toxicology to pharmacology. Adv Drug Deliv Rev 2006; 58(14): 1460-70. [http://dx.doi.org/10.1016/j.addr.2006.09.015] [PMID: 17113677]

[53] Takanashi S, Hara K, Aoki K, *et al.* Carcinogenicity evaluation for the application of carbon nanotubes as biomaterials in rasH2 mice. Sci Rep 2012; 2(498): 498. [http://dx.doi.org/10.1038/srep00498] [PMID: 22787556]

[54] Donaldson K, Aitken R, Tran L, *et al.* Carbon nanotubes: a review of their properties in relation to pulmonary toxicology and workplace safety. Toxicol Sci 2006; 92(1): 5-22. [http://dx.doi.org/10.1093/toxsci/kfj130] [PMID: 16484287]

[55] Poland CA, Duffin R, Kinloch I, *et al.* Carbon nanotubes introduced into the abdominal cavity of mice show asbestos-like pathogenicity in a pilot study. Nat Nanotechnol 2008; 3(7): 423-8. [http://dx.doi.org/10.1038/nnano.2008.111] [PMID: 18654567]

[56] Ali-Boucetta H, Nunes A, Sainz R, *et al.* Asbestos-like pathogenicity of long carbon nanotubes alleviated by chemical functionalization. Angew Chem Int Ed Engl 2013; 52(8): 2274-8. [http://dx.doi.org/10.1002/anie.201207664] [PMID: 23319294]

[57] Yang ST, Luo J, Zhou Q, Wang H. Pharmacokinetics, metabolism and toxicity of carbon nanotubes for biomedical purposes. Theranostics 2012; (2): 271-82. [http://dx.doi.org/10.7150/thno.3618]

[58] Riviere JE. Pharmacokinetics of nanomaterials: an overview of carbon nanotubes, fullerenes and quantum dots. Wiley Interdiscip Rev Nanomed Nanobiotechnol 2009; 1(1): 26-34. [http://dx.doi.org/10.1002/wnan.24] [PMID: 20049776]

[59] Singh R, Pantarotto D, Lacerda L, *et al.* Tissue biodistribution and blood clearance rates of intravenously administered carbon nanotube radiotracers. Proc Natl Acad Sci USA 2006; 103(9): 3357-62. [http://dx.doi.org/10.1073/pnas.0509009103] [PMID: 16492781]

[60] Dutta D, Sundaram SK, Teeguarden JG, *et al.* Adsorbed proteins influence the biological activity and molecular targeting of nanomaterials. Toxicol Sci 2007; 100(1): 303-15. [http://dx.doi.org/10.1093/toxsci/kfm217] [PMID: 17709331]

[61] Ali-Boucetta H, Al-Jamal KT, McCarthy D, Prato M, Bianco A, Kostarelos K. Multiwalled carbon nanotube-doxorubicin supramolecular complexes for cancer therapeutics. Chem Commun (Camb) 2008; 4(4): 459-61. [http://dx.doi.org/10.1039/B712350G] [PMID: 18188467]

[62] Bianco A, Kostarelos K, Partidos CD, Prato M. Biomedical applications of functionalised carbon nanotubes. Chem Commun (Camb) 2005; 5(5): 571-7. [http://dx.doi.org/10.1039/b410943k] [PMID: 15672140]

[63] Kostarelos K, Bianco A, Prato M. Promises, facts and challenges for carbon nanotubes in imaging and therapeutics. Nat Nanotechnol 2009; 4(10): 627-33. [http://dx.doi.org/10.1038/nnano.2009.241] [PMID: 19809452]

[64] Lacerda L, Herrero MA, Venner K, Bianco A, Prato M, Kostarelos K. Carbon-nanotube shape and individualization critical for renal excretion. Small 2008; 4(8): 1130-2. [http://dx.doi.org/10.1002/smll.200800323] [PMID: 18666166]

[65] Al-Jamal KT, Nunes A, Methven L, *et al.* Degree of chemical functionalization of carbon nanotubes determines tissue distribution and excretion profile. Angew Chem Int Ed Engl 2012; 51(26): 6389-93. [http://dx.doi.org/10.1002/anie.201201991] [PMID: 22623333]

[66] McDevitt MR, Chattopadhyay D, Jaggi JS, *et al.* PET imaging of soluble yttrium-86-labeled carbon nanotubes in mice. PLoS One 2007; 2(9): e907. [http://dx.doi.org/10.1371/journal.pone.0000907] [PMID: 17878942]

[67] McDevitt MR, Chattopadhyay D, Kappel BJ, *et al.* Tumor targeting with antibody-functionalized, radiolabeled carbon nanotubes. J Nucl Med 2007; 48(7): 1180-9. [http://dx.doi.org/10.2967/jnumed.106.039131] [PMID: 17607040]

[68] Liu Z, Davis C, Cai W, He L, Chen X, Dai H. Circulation and long-term fate of functionalized, biocompatible single-walled carbon nanotubes in mice probed by Raman spectroscopy. Proc Natl Acad Sci USA 2008; 105(5): 1410-5. [http://dx.doi.org/10.1073/pnas.0707654105] [PMID: 18230737]

[69] Bianco A, Kostarelos K, Prato M. Making carbon nanotubes biocompatible and biodegradable. Chem Commun (Camb) 2011; 47(37): 10182-8. [http://dx.doi.org/10.1039/c1cc13011k] [PMID: 21776531]

[70] Samorì C, Ali-Boucetta H, Sainz R, *et al.* Enhanced anticancer activity of multi-walled carbon nanotube-methotrexate conjugates using cleavable linkers. Chem Commun (Camb) 2010; 46(9): 1494-6. [http://dx.doi.org/10.1039/B923560D] [PMID: 20162159]

[71] Kam NW, O'Connell M, Wisdom JA, Dai H. Carbon nanotubes as multifunctional biological transporters and near-infrared agents for selective cancer cell destruction. Proc Natl Acad Sci USA 2005; 102(33): 11600-5. [http://dx.doi.org/10.1073/pnas.0502680102] [PMID: 16087878]

[72] Cherukuri P, Gannon CJ, Leeuw TK, *et al.* Mammalian pharmacokinetics of carbon nanotubes using intrinsic near-infrared fluorescence. Proc Natl Acad Sci USA 2006; 103(50): 18882-6. [http://dx.doi.org/10.1073/pnas.0609265103] [PMID: 17135351]

[73] Yang S, Guo W, Lin Y, *et al.* Biodistribution of pristine single-walled carbon nanotubes *in vivo*. Phys J Chem C 2007; 111: 17761-4. [http://dx.doi.org/10.1021/jp070712c]

[74] Liu Z, Cai W, He L, *et al. In vivo* biodistribution and highly efficient tumour targeting of carbon nanotubes in mice. Nat Nanotechnol 2007; 2(1): 47-52. [http://dx.doi.org/10.1038/nnano.2006.170] [PMID: 18654207]

[75] Prencipe G, Tabakman SM, Welsher K, *et al.* PEG branched polymer for functionalization of nanomaterials with ultralong blood circulation. J Am Chem Soc 2009; 131(13): 4783-7. [http://dx.doi.org/10.1021/ja809086q] [PMID: 19173646]

[76] Liu X, Tao H, Yang K, Zhang S, Lee S-T, Liu Z. Optimization of surface chemistry on single-walled carbon nanotubes for in vivo photothermal ablation of tumors. Biomaterials 2011; 32(1): 144-51. [http://dx.doi.org/10.1016/j.biomaterials.2010.08.096] [PMID: 20888630]

[77] Singh R, Pantarotto D, Lacerda L, *et al.* Tissue biodistribution and blood clearance rates of intravenously administered carbon nanotube radiotracers. Proc Natl Acad Sci USA 2006; 103(9): 3357-62. [http://dx.doi.org/10.1073/pnas.0509009103] [PMID: 16492781]

[78] Guo J, Zhang X, Li Q, Li W. Biodistribution of functionalized multiwall carbon nanotubes in mice. Nucl Med Biol 2007; 34(5): 579-83. [http://dx.doi.org/10.1016/j.nucmedbio.2007.03.003] [PMID: 17591558]

[79] Wang H, Wang J, Deng X, *et al.* Biodistribution of carbon single-wall carbon nanotubes in mice. J Nanosci Nanotechnol 2004; 4(8): 1019-24. [http://dx.doi.org/10.1166/jnn.2004.146] [PMID: 15656196]

[80] Deng X, Jia G, Wang H, *et al.* Translocation and fate of multi-walled carbon nanotubes *in vivo*. Carbon 2007; 45: 1419-24.

[http://dx.doi.org/10.1016/j.carbon.2007.03.035]

[81] Wang J, Deng X, Yang S, Wang H, Zhao Y, Liu Y. Rapid translocation and pharmacokinetics of hydroxylated single-walled carbon nanotubes in mice. Nanotoxicology 2008; 2: 28-32. [http://dx.doi.org/10.1080/17435390701851747]

[82] Deng X, Yang S. A generally adoptable radio tracing method for tracking carbon nanotubes in animals. Nanotechnology 2008; 19: 75-101. [http://dx.doi.org/http://dx.doi.org/10.1088/0957-4484/19/7/075101] [PMID: 21817626]

[83] Yang ST, Fernando KA, Liu J-H, *et al.* Covalently PEGylated carbon nanotubes with stealth character *in vivo.* Small 2008; 4(7): 940-4. [http://dx.doi.org/10.1002/smll.200700714] [PMID: 18574799]

[84] Lacerda L, Soundararajan A, Pastorin G, *et al.* Dynamic imaging of functionalized multi-walled carbon nanotube systemic circulation & urinary excretion. Adv Mater 2008; 20(2): 225-30. [http://dx.doi.org/10.1002/adma.200702334]

[85] Lacerda L, Ali-Boucetta H, Herrero MA, *et al.* Tissue histology and physiology following intravenous administration of different types of functionalized multiwalled carbon nanotubes. Nanomedicine (Lond) 2008; 3(2): 149-61. [http://dx.doi.org/10.2217/17435889.3.2.149] [PMID: 18373422]

[86] Ruggiero A, Villa CH, Bander E, *et al.* Paradoxical glomerular filtration of carbon nanotubes. Proc Natl Acad Sci USA 2010; 107(27): 12369-74. [http://dx.doi.org/10.1073/pnas.0913667107] [PMID: 20566862]

[87] Jain S, Thakare VS, Das M, *et al.* Toxicity of multiwalled carbon nanotubes with end defects critically depends on their functionalization density. Chem Res Toxicol 2011; 24(11): 2028-39. [http://dx.doi.org/10.1021/tx2003728] [PMID: 21978239]

[88] Jokerst JV, Lobovkina T, Zare RN, Gambhir SS. Nanoparticle PEGylation for imaging and therapy. Nanomedicine (Lond) 2011; 6(4): 715-28. [http://dx.doi.org/10.2217/nnm.11.19] [PMID: 21718180]

[89] Liu Z, Chen K, Davis C, *et al.* Drug delivery with carbon nanotubes for *in vivo* cancer treatment. Cancer Res 2008; 68(16): 6652-60. [http://dx.doi.org/10.1158/0008-5472.CAN-08-1468] [PMID: 18701489]

[90] Liu Z, Sun X, Nakayama-Ratchford N, Dai H. Supramolecular chemistry on water-soluble carbon nanotubes for drug loading and delivery. ACS Nano 2007; 1(1): 50-6. [http://dx.doi.org/10.1021/nn700040t] [PMID: 19203129]

[91] Bhirde AA, Patel S, Sousa AA, *et al.* Distribution and clearance of PEG-single-walled carbon nanotube cancer drug delivery vehicles in mice. Nanomedicine (Lond) 2010; 5(10): 1535-46. [http://dx.doi.org/10.2217/nnm.10.90] [PMID: 21143032]

[92] Zhang D, Deng X, Ji Z, *et al.* Long-term hepatotoxicity of polyethylene-glycol functionalized multi-walled carbon nanotubes in mice. Nanotechnology 2010; 21(17): 175101. [http://dx.doi.org/10.1088/0957-4484/21/17/175101] [PMID: 20357413]

[93] Yang S-T, Wang Y-W, Liu J-H, Wang H. Biodistribution of multi-walled carbon nanotubes functionalized by hydroxyl terminated poly(ethylene glycol) in mice. J Radioanal Nucl Chem 2013; 295: 1181-6. [http://dx.doi.org/10.1007/s10967-012-1901-0]

[94] Yang S-T, Fernando KA, Liu J-H, *et al.* Covalently PEGylated carbon nanotubes with stealth character *in vivo.* Small 2008; 4(7): 940-4. [http://dx.doi.org/10.1002/smll.200700714] [PMID: 18574799]

[95] Faure A-C, Dufort S, Josserand V, *et al.* Control of the *in vivo* biodistribution of hybrid nanoparticles with different poly(ethylene glycol) coatings. Small 2009; 5(22): 2565-75. [http://dx.doi.org/10.1002/smll.200900563] [PMID: 19768700]

[96] Murphy FA, Poland CA, Duffin R, *et al.* Length-dependent retention of carbon nanotubes in the pleural space of mice initiates sustained inflammation and progressive fibrosis on the parietal pleura. Am J Pathol 2011; 178(6): 2587-600.
[http://dx.doi.org/10.1016/j.ajpath.2011.02.040] [PMID: 21641383]

[97] Torchilin VP. Passive and active drug targeting: drug delivery to tumors as an example. Handb Exp Pharmacol 2010; 197(197): 3-53.
[http://dx.doi.org/10.1007/978-3-642-00477-3_1] [PMID: 20217525]

[98] Wang X, Wang Y, Chen ZG, Shin DM. Advances of cancer therapy by nanotechnology. Cancer Res Treat 2009; 41(1): 1-11.
[http://dx.doi.org/10.4143/crt.2009.41.1.1] [PMID: 19688065]

[99] Torchilin VP. Multifunctional, stimuli-sensitive nanoparticulate systems for drug delivery. Nat Rev Drug Discov 2014; 13(11): 813-27.
[http://dx.doi.org/10.1038/nrd4333] [PMID: 25287120]

[100] Kelley EG, Albert JN, Sullivan MO, Epps TH III. Stimuli-responsive copolymer solution and surface assemblies for biomedical applications. Chem Soc Rev 2013; 42(17): 7057-71.
[http://dx.doi.org/10.1039/c3cs35512h] [PMID: 23403471]

[101] Shen Y, Fu X, Fu W, Li Z. Biodegradable stimuli-responsive polypeptide materials prepared by ring opening polymerization. Chem Soc Rev 2015; 44(3): 612-22.
[http://dx.doi.org/10.1039/C4CS00271G] [PMID: 25335988]

[102] Yang D, Yang F, Hu J, *et al.* Hydrophilic multi-walled carbon nanotubes decorated with magnetite nanoparticles as lymphatic targeted drug delivery vehicles. Chem Commun (Camb) 2009; 29(29): 4447-9.
[http://dx.doi.org/10.1039/b908012k] [PMID: 19597621]

[103] Liu Z, Chen K, Davis C, *et al.* Drug delivery with carbon nanotubes for *in vivo* cancer treatment. Cancer Res 2008; 68(16): 6652-60.
[http://dx.doi.org/10.1158/0008-5472.CAN-08-1468] [PMID: 18701489]

[104] Wu W, Li R, Bian X, *et al.* Covalently combining carbon nanotubes with anticancer agent: preparation and antitumor activity. ACS Nano 2009; 3(9): 2740-50.
[http://dx.doi.org/10.1021/nn9005686] [PMID: 19702292]

[105] Bhirde AA, Patel V, Gavard J, *et al.* Targeted killing of cancer cells *in vivo* and *in vitro* with EGF-directed carbon nanotube-based drug delivery. ACS Nano 2009; 3(2): 307-16.
[http://dx.doi.org/10.1021/nn800551s] [PMID: 19236065]

[106] Yang R, Yang X, Zhang Z, *et al.* Single-walled carbon nanotubes-mediated *in vivo* and *in vitro* delivery of *si*RNA into antigen-presenting cells. Gene Ther 2006; 13(24): 1714-23.
[http://dx.doi.org/10.1038/sj.gt.3302808] [PMID: 16838032]

[107] Zhang Z, Yang X, Zhang Y, *et al.* Delivery of telomerase reverse transcriptase small interfering RNA in complex with positively charged single-walled carbon nanotubes suppresses tumor growth. Clin Cancer Res 2006; 12(16): 4933-9.
[http://dx.doi.org/10.1158/1078-0432.CCR-05-2831] [PMID: 16914582]

[108] Bartholomeusz G, Cherukuri P, Kingston J, *et al. In vivo* therapeutic silencing of hypoxia-inducible factor 1 a (HIF-1a) using single-walled carbon nanotubes noncovalently coated with *si*-RNA. Nano Res 2009; 2(4): 279-91.
[http://dx.doi.org/10.1007/s12274-009-9026-7] [PMID: 20052401]

[109] GabayT.JakobsE.Ben-JacobE.HaneinY.Engineered self-organization of neural networks using carbon nanotube clusters. Physica A: Statist Mech Appl 2005; 350: 611-21.

[110] Hu H, Ni Y, Montana V, Haddon RC, Parpura V. Chemically functionalized carbon nanotubes as substrates for neuronal growth. Nano Lett 2004; 4(3): 507-11.
[http://dx.doi.org/10.1021/nl035193d] [PMID: 21394241]

[111] Webster TJ, Waid MC, McKenzie JL, Price RL, Ejiofor JU. Nano-biotechnology: carbon nanofibres as improved neural and orthopaedic implants. Nanotechnology 2004; 15: 48-54. [http://dx.doi.org/10.1088/0957-4484/15/1/009]

[112] Lovat V, Pantarotto D, Lagostena L, *et al.* Carbon nanotube substrates boost neuronal electrical signaling. Nano Lett 2005; 5(6): 1107-10. [http://dx.doi.org/10.1021/nl050637m] [PMID: 15943451]

[113] Cheung W, Pontoriero F, Taratula O, Chen AM, He H. DNA and carbon nanotubes as medicine. Adv Drug Deliv Rev 2010; 62(6): 633-49. [http://dx.doi.org/10.1016/j.addr.2010.03.007] [PMID: 20338203]

[114] Cai D, Mataraza JM, Qin ZH, *et al.* Highly efficient molecular delivery into mammalian cells using carbon nanotube spearing. Nat Methods 2005; 2(6): 449-54. [http://dx.doi.org/10.1038/nmeth761] [PMID: 15908924]

[115] Shi Kam NW, Jessop TC, Wender PA, Dai H. Nanotube molecular transporters: internalization of carbon nanotube-protein conjugates into Mammalian cells. J Am Chem Soc 2004; 126(22): 6850-1. [http://dx.doi.org/10.1021/ja0486059] [PMID: 15174838]

[116] McDevitt MR, Chattopadhyay D, Kappel BJ, *et al.* Tumor targeting with antibody-functionalized, radiolabeled carbon nanotubes. J Nucl Med 2007; 48(7): 1180-9. [http://dx.doi.org/10.2967/jnumed.106.039131] [PMID: 17607040]

[117] Meng J, Meng J, Duan J, *et al.* Carbon nanotubes conjugated to tumor lysate protein enhance the efficacy of an antitumor immunotherapy. Small 2008; 4(9): 1364-70. [http://dx.doi.org/10.1002/smll.200701059] [PMID: 18720440]

[118] Ruggiero A, Villa CH, Holland JP, *et al.* Imaging and treating tumor vasculature with targeted radiolabeled carbon nanotubes. Int J Nanomedicine 2010; 5: 783-802. [http://dx.doi.org/10.2147/IJN.S13300] [PMID: 21042424]

[119] Pantarotto D, Singh R, McCarthy D, *et al.* Functionalized carbon nanotubes for plasmid DNA gene delivery. Angew Chem Int Ed Engl 2004; 43(39): 5242-6. [http://dx.doi.org/10.1002/anie.200460437] [PMID: 15455428]

[120] Pruthi J, Mehra NK, Jain NK. Macrophages targeting of amphotericin B through mannosylated multiwalled carbon nanotubes. J Drug Target 2012; 20(7): 593-604. [http://dx.doi.org/10.3109/1061186X.2012.697168] [PMID: 22690657]

[121] Prajapati VK, Awasthi K, Gautam S, *et al.* Targeted killing of Leishmania donovani *in vivo* and *in vitro* with amphotericin B attached to functionalized carbon nanotubes. J Antimicrob Chemother 2011; 66(4): 874-9. [http://dx.doi.org/10.1093/jac/dkr002] [PMID: 21393222]

[122] Li Z, de Barros AL, Soares DC, Moss SN, Alisaraie L. Functionalized single-walled carbon nanotubes: cellular uptake, biodistribution and applications in drug delivery. Int J Pharm 2017; 524(1-2): 41-54. [http://dx.doi.org/10.1016/j.ijpharm.2017.03.017] [PMID: 28300630]

[123] Bianco A, Kostarelos K, Prato M. Applications of carbon nanotubes in drug delivery. Curr Opin Chem Biol 2005; 9(6): 674-9. [http://dx.doi.org/10.1016/j.cbpa.2005.10.005] [PMID: 16233988]

[124] Erol O, Uyan I, Hatip M, Yilmaz C, Tekinay AB, Guler MO. Recent advances in bioactive 1D and 2D carbon nanomaterials for biomedical applications. Nanomedicine: NBM 2017. in press.

[125] Eatemadi A, Daraee H, Karimkhanloo H, *et al.* Carbon nanotubes: properties, synthesis, purification, and medical applications. Nanoscale Res Lett 2014; 9(1): 393. [http://dx.doi.org/10.1186/1556-276X-9-393] [PMID: 25170330]

[126] Foldvari M, Bagonluri M. Carbon nanotubes as functional excipients for nanomedicines: II. Drug

delivery and biocompatibility issues. Nanomedicine (Lond) 2008; 4(3): 183-200. [http://dx.doi.org/10.1016/j.nano.2008.04.003] [PMID: 18550450]

[127] Sitharaman B, Shi X, Tran LA, *et al.* Injectable in situ cross-linkable nanocomposites of biodegradable polymers and carbon nanostructures for bone tissue engineering. J Biomater Sci Polym Ed 2007; 18(6): 655-71. [http://dx.doi.org/10.1163/156856207781034133] [PMID: 17623549]

[128] Shityakov S, Salvador E, Pastorin G, Förster C. Blood-brain barrier transport studies, aggregation, and molecular dynamics simulation of multiwalled carbon nanotube functionalized with fluorescein isothiocyanate. Int J Nanomedicine 2015; 10: 1703-13. [http://dx.doi.org/10.2147/IJN.S68429] [PMID: 25784800]

[129] Yang Z, Zhang Y, Yang Y, *et al.* Pharmacological and toxicological target organelles and safe use of single-walled carbon nanotubes as drug carriers in treating Alzheimer disease. Nanomedicine (Lond) 2010; 6(3): 427-41. [http://dx.doi.org/10.1016/j.nano.2009.11.007] [PMID: 20056170]

[130] Lee HJ, Park J, Yoon OJ, *et al.* Amine-modified single-walled carbon nanotubes protect neurons from injury in a rat stroke model. Nat Nanotechnol 2011; 6(2): 121-5. [http://dx.doi.org/10.1038/nnano.2010.281] [PMID: 21278749]

[131] Ren J, Shen S, Wang D, *et al.* The targeted delivery of anticancer drugs to brain glioma by PEGylated oxidized multi-walled carbon nanotubes modified with angiopep-2. Biomaterials 2012; 33(11): 3324-33. [http://dx.doi.org/10.1016/j.biomaterials.2012.01.025] [PMID: 22281423]

[132] Kafa H, Wang JT, Rubio N, *et al.* The interaction of carbon nanotubes with an in vitro blood-brain barrier model and mouse brain in vivo. Biomaterials 2015; 53: 437-52. [http://dx.doi.org/10.1016/j.biomaterials.2015.02.083] [PMID: 25890741]

[133] Ye Y, Ahn C, Witham C, *et al.* Hydrogen adsorption and cohesive energy of single-walled carbon nanotubes. Appl Phys Lett 1999; 74: 2307-9. [http://dx.doi.org/10.1063/1.123833]

CHAPTER 6

Tautomycetin, Protein Phosphatase 1 Specific Inhibitor, opened the Door for understanding the Role of PP1 in Minkowski Space

Makoto Ubukata*

Research Faculty of Agriculture, Hokkaido University, Sapporo, Japan

Abstract: When the structure of tautomycin (TM) was determined, it was pointed out that the molecule must have potent inhibitory effects against protein phosphatases (PPs). The hint was brought from the structure of okadaic acid (OA), which was just disclosed to be a potent inhibitor of PPs. Protein phosphatase type 2A (PP2A) was the primary molecular target of OA, while TM was found to be a dual inhibitor for protein phosphatase type 1 (PP1) and PP2A, with partial selectivity to PP1. The rigid structure of OA, however, exhibited the biological activity as a tumor promoter, whereas the flexible structure of TM did not induce any tumor promotion *in vivo*. Tautomycetin (TC), having very similar structure to TM, was found to be the only inhibitor specific to PP1 at low concentrations, and the discovery opened the door to therapeutic strategies for immune disorders, cancer or neurological disorders involving PP1 and to understanding the distinguishable roles of PP1 and PP2A, two major Ser/Thr PPs in human cells. Sephin1, a selective inhibitor of PP1 holoenzyme containing growth arrest and DNA damage-inducible protein (GADD34), attenuated expression of stress-inducible gene products. The approach was one of the several challenges for developing PP1-targeted therapeutics for neurological disorders to which circadian rhythm would be related. Thus appropriately modulating PP1 activity could lead to new treatments for neurological disorders in Minkowski space, a combination of three-dimensional space and one dimension of time.

Keywords: Anti-cancer agents, Circadian clock, Immunosuppressant, Okadaic acid (OA), PP1/PP2A balance, PP1-target therapeutics, Protein phosphatase type 1 (PP1), Protein phosphatase type 2A (PP2A), Tautomycetin (TC), Tautomycin (TM), Tumor promoter.

INTRODUCTION

The phosphorylation reactions of proteins govern the regulations of various cellular functions such as cell proliferation, differentiation, and apoptosis. Protein

* **Corresponding author Makoto Ubukata:** Research Faculty of Agriculture, Hokkaido University, Japan; Tel/Fax: +81-11-737-3243; E-mail: m-ub@for.agr.hokudai.ac.jp

Atta-ur-Rahman & M. Iqbal Choudhary (Eds.)

kinase and protein phosphatase function as the enzymes, that phosphorylate and dephosphorylate protein substrates, respectively, as shown in Fig. (**1**).

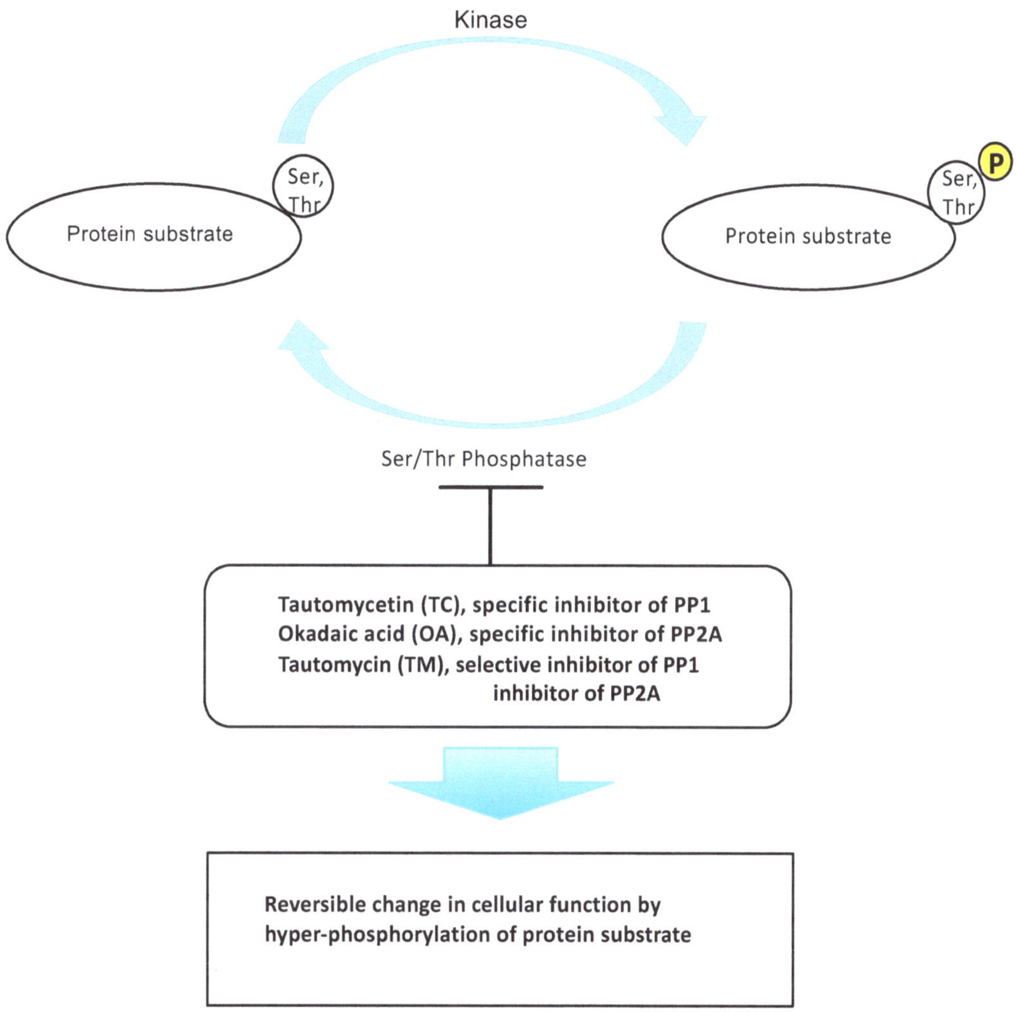

Fig. (1). General mechanism of action of Ser/Thr protein phosphatase inhibitors, tautomycetin (TC), okadaic acid (OA), and tautomycin (TM).

Phosphorylation by kinase and dephosphorylation by phosphatase rapidly change the local charge distribution and the local three-dimensional structure of substrate protein whose function can be turned on and off by using the molecular switch, kinase inhibitor or phosphatase inhibitor. Kinases have been established as therapeutic targets for various diseases, and the approval of imatinib (Gleevec,

Novartis) by the FDA in 2001 for the treatment of patients with chronic myeloid leukemia (CML) paved the way for development of many other kinase inhibitors for cancer, rheumatoid arthritis, idiopathic pulmonary fibrosis, and so on [1].

Phosphorylation of protein mainly occurs on serine, threonine, or tyrosine residue, and protein serine/threonine phosphatases (PPs) remove phosphate groups from phospho-serine/threonine residue in substrate proteins, while protein tyrosine phosphatases (PTPs) remove phosphate groups from phospho-tyrosine residues. Dual specificity phosphatase (DUSP) can act on tyrosine or serine/threonine residues.

Protein phosphatase type 1 (PP1), PP2A, PP2B, and PP2C are 4 major PPs (also known as PPP1, PPP2, PPP3, and PPP2C, respectively), while PTPs estimated from genome data analyses are over 100 types in human. Cdc25 is a dual-specificity phosphatase, which activates cyclin dependent kinases (Cdks) by removing inhibitory phosphate groups from the Cdk active site. Protein phosphatases have been shown to be the extremely important molecular switches for regulating various cell functions [2 - 7]. A number of studies using individual protein phosphatase-inhibitors contributed to discovering the physiological significance of protein dephosphorylation and the mechanisms of actions of these protein phosphatases in cells.

Tautomycetin (TC), PP1 specific inhibitor, okadaic acid (OA), PP2A specific inhibitor are useful chemical probes for dissecting the physiological roles of two major protein phosphatases, PP1 and PP2A [8]. The use of TC led to the possibility of PP1 as a therapeutic target for treating various intractable diseases. TC might be useful for the development of a new class of anti-cancer agents [8, 9] or a new class of immunosuppressant [10]. The PP1 holoenzymes consist of a catalytic subunit (PP1C) and various targeting/regulatory subunits. One of the successful inhibitors for PP1 holoenzymes is sephin1 (selective inhibitor of a holophosphatase) that selectively inhibits a regulatory subunit (GADD34, another name is PPP1R15A) of PP1 *in vivo*. The compound safely prevented the motor, morphological, and molecular defects of two otherwise unrelated protein-misfolding diseases in mice, Charcot-Marie-Tooth 1B, and amyotrophic lateral sclerosis (ALS) [11]. Nevertheless, TC, the specific inhibitor of pan PP1 provided the first recognition of the potential therapeutic importance of PP1.

Discoveries of Tautomycin (TM) and Tautomycetin (TC) as Protein Phosphatase Inhibitors

Tautomycin (TM) and TC were screened as the new antifungal antibiotics against *Sclerotinia sclerotiorum* for the purpose of protecting oilseed rape from sclerotial disease in RIKEN, the Institute of Physical and Chemical Research [12, 13]. The structure of TM isolated from a culture broth of *Streptomyces spiroverticillatus* and that of TC from *Streptomyces griseochromogenes* were determined to be a new class of dialkylmaleic anhydride antibiotics [14 - 16]. The stability of TM was much better than that of TC, thus the absolute configuration of TM was determined as shown in Fig. (**2**) [17]. Ubukata who determined these structures pointed out for the first time that the molecular size and arrangement of functional groups of TM are similar to those of okadaic acid (OA) and it must have potent inhibitory activity against PPs. OA was found to be a toxin produced by several species of dinoflagellates in marine sponges and shellfish. The complex structure of OA includes multiple spiroketals and fused ether rings [18], and PP2A can be specifically inhibited by OA [19].

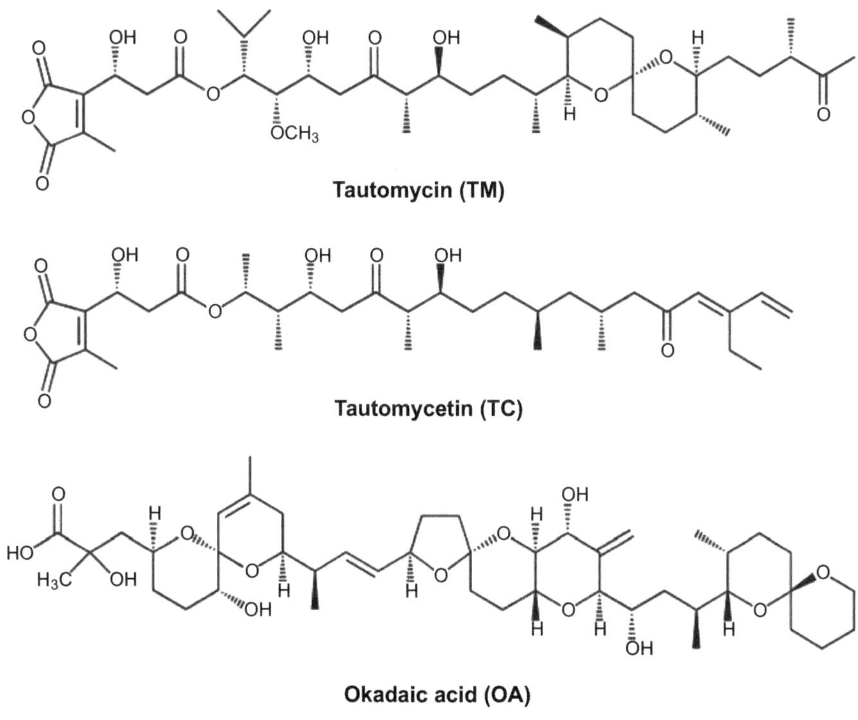

Fig. (2). Structures of tautomycin (TM), tautomycetin (TC), and okadaic acid (OA).

Ubukata gave a presentation on the structure and biological properties of TM at 27th Interscience Conference on Antimicrobial Agents and Chemotherapy (ICAAC) held in Houston, USA in 1989 [20]. In 1990, MacKintosh and Klumpp published a paper that TM was an inhibitor of PP1 and PP2A [21]. They described "When drawn following its structural formula, the backbone of tautomycin matches that of okadaic acid in the region from carbon-1 to carbon-16" and "tautomycin, like okadaic acid, might be a protein phosphatase inhibitor." The structure of TM was shown in the opposite direction in their paper suggesting that the right-half moiety of TM bares a resemblance to the left-half moiety of OA. The ostensible similarity exists between methyl ketone - spiroketal moiety in the right-half moiety of TM and carboxylic acid - spiroketal moiety in the left-half moiety of OA [21]. That idea was different from Ubukata's idea that the carboxylic acid group of OA mimics phosphoric acid of substrates in PP1 or PP2A, and di-carboxylic acid group of TM easily generated from the maleic anhydride moiety in the presence of water mimics the phosphate group of substrate in PP1 or PP2A as will be shown later in Fig. (**4**). Therefore, the structures of TM and OA should be depicted to show molecular similarity between TM and OA as shown in Fig. (**2**). Nevertheless, TM inhibited the catalytic subunits of PP1 (K_{iapp} = 0.16 nM) more potently than PP2A (K_{iapp} = 0.4 nM), while PP2B was inhibited 10,000-fold less potently [21]. Various IC_{50} values of TM for the catalytic subunits of PP1 and PP2A have been reported: 0.16 and 0.40 nM [21], 22 and 32 nM [22], and 0.7 and 0.65 nM [23]. A non-ionic detergent, Brij-35 lowers the IC_{50} values for PP1 from 0.20 to 0.020 nM and for PP2A from 0.70 to 0.075 nM, suggesting that the wide variation in the IC_{50} values was due to differences in the assay protocols [24]. Structure-activity relationships of TM [24 - 26] suggests that dialkyl maleic anhydride moiety is essential for the potent inhibitory effect of TM and the right-half moiety of TM is involved in the specificity for PP1.

The left-half moiety of TC is very similar to that of TM, whereas the right-half moiety of TC is different from that of TM. The PP1 inhibitory activities of 22-epi-tautomycin and TM-dimethyl ester were three-order and four-order magnitude less active, respectively, than TM, whereas their inhibitory selectivities for PP1 over PP2A were comparable to TM. Although anhydrodeacy-TM lacking C1'-C7' in the left-half moiety of TM showed only poor inhibitory effects against PP1 and PP2A, it possessed potent apoptosis-inducing activities comparable to TM. These observations suggest that the spiroketal structure in TM is not essential for inhibitory activity against phosphatase, while it is involved in apoptosis induction. TC was found to be the most specific PP1 inhibitor out of over 40 phosphatase inhibitors reported. Table **1** lists the inhibitory effects of TC and TM [27].

TM inhibited PP1 more potently than PP2A, with the PP2A/PP1 ratio of IC_{50} being 4.5 within purified enzymes. The inhibitory effect of TC against PP1 was 39-fold more potent than that against PP2A in both the purified form. TC had no effect on purified PP2B up to 1 μM [27]. PP1 acts *in vivo* as multiple holoenzymes composed of catalytic subunit, PP1C and targeting or regulatory subunits [28, 29]. Three genes in mammals encode the isoforms PP1Cα, PP1Cγ1, PP1Cγ2, and PP1Cβ/δ. All PP1Cs except PP1Cγ2 are ubiquitously expressed [2, 30 - 32]. The effects of TC and TM on endogenous holoenzymes of PP1 and

PP2A were analyzed by using crude enzymes. TC and TM inhibited PP1 and PP2A as shown in Table **1**, while PP2B activity was not inhibited under these conditions [27].

Table 1. Inhibition of PP1 and PP2A by TC and TM [27]

Table 1. Inhibition of PP1 and PP2A by TC and TM [27].

		IC_{50} (nM)		ratio
Inhibitor	Enzyme source	PP1	PP2A	(PP2A/PP1)
TC	Purified	1.6	62	39
	Extract	2.0	46	23
TM	Purified	0.2	0.9	4.5
	Extract	0.4	1.3	3

The IC_{50} values of TC for PP1 activities with phosphorylase, myelin basic protein (MBP) and 20 kDa smooth muscle myosin light chain (LC20) were 2, 0.6, and 1.7 nM, respectively. These comparable IC_{50} values suggest that TC inhibits PP1 *via* a similar molecular mechanism regardless of substrate. Inhibitor 2 (I-2), one of the endogenous PP1 inhibitor proteins has been often used in a differential assay for PP1 *in vitro*. In sharp contrast to TC, I-2 potently inhibited PP1 activity for phosphorylase (IC_{50}: 1.8 nM), but did not inhibited PP1 activity for MBP and LC20. These results indicate that TC is more useful than I-2 to suppress most of PP1 holoenzymes in cells [27].

As PP1C is encoded by three independent genes and is associated with numerous regulatory subunits, it may not be easy to detect the phenotype of loss-of-function in mammalian cells by gene disruption or suppression [27].

Usage of tautomycetin (TC) as specific PP1 inhibitor

TC can be used as a powerful chemical probe to elucidate the physiological roles of PP1 in various biological events. For instance, treatment of COS-7 or HEK293T cells with 5 μM TC causes a complete inhibition of PP1 activity without affecting PP2A activity [8]. Tautomycin (TM) has been used to analyze PP1 *in vivo* because of its relatively high affinity to PP1. However, 10 μM TM, which completely inhibited PP1, also inhibited PP2A by 50% [33]. Therefore, it is great advantage for *in vivo* analysis that 5 μM TC and 100 nM OA make it possible to differentiate between the roles of PP1 and PP2A, respectively, in cells [8].

Activations of mitogen-activated protein kinase (MAPK) cascades play critical roles in transducing extracellular signals to nucleus [34 - 36]. The MAPK superfamily includes extracellular signal-regulated kinase (ERK), c-Jun N-terminal kinase (JNK), and p38 MAPK. The ERK pathway is activated by such mitogenic stimuli as growth factor, cytokines, and phorbol esters. The stress-activated MAPK pathways, JNK pathway and p38 pathway, are activated by UV light, heat shock, osmotic stress, and inflammatory cytokines [34 - 36]. Numerous observations suggested that PP2A plays a major role in the down-regulation of JNK, mitogen-activated protein kinase kinase (MEK, also known as MAP2K, MAPKK), and ERK activities, and highlighted the role of PP2A as a tumor suppressor [37]. This was first suggested in experiments showing that OA promoted tumor growth in mice [19, 38, 39]. Given the specific inhibitor of PP1, the involvement of PP1 in MAPK pathways was revealed. Treatment of COS-7 cells with 5 μM TC selectively inhibited PP1 activity. TC specifically inhibited activation of ERK among three MAPKs (ERK, JNK, and p38) in the cells treated with TPA or EGF, and suppressed activation of Raf-1 activity, which resulted in inactivation of the MEK-ERK pathway. Over expression of PP1C activated Raf-1, and PP1C directly interacted with Raf-1 in COS-7 cells. Thus, it was proposed that PP1 is a positive regulator of Raf-1, whereas PP2A functions as a positive regulator of Raf-1 but a negative regulator of MEK and ERK in EGF/ERK pathway as shown in Fig. (**3**). The up-regulation of Raf-MEK-ERK by PP1 was observed not only in COS-7 but also in HEK293T, HeLa, and HepG2 cells [8].

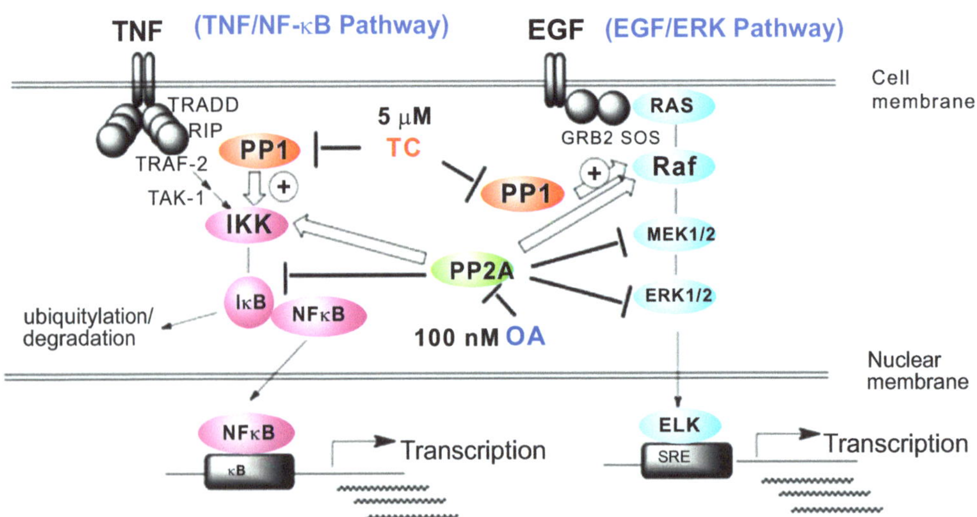

Fig. (3). Five μM tautomycetin (TC) inhibits both EGF/ERK pathway and TNF/NF-κB pathway, whereas 100 nM okadaic acid (OA) activates both pathways. Proposed mechanisms show that PP1 is a positive regulator of Raf and IKK, while PP2A is a negative regulator of MEK/ERK and IκB pathways [8, 40].

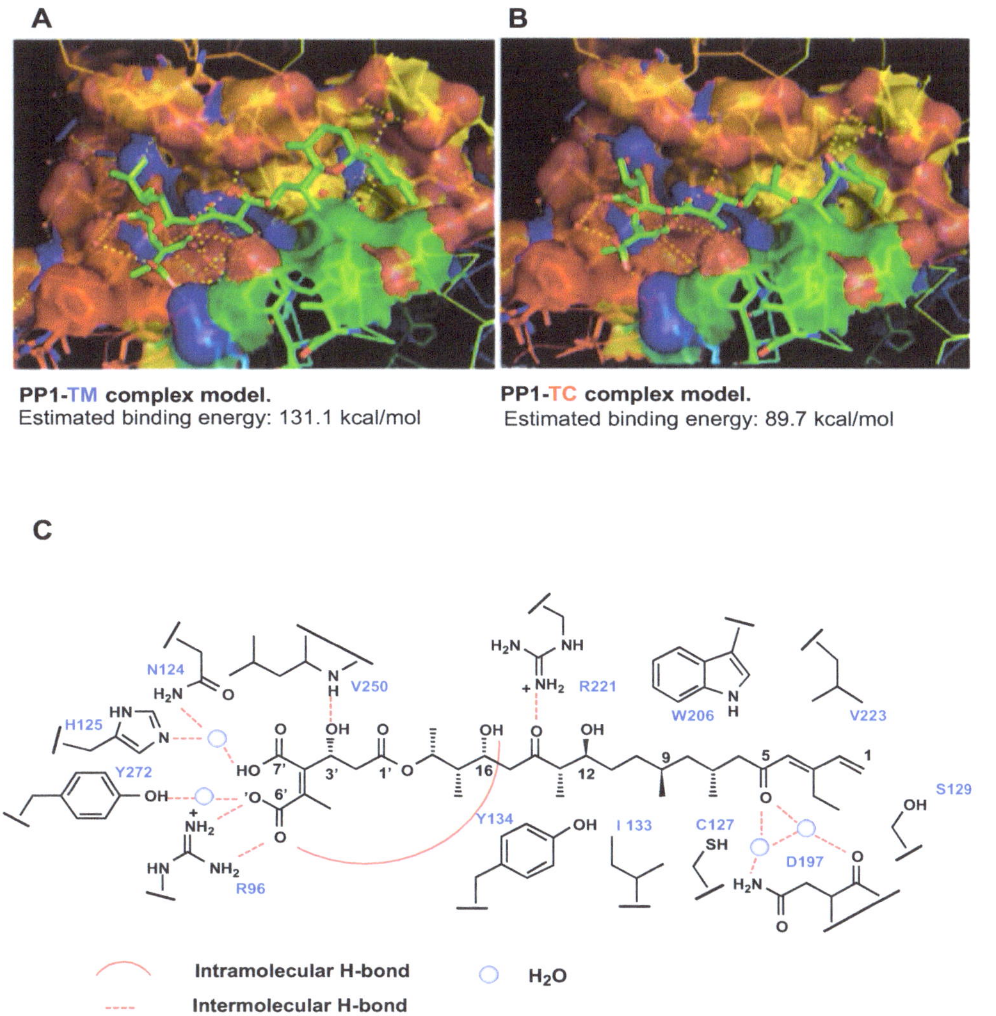

Fig. (4). Tautomycin (TM) and tautomycetin (TC) bind to PP1. **A**: PP1-TM complex model. **B**: PP1-TC complex model. **C**: Interaction between TC and amino acid residues in the active site in the catalytic subunit of PP1 [57].

A proposed mechanism for the NF-κB pathway was deduced from TNFα-induced phosphorylation using 5 μM TC and 100 nM OA in HEK293T cells as shown in Fig. (**3**) [40]. To determine whether PP1 functions in TNFα-induced NF-κB activation, the effect of TC or OA was evaluated by a reporter gene assay in HEK293T cells transfected with pNF-κB-Luc and pCMV-β-galactosidase. As reported, 100 nM OA enhanced basal and TNFα-induced NF-κB-dependent luciferase activity in the reporter gene assay. In contrast, 5 μM TC dramatically decreased TNFα-induced NF-κB-dependent luciferase activity. Neither TC nor OA had effect on caspase-3/7 activation or caspase-8 in HeLa cells. These results suggest that the targets of PP1 and PP2A in TNFα signaling are not downstream of complex II-mediated caspase activation pathways but rather downstream of complex I-mediated IKK/NF-κB pathway. Since downstream of complex I bifurcate into NF-κB and JNK pathways, the effects of OA and TC were compared on phosphorylation and degradation of IκBα and activation of JNK in HEK293T cells. As reported, 100 nM OA induced degradation of IκBα and hyper-phosphorylation of JNK, confirming that PP2A is a negative regulator of NF-κB and JNK activation [41, 42]. Under the same conditions, 5 μM TC dramatically reduced IκBα degradation but had no effect on JNK activation. Since IκBα is phosphorylated by IKK, the effects of TC and OA were analyzed on IKK activation stimulated with TNFα. IKK activity was increased to a peak at 10-15 min and subsided by 30 min after TNFα stimulation, whereas 5 μM TC clearly suppressed IKK activation within 30 min after TNFα stimulation. Inhibition of PP2A with OA resulted in significant suppression of IKK activation, too. Phosphorylation at Ser-180 in T-loop of IKKα and Ser-181 in T-loop of IKKβ appeared at 10 to 25 min and IκBα started to degrade at 10 min and completely disappeared at 25 min after stimulation with TNFα in HEK293T cells. However, 5 μM TC completely inhibited both phosphorylation at these sites and degradation of IκBα. Overexpressed Myc-PP1C interacted with Flag-IKKα and Flag-IKKβ. Endogenous PP1C interacted with the endogenous IKK complex, too. The spatiotemporal profile of IKK activation triggered with TNFα was very similar to those of PP1C, RIP, TAK1, IKKα, and IKKβ amounts recruited to TNFR1. The recruited amounts of these signaling molecules increased to a peak at 10 min and subsided by 60 min after TNFα stimulation. Five μM TC inhibited TNFα-induced IkBα phosphorylation and degradation in HeLa, COS-7, HepG2, Jurkat cells, too [40].

Constitutive activation of NF-κB and IKK are related to a variety of tumor malignancies [43 - 46], and regulates the expression of genes involved in cancer such as c-myc, cyclin D1, matrix metalloproteinases, Bcl-xL, c-FLIP, and VEGF. Therefore, inhibition of NF-κB alone or in combination with cancer therapies leads to effective tumor cell death or growth inhibition of tumor [43].

Relationship Between ATM Kinase and Protein Phosphatase 1 (PP1) *via* Ribosomal Protein S6 (RPS6) Revealed by Tautomycetin (TC)

The cellular responses to DNA damage are also regulated by kinase and phosphatase [47 - 50]. ATM kinase (ATM) is a serine/threonine kinase that is a sensor protein activated by DNA double-strand breaks. It phosphorylates several proteins that initiate activation of the DNA damage checkpoint leading to cell cycle arrest, DNA repair, or apoptosis [47].

Although a number of downstream targets of ATM have been identified [47, 51], the detailed functions of ATM and its pathways have not been fully understood. Ser/Thr protein phosphatases have been found to play important roles in regulating the DNA damage response during recovery from cell cycle arrest [49]. PP2A has been reported to regulate ATM autophosphorylation at Ser1981 after ionizing radiation-induced DNA damage [48]. A member of PP2Cδ, Wip1 dephosphorylates multiple phospho-(S/T)Q sites targeted by ATM, including those in ATM itself and its substrates, *in vitro* [50, 52]. Despite the fact that PP1 is an essential mediator in ATM-induced signal transduction, little is known about the substrates of PP1 and its role in this pathway.

A Ser/Thr kinase, mammalian target of rapamycin (mTOR) directly phosphorylates and activates S6 Ser/Thr kinase (S6K) [53]. Ribosomal protein S6 (RPS6) is the first identified substrate of S6K, and the phosphorylation of RPS6 is involved in regulating mRNA translation [54]. The phosphorylation sites in RPS6 by S6K have been mapped to the five clustered residues, Ser235, Ser236, Ser240, Ser244, and Ser247 located at the C terminus [55]. The phosphorylation of the five clustered serines implies the affinity of RPS6 for the 5'-cap, and it is strongly suggested that RPS6 phosphorylation facilitates the initiation of mRNA translation [56]. The S6 phosphatase was sensitive to I-2 and specifically recognized by an antibody to PP1 [56].

RPS6 was identified as a common substrate of PP1 and ATM by specific antibody to the phospho-(Ser/Thr) ATM substrate and 5μM TC, specific PP1 inhibitor [57]. Although the crystal structure of PP1 bound to TM was determined [58], the co-crystal structure of PP1-TC has not yet been obtained. A conformational analysis of the PP1-TC complex was therefore performed using the crystal structure of the PP1-TM complex as a starting point as indicated in Fig. (**4A**).

The energy minimized structure shown in Fig. (**4B**) indicated that the C-6' carboxylic acid was directly bound to Arg96 and indirectly to Tyr272 *via* a water molecule. The C-16 hydroxy group was involved in the intramolecular hydrogen bond to the C-6' carboxylic acid, the 3'-hydroxy group was bound to Val250, and the C-14 carbonyl oxygen was bound to Asn124 and His125 *via* a water molecule [57]. The three-dimensional structure of the left-half moiety of TC was superimposed on that of TM to fit in the active site of PP1 as shown in Fig. (**4A** and **B**). The C-1-C-11 moiety of TC fitted in the amino acid residues Trp206, Val223, Ile133, Gly222, Ser129, and Cys127, and by the C-5 carbonyl oxygen which was fixed due to harmonic hydrogen bonds with Asp197 *via* two water molecules as indicated in Fig. (**4C**) [57]. The estimated binding energy of PP1/TC complex, 89.7 kcal/mol is extremely low in comparison with that of PP1/TM complex, 131.1 kcal/mol as indicated in Fig. (**4A** and **B**). Thus the potent inhibitory activity of TC against PP1 was verified by the theoretical calculations, too.

To search for common novel substrates of PP1 and ATM, the extract of HEK293T cells treated with TC were analyzed by immunoblotting with the α-phospho-ATM substrate antibody. Among several enhanced signals as candidate substrates detected with the TC treatment, the analysis was focused on a major band of 30 kDa. It is well known that S6K phosphorylates Ser235, Ser236, Ser240, Ser244, and Ser247. Importantly, Ser247 was present in the ATM recognition motif (SESSQK), and the two-dimensional gel electrophoretic analysis suggested that the Ser247 residue of RPS6 was a common phosphorylation site of the substrate for both PP1 and ATM [57].

Ubukata *et al*. proposed that intracellular ATM directly phosphorylates Ser247 of RPS6 as a DNA damage response and PP1 dephosphorylates RPS6 at phospho-Ser247 in response to UV irradiation. The modulation mechanism of RPS6 by switching the functional balance of PP1 and ATM is shown in Fig. (**5**).

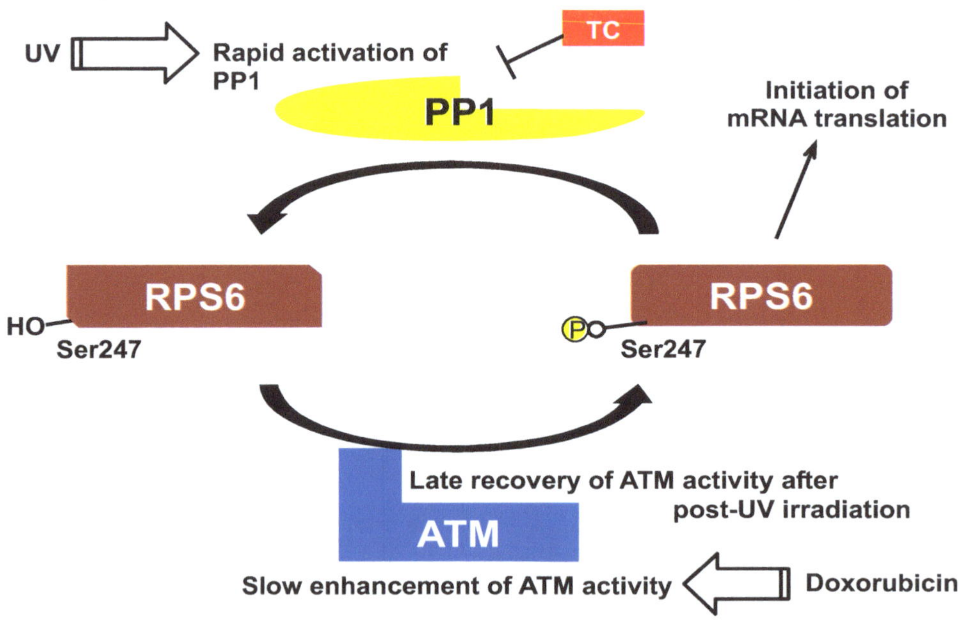

Fig. (5). Modulation mechanism of RPS6 by switching the functional balance of PP1 and ATM. Tautomycetin (TC) inhibits the rapid activation of PP1 by attenuating the impact of dephosphorylation of RPS6 at Ser247 post-UV irradiation. The genotoxic drug, doxorubicin slowly enhances the ATM activity leading to an accumulation of phopho-RPS6 at Ser247 [57].

It is noteworthy that the emergent activity of PP1 as a S6 phosphatase is stimulated *via* an ATM-independent pathway in the UV irradiation-mediated DNA damage response [57].

Tautomycetin (TC) as a Novel Immunosuppressant

Lee *et al.* identified TC as an effective immunosuppressant of activated T cells. In TC-treated rats transplanted with heterotopic cardiac allografts, the grafts survived more than 160 days as shown in Table **2** [10].

Table 2. ***In vivo*** **immunosuppressive effect of tautomycetin (TC) on graft survival after heterocardiac transplantation in rat [10].**

Drug	Dose (mg/ml)	Injection	Survival days	No. of rats
Cremophor-EL	5	i.p.	9	4
only	5	i.p.	10	3
CsA/cremophor-EL	5	i.p.	>100	2
TM/PBS	0.05	i.p.	10	3
TC/PBS	0.05	i.p.	>160	12
TC/ME	0.03	i.v.	>160	8

It was also observed that biochemical analysis of glutamic-oxaloacetic transaminase (GOT), glutamic-pyruvic transaminase (GPT), glucose, and creatinine levels in blood after administration of TC up to 5 mg/kg in rats for 30 days did not show any significant toxicity in liver or kidney. In contrast to TC, TM did not show any immunosuppressive effect in rat heart allograft. The immunosuppressive effect of TC was 100-fold more potent than cyclosporin A (CsA), increasing graft survival up to more than 160 days [10]. The combined use of TC as PP1 inhibitor and CsA as calcineurin inhibitor showed a remarkable potentiation in islet transplantation, a strong rejection model compared with cardiac organ transplantation [59]. Two papers [60, 61] demonstrated that PP1 was targeted to microtubules by the microtubule-associated protein τ, and associated with and activated the actin-binding protein cofilin in human T lymphocytes, and regulated actin polymerization through phosphorylation. Phosphorylation of cytoskeleton or its associated protein has been recognized to play an important role in T cell activation. Thus, these results strongly supported the inhibitory effect of TC on T cell activation, and indicated that TC might act on a PP1 holoenzyme with unknown interacting protein in T cell-specific manner [10].

A report suggested that Src-homology 2 domain-containing protein phosphatase 2 (SHP2) was a potential target for the immunosuppressive activity of TC [62]. TC and its analog TMC-D1 inhibited SHP2 with a similar IC_{50} values (TC: 2.9 μM, TMC-D1: 4.4 μM), whereas TMC-D2, another analog of TC, did not exhibit significant inhibitory activity against SHP2 [62, 63]. Inconsistent with this hypothesis, TC and TMC-D2 inhibited IL2 production at 0.2 μM and 20 μM, respectively, while TMC-D1 did not inhibit IL2 production in Jurkat T cells at 20 μM and did not show cytotoxicity against Jurkat T cells at 50 μM. These observations suggest that SHP2 is not a primary target of TC for immunosuppressive effect [63]. However, inhibition of SHP2 by TC might synergistically contribute to the remarkable effect of TC as immunosuppressant, because PP1 blocks the lysophosphatidic acid (LPA)-stimulated activity of SHP2, SHP2 tyrosine phosphorylation, and SHP2 association with growth factor receptor-bound protein 2 (Grb2) [64].

In contrast to TC, TM did not show any immunosuppressive activity in rat heart allograft [10]. This observation suggests that pharmacological modulation of PP1/PP2A balance is critical for the remarkable biological activities of TC *in vivo*. The right amount of TC specifically inhibits PP1 without complete inhibition of PP2A, whereas the right amount of OA specifically inhibits PP2A without complete inhibition of PP1 *in vivo* [8]. PP2A has been recognized to be a tumor suppressor in human cancers [65 - 70]. PP1, therefore, seems to be an attractive therapeutic target in comparison with PP2A for treating various human diseases [2, 3, 71 - 73].

Low Concentration TC Selectively Inhibits Growth of Colorectal Cancer Cells *via* $p21^{cip1}$ Induction

Choi *et al.* identified the efficacy of TC on cellular proliferation and transformation of colorectal cancer cells [9]. The proliferation and anchorage-independent growth of HCT-15, HT-29, and DLD-1 colorectal cancer cells were efficiently inhibited without induction of apoptosis by 150 nM of TC. These growth inhibitory effects were dependent on cyclin-dependent kinase inhibitor 1 ($p21^{Cip1}$) induction *via* the extracellular signal-regulated kinase pathway, and 150 nM TC did not inhibit cell proliferation of HCT-116 colon cancer, COS-7, HEK293, L929, NIH3T3, Chang, HeLa, MCF-7, and HepG2 cells [9]. The growth inhibitory effect of TC can be acquired by regulation of Raf-1 activity through inhibition of PP1 [8, 9].

Transcriptional factor Sp3 was shown to promote the binding of PP1 to the p21[Cip1] promoter, reducing phosphorylation of histone H3 at serine10 (H3S10). A finding was consistent with Sp3-dependent regulation of the local balance between kinase and phosphatase activities [74]. It was indicated that histone H3 phosphorylations at Ser10 (p-H3S10) and Ser28 (p-H3S28) are important for stress-induced responses and closely related to the transcriptional activation of immediate-early genes, such as *c-jun* and *c-fos*. 12-*O*-Tetradecanoylphorbol-13-acetate (TPA) induced histone H3 phosphorylation at Ser10 and Ser28 in HaCaT human keratinocyte cells. Activation of PI3K/Akt1 pathway with TPA contributes to induction of c-Jun and c-Fos: TPA directly induces phosphorylation of heterochromatin protein 1γ (HP1γ) and attenuates its suppressive function, permitting the binding of 14-3-3 protein to phosphorylated histone H3 [75].

Although ribosomal S6 kinase 2 (Rsk2), mitogen-and stress-activated kinase 1 (Msk1), IkB kinase α (IKKα), and aurora B kinase have been reported to be involved in phosphorylation of H3S10 [76 - 79], little is known about the role of PP1 and/or PP2A in dephosphorylation of H3S10. Ser10 is phosphorylated by aurora B in mitosis and dephosphorylated by PP1 that can dephosphorylate and inactivate aurora B [80 - 82]. The careful experiments using low concentration TC (specific inhibitor of PP1, 150 nM - 5μM) might provide further insight into the mechanism at the transcriptional level, because low concentration TC (PP1 specific inhibitor) or low concentration OA (PP2A specific inhibitor) can differentiate the function of two major serine/threonine phosphatases, PP1 or PP2A *in vivo* [8, 83], and a report showed that aurora B was activated by high concentration OA (400nM), whereas low concentration OA (100 nM) did not activate aurora B [82].

Tautomycetin (TC) Modulates LPS-induced Toll-like Receptor (TLR) Response *via* Inhibition of PP1/GADDA34 Holoenzyme

The role of PP1 has remained unclear in Toll-like receptor (TLR)-triggered immune response. Mice pretreated with 0.05 mg/kg of TC significantly enhanced the expression level of TNF-α in serum measured at 1.5 h after sublethal LPS injection. Consequently, all pretreated mice died within a week, whereas 62.5% of control mice survived, suggesting that inhibition of PP1 enhanced LPS-induced endotoxin shock [84]. PP1 inhibited the activation of MAPK and NF-κB pathways and production of TNF-α, IL-6 in macrophages triggered by TLR3, TLR4, and TLR9. PP1 knockdown increased TLR-mediated signaling and proinflammatory cytokine production. It seems that holoenzyme PP1/GADD34 tightly controls phosphorylation of TLR-induced TAK1 at Ser412 preventing excess activation of inflammatory responses [84]. Growth arrest and DNA damage-inducible protein GADD34 is also known as protein phosphatase 1 regulatory subunit 15A (PPP1R15A).

Selective Inhibitor of PP1/GADD34 Holoenzyme can Ameliorate Protein-misfolding Diseases, Amyotrophic Lateral Sclerosis (ALS) and Multiple Sclerosis (MS)

Amyotrophic lateral sclerosis (ALS) and multiple sclerosis (MS) are chronic neurodegenerating diseases. A selective inhibitor of PP1/GADD34, guanabenz, which inhibited dephosphorylation of eukaryotic initiation factor 2 (eIF2α), increased cellular protection from apoptosis caused by overwhelmed stress response. Though its FDA-approved status as antihypertensive drug would significantly accelerate the use of guanabenz in patient, sephin 1, another PP1/GADD34 inhibitor minimized its anti-hypertension side effects *in vivo*. Thus second-generation small molecular inhibitors of eIFα dephosphorylation would be tested in these inflammatory disorders [11, 85, 86].

Despite inhibiting global protein synthesis, p-eIFα preferentially stimulated translation of activating transcription factor 4 (ATF4). Induction of ATF4 increased the expression of certain cytoprotective genes. ATF4 also induced expression of CAAT/enhancer-binding protein homologous protein (CHOP), which in turn induced GADD34 expression. PP1/GADD34 holoenzyme specifically dephosphorylated p-eIF2α as indicated in Fig. (**6**) [86].

Sephin1, selective inhibitor of PP1/GADD34, specifically disrupted the PP1/GADD34 (also known as PPP1R15A) complex but did not affect the highly related PP1/PPP1R15B complex. As a result, sephin1 prolonged eIF2α phosphorylation after endoplasmic reticulum (ER) stress, delaying translation recovery. Consequently, sephin1 attenuated expression of stress-inducible gene products such as CHOP, a pro-apoptotic protein. ATF4 is selectively translated when eIF2α is phosphorylated [11], and this was prolonged in sephin1-treated cells. The inhibitor did not affect eIF2α signaling under non-stress conditions, probably because GADD34 is preferentially expressed under stress conditions. Sephin1 is a selective PP1/GADD34 inhibitor and has no adverse effects *in vivo*. Sephin1-treated mice (1–5 mg/kg) continued to run for 45 min in the rotarod performance test without any adverse effect on general health or memory [11]. Further analysis on potential side effect of PP1/GADD34 inhibitor might be necessary related to TLR-triggered immune response [84].

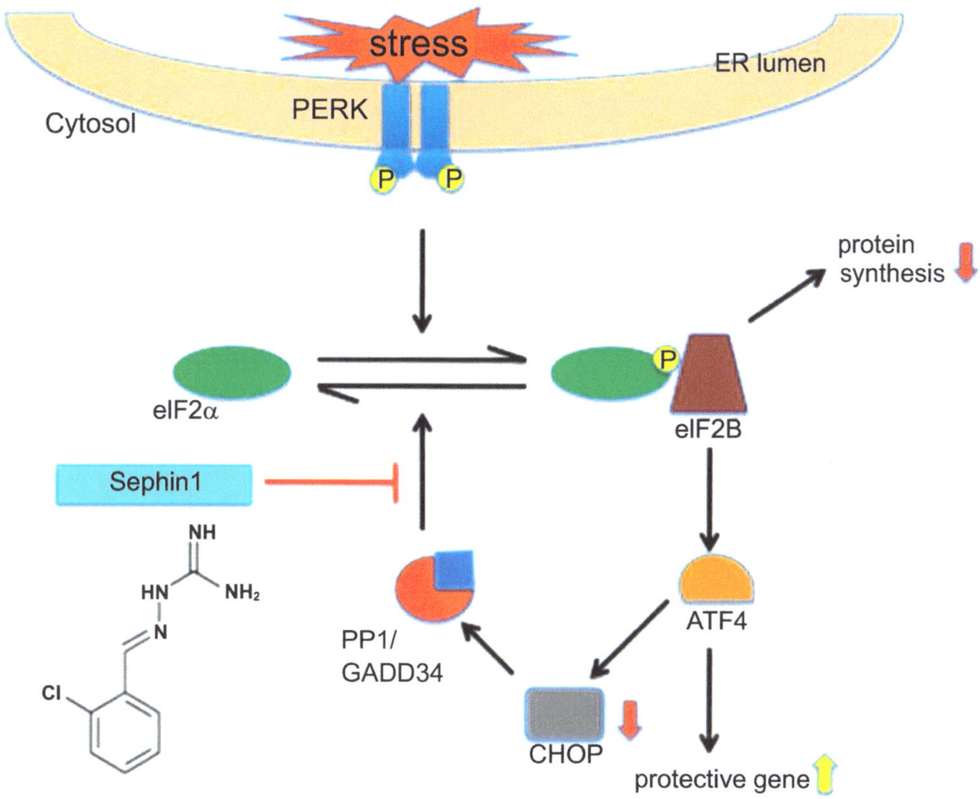

Fig. (6). Selective inhibitor of PP1/GADD34 modulates stress-induced PERK pathway. Sephin1 is an analog of guanabenz, which enhances the unfolded protein response and ameliorates mutant SOD-induced amyotrophic lateral sclerosis. Sephin1 selectively binds to protein phosphatase 1 regulatory subunit 15A, GADD34 [11, 86].

Is Protein Phosphatase 1 (PP1) a Regulator of Mammalian Circadian Clock?

ATF4 is a central transcription-factor of protein kinase RNA-like endoplasmic reticulum kinase (PERK) signaling pathway, and it is transcriptionally regulated by circadian locomotor output cycles kaput (CLOCK) [87]. Among the ATF/c-AMP-response element (CRE)-binding (CREB) family protein, ATF4 is essential for the circadian expression of the *Period2* (*Per2*) gene, a key component of the circadian clock [87]. The mammalian circadian clock system is organized hierarchically and the master pacemaker in the suprachiasmatic nucleus (SCN), which is responsible for controlling circadian rhythms in other brain regions and peripheral tissues [88, 89]. Transfection of *Atf4* siRNA into the SCN slices resulted in the reduction of *Atf4* mRNA decreasing *Per2* and increasing *Bmal1* at the mRNA levels. Bioluminescence oscillation driven by *Per2::Luc* in the SCN slice culture prepared from *Per2::Luc* knock-in mice was attenuated by transfection of *Atf4* siRNA. The amplitude of bioluminescence oscillation of Per2::Luc after transfection with *Atf4* siRNA was significantly lower than control siRNA treatment. These findings suggest that ATF4 acts as a component of mammalian circadian oscillator not only in peripheral cells but also in the center of mammalian circadian clock [87].

Transcription of the *Atf4* gene was controlled by the molecular components of the circadian clock. CLOCK/brain and muscle Arnt-like protein 1 (BMAL1) heterodimers transactivated the *Atf4* gene through the proximal E-box located near the transcriptional start site. CLOCK/BMAL1-mediated transactivation is repressed by PER and CRY proteins. PER and CER proteins form complexes and re-enter the nucleus, where they inhibit their own transcription as shown in Fig. (**7**). The time-dependent increase and decrease in *Atf4* mRNA levels seems to cause the circadian fluctuation of its protein expression, because ATF4 protein has a half-life of only 30 – 60 min [87].

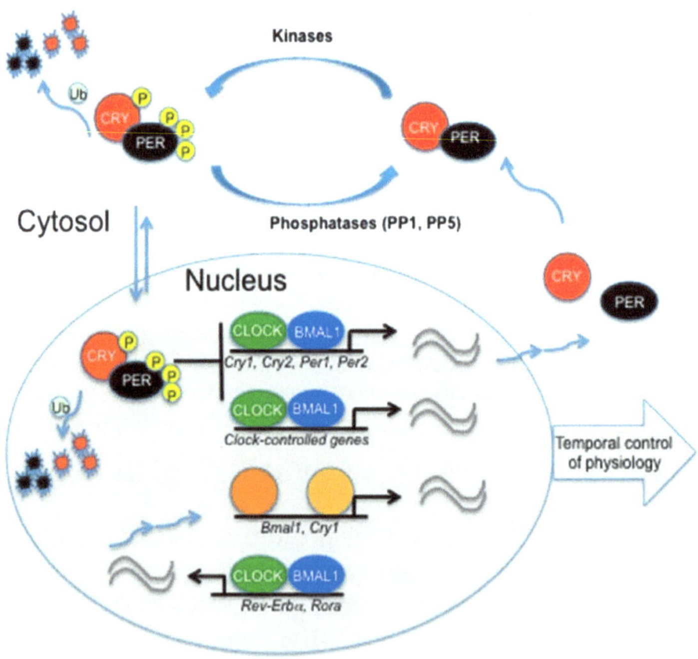

Fig. (7). Temporal control of physiology *via* integral clock proteins: two activators, CLOCK and BMAL1, and two repressors, PER and CRY. Abbreviations: CLOCK, circadian locomotor output cycles kaput; BMAL1, brain and muscle ARNT-like 1; CRY, cryptochrome; PER, period; PP, protein phosphatase [91].

Little is known about the role of PP1 on circadian rhythms, although PP1 appears to be implicated in the regulation of the subcellular localization of the PER2 protein. To study the function of PP1, PP1 activity was modulated in cells and in the brain of transgenic mice by inducible expression of the regulatory protein-1 (I-1) or nuclear inhibitor of PP1 (NIPP1). The experiments suggest that PP1 is a post-translational regulator of the mammalian circadian clock *in vivo* [90].

The casein kinase CKIδ/ε plays an important role in determining the intrinsic period of the clock by controlling the rate at which the PER/CRY complex is either degraded or enter the nucleus, and their activities are either counteracted or regulated by PP1 [92]. Calyculin A (CA), which inhibits PP1 and PP2A with almost equal potency [93], was used to determine the role of PP1 in the experiment. All CA treatment conditions that significantly shortened the circadian period also caused cell death and allowed only one cycle after the CA treatment in fibroblast and lung explants [92].

Instead of CA or OA, the use of low concentration of TC, pan-PP1 specific inhibitor, might provide further insight on the total function of PP1 in the mammalian circadian clock or other homeostasis *in vivo*, because substrate specificity and intracellular localization of PP1 are tightly regulated by the interaction with over 200 targeting and regulatory subunits [3].

Rescue of amyotrophic lateral sclerosis (ALS) phenotype has been realized by intravenous adeno-associated virus (AAV9)-adenosine deaminase acting on RNA 2 (ADAR2) delivery to motor neurons [94]. ALS is the most common adult-onset motor neuron disease, lacking effective therapy results in inevitable death within a few years of onset. Failure of GluA2 RNA editing resulting from down regulation of ADAR2 is observed in the majority of ALS cases, and it causes the death of motor neurons *via* Ca^{2+}-permeable AMPA receptor-mediated mechanism [94]. ADAR2 (also known as ADARB1) catalyzes circadian A-to-I editing and regulates RNA rhythm. *Adar2*-knockout mice exhibited short-period rhythms in locomotor activity and gene expression. These phenotypes were associated with abnormal accumulation of CRY2. Thus, A-to-I RNA editing was proposed to be a key mechanism of post-transcription in the circadian clockwork [95].

Protein phosphatase 1 (PP1) might be a therapeutic target for treating various diseases related to abnormality of mammalian circadian rhythms. Although current understanding of post-translational modification of the histone H3 tail suggests that phosphorylation (Ser10), acetylation (Lys9, Lys14), and methylation (Lys4, Lys27) are associated with circadian translation [96], little is known about the role to PP1 in this sense [80, 81].

Inhibition of nuclear PP1 in the mouse brain, or training on an object recognition task similarly increases expression of a cluster of microRNA (miRNA) comprising miR-183/96/182 in the hippocampus. Mimicking this increase by overexpression of miR-183/96/182 enhanced object memory, while knockdown of endogenous miR-183/96/182 impaired it. This effect involves the modulation of several plasticity-related genes, with *histone deacetylase 9* (*HDAC9*) identified as an important functional target of the miRNA. PP1 controls miR-183/96/182 in a transcription-independent manner *via* processing their precursors. These findings provided novel evidence for a role of miRNAs in memory formation and suggested participation of PP1 in miRNA processing in the adult brain [97].

Both miR-182 and HDAC9 have been implicated in schizophrenia, and *HDAC9* is one of the few genes with rare copy number variation in schizophrenia patients [98]. Further, disruption of hippocampal miR-182 signaling caused changes in gene expression observed in schizophrenia and other mental illness [99]. PP1 act as memory suppressors [100], and can regulate the expression of genes important for memory formation by influencing the epigenetic state of these genes *via* post-translational modifications of histone proteins [101 - 104]. PP1 is also known as an important regulator of gene transcription and ribonucleic acid (RNA) processing [32, 105]. Thus, spatiotemporal modulation of PP1 as a regulator of circadian clock might provide a new therapeutic option for treating some types of mental illness and chronic neurodegenerating diseases.

Table **3** summarizes selectivity, potency, and usefulness as chemical probe or potential for drug development of TC, OA, TM, CA, and sephin1, PP1 or PP2A inhibitors appeared in this chapter. Sephin1 is an interesting compound relating to an effective therapeutic approach for ALS or MS. Discoveries of the other PP1-holoenzyme inhibitors would be important for further understanding of the role of PP1 and the development of PP1-targeted therapeutics for neurological disorders. Tautomycetin (TC) is useful as an efficient positive control for PP1 inhibition in such developmental experiments. Reverse chemical genetics using TC, OA, and sephin1 have several unique advantages complementing genetic techniques. Unique properties of these chemical probes involve high rate of phenotypic change, high tunability, low cost of individual experiment, reversibility, and high developmental potential [106].

CONCLUSION

Tautomycetin (TC) has specific inhibitory activity against PP1 in two major Ser/Thr phosphatases (PP1, PP2A) and tyrosine phosphatases, and shows many interesting phenotypic changes *in vivo*. Even though TC or its analog has not been used yet as a therapeutic drug, low concentration TC may be a useful chemical probe for searching therapeutic strategies in epigenetic levels, which mean treating individual patients in Minkowski space, a combination of three-dimensional Euclidean space (x, y, z) and one dimension of time.

Table 3. Properties of PP1 or PP2A inhibitors appeared in this chapter.

Inhibitor	Selectivity	Potency	Usefulness
TC	++++	+++	++++ (chemical probe)
	(PP1)		++ (drug development)
OA	+++	++++	+++ (chemical probe)
	(PP2A)		– (drug development)
TM	+	+++	+ (chemical probe)
	(PP1>PP2A)		+ (drug development)
CA	–	++++	+ (chemical probe)
	(PP1, PP2A)		– (drug development)
Sephin1	+++	++	++ (chemical probe)
	(PP1/GADD34)		+++ (drug development)

+: positive, –: negative.

APPENDIX

Tautomycetin (TC): Specific inhibitor of PP1 at 150 nM [9] to 5 μM [8,40], [$C_{33}H_{50}O_{10}$; molecular weight (MW) 606]. Principal source: *Streptomyces griseochromogenes*, *Streptomyces* sp. TC is a polyketide small molecule obtained as a colorless liquid. Lyophilization may cause insolubilization of TC, because of the presence of conjugated dienone moiety in the molecule. It is readily soluble in dimethyl sulfoxide (DMSO), DMF, EtOH and many other organic solvents. Repackage into small sizes and desiccate at –20°C. Nitrogen or argon replacement might be useful for long-term storage. TC can be purchased from Tocris Bioscience (EI-243), WAKO Chemicals (209-12041), or R&D systems (2305).

Okadaic acid (OA): Specific inhibitor of PP2A at 2.5 nM [107] to 100 nM [8,40], [$C_{44}H_{68}O_{13}$; molecular weight (MW) 804]. Principal source: marine dinoflagellates, *Prorocentrum lima*, *Prorocentrum elegans*, and *Prorocentrum concavum*; also produced by several other species of *Dinophysis*. OA is a complex hydrophobic polyether obtained as a white crystalline solid. It is readily soluble in DMSO and a number of organic solvents, the recommended solvent is *N,N*-

dimethylformamide (DMF). When stored in DMF at –20°C, OA is stable for years (>5). OA can be purchased from Sigma-Aldrich (O8010, O4511, O7760), Abcam (ab120375), Santa Cruz Biotechnology (CAS 78111-17-8), or Cell Signaling Technology (#5934).

Tautomycin (TM): Inhibitor of PP1 and PP2A, [$C_{41}H_{65}O_{13}$; M.W. 766]. Principal source: *Streptomyces spiroverticillatus*. TM is a polyketide small molecule obtained as a lyophilized solid. It is soluble in DMSO, DMF, EtOH and many other organic solvents. TM is fairly stable and can be stored at –20°C as a lyophilized solid or in a concentrated DMSO or DMF solution. Dilute solutions in DMSO or DMF can be stored at –20°C. Repackage into small sizes and desiccate at –20°C. Nitrogen or argon replacement might need for long-term storage. TM can be purchased from BIOMOL (EI-243), WAKO Chemicals (209-12041), or Calbiochem (580551).

CONSENT FOR PUBLICATION

Not applicable.

CONFLICT OF INTEREST

The author (editor) declares no conflict of interest, financial or otherwise.

ACKNOWLEDGEMENT

The author would like to express the deepest appreciation to research collaborators in the Institute of Physical and Chemical Research (RIKEN), Toyama Prefectural University, Hokkaido University, and other institutions. The author would also like to express gratitude to The Japan Society for the Promotion of Science (JSPS), and the Ministry Education, Culture, Sports, Science and Technology (MEXT) for their financial support.

REFERENCES

[1] Wu P, Nielsen TE, Clausen MH. Small-molecule kinase inhibitors: an analysis of FDA-approved drugs. Drug Discov Today 2016; 21(1): 5-10. [http://dx.doi.org/10.1016/j.drudis.2015.07.008] [PMID: 26210956]

[2] Ceulemans H, Bollen M. Functional diversity of protein phosphatase-1, a cellular economizer and reset button. Physiol Rev 2004; 84(1): 1-39. [http://dx.doi.org/10.1152/physrev.00013.2003] [PMID: 14715909]

[3] Peti W, Nairn AC, Page R. Structural basis for protein phosphatase 1 regulation and specificity. FEBS J 2013; 280(2): 596-611. [http://dx.doi.org/10.1111/j.1742-4658.2012.08509.x] [PMID: 22284538]

[4] Janssens V, Goris J. Protein phosphatase 2A: a highly regulated family of serine/threonine phosphatases implicated in cell growth and signalling. Biochem J 2001; 353(Pt 3): 417-39. [http://dx.doi.org/10.1042/bj3530417] [PMID: 11171037]

[5] Aramburu J, Heitman J, Crabtree GR. Calcineurin: a central controller of signalling in eukaryotes. EMBO Rep 2004; 5(4): 343-8.
[http://dx.doi.org/10.1038/sj.embor.7400133] [PMID: 15060569]

[6] Alonso A, Sasin J, Bottini N, *et al.* Protein tyrosine phosphatases in the human genome. Cell 2004; 117(6): 699-711.
[http://dx.doi.org/10.1016/j.cell.2004.05.018] [PMID: 15186772]

[7] Boutros R, Dozier C, Ducommun B. The when and wheres of CDC25 phosphatases. Curr Opin Cell Biol 2006; 18(2): 185-91.
[http://dx.doi.org/10.1016/j.ceb.2006.02.003] [PMID: 16488126]

[8] Mitsuhashi S, Shima H, Tanuma N, *et al.* Usage of tautomycetin, a novel inhibitor of protein phosphatase 1 (PP1), reveals that PP1 is a positive regulator of Raf-1 *in vivo*. J Biol Chem 2003; 278(1): 82-8.
[http://dx.doi.org/10.1074/jbc.M208888200] [PMID: 12374792]

[9] Lee JH, Lee JS, Kim SE, *et al.* Tautomycetin inhibits growth of colorectal cancer cells through p21cip/WAF1 induction *via* the extracellular signal-regulated kinase pathway. Mol Cancer Ther 2006; 5(12): 3222-31.
[http://dx.doi.org/10.1158/1535-7163.MCT-06-0455] [PMID: 17172426]

[10] Shim JH, Lee HK, Chang EJ, *et al.* Immunosuppressive effects of tautomycetin *in vivo* and *in vitro via* T cell-specific apoptosis induction. Proc Natl Acad Sci USA 2002; 99(16): 10617-22.
[http://dx.doi.org/10.1073/pnas.162522099] [PMID: 12149481]

[11] Das I, Krzyzosiak A, Schneider K, *et al.* Preventing proteostasis diseases by selective inhibition of a phosphatase regulatory subunit. Science 2015; 348(6231): 239-42.
[http://dx.doi.org/10.1126/science.aaa4484] [PMID: 25859045]

[12] Cheng XC, Kihara T, Kusakabe H, *et al.* A new antibiotic, tautomycin. J Antibiot 1987; 40(6): 907-9.
[http://dx.doi.org/10.7164/antibiotics.40.907] [PMID: 3610839]

[13] Cheng XC, Kihara T, Ying X, *et al.* A new antibiotic, tautomycetin. J Antibiot 1989; 42(1): 141-4.
[http://dx.doi.org/10.7164/antibiotics.42.141] [PMID: 2921220]

[14] Ubukata M, Cheng XC, Isono K. The structure of tautomycin, a regulator of eukaryotic cell growth. J Chem Soc Chem Commun 1990; 244-6.
[http://dx.doi.org/10.1039/c39900000244]

[15] Cheng XC, Ubukata M, Isono K. The structure of tautomycin, a dialkylmaleic anhydride antibiotic. J Antibiot 1990; 43(7): 809-19.
[http://dx.doi.org/10.7164/antibiotics.43.809] [PMID: 2387775]

[16] Cheng XC, Ubukata M, Isono K. The structure of tautomycetin, a dialkylmaleic anhydride antibiotic. J Antibiot 1990; 43(7): 890-6.
[http://dx.doi.org/10.7164/antibiotics.43.890] [PMID: 2387780]

[17] Ubukata M, Cheng XC, Isobe M, Isono K. Absolute configuration of tautomycin, a protein phosphatase inhibitor from a Streptomycete. J Chem Soc, Perkin Trans 1 1993; 617-24.
[http://dx.doi.org/10.1039/p19930000617]

[18] Tachibana K, Scheuer PJ. Tsukitani, Y.; Kikuchi, H.; Engen, D.V.; Clardy, J.; Gopichand, Y.; Schmitz, F.J. Okadaic acid, a cytotoxic polyether from two marine sponges of the genus Halichondria. J Am Chem Soc 1981; 103: 2469-71.
[http://dx.doi.org/10.1021/ja00399a082]

[19] Bialojan C, Takai A. Inhibitory effect of a marine-sponge toxin, okadaic acid, on protein phosphatases. Specificity and kinetics. Biochem J 1988; 256(1): 283-90.
[http://dx.doi.org/10.1042/bj2560283] [PMID: 2851982]

[20] Ubukata M, Cheng XC, Magae J, Yamazaki M, Isono K. Tautomycin, a new antibiotic which

regulates eukaryotic cell growth. 27th Interscience Congress on Antimicrobial Agents and Chemotherapy. Houston. 1989.

[21] MacKintosh C, Klumpp S. Tautomycin from the bacterium Streptomyces verticillatus. Another potent and specific inhibitor of protein phosphatases 1 and 2A. FEBS Lett 1990; 277(1-2): 137-40. [http://dx.doi.org/10.1016/0014-5793(90)80828-7] [PMID: 2176611]

[22] Hori M, Magae J, Han YG, Hartshorne DJ, Karaki H. A novel protein phosphatase inhibitor, tautomycin. Effect on smooth muscle. FEBS Lett 1991; 285(1): 145-8. [http://dx.doi.org/10.1016/0014-5793(91)80745-O] [PMID: 1648511]

[23] Suganuma M, Okabe S, Sueoka E, *et al.* Tautomycin: an inhibitor of protein phosphatases 1 and 2A but not a tumor promoter on mouse skin and in rat glandular stomach. J Cancer Res Clin Oncol 1995; 121(9-10): 621-7. [http://dx.doi.org/10.1007/BF01197780] [PMID: 7559747]

[24] Kawamura T, Matsuzawa S, Mizuno Y, *et al.* Different moieties of tautomycin involved in protein phosphatase inhibition and induction of apoptosis. Biochem Pharmacol 1998; 55(7): 995-1003. [http://dx.doi.org/10.1016/S0006-2952(97)00539-X] [PMID: 9605423]

[25] Nishiyama U, Ubukata M, Magae J, *et al.* Structure-activity relationship within a series of degradation products of tautomycin. Biosci Biotechnol Biochem 1996; 60(1): 103-7. [http://dx.doi.org/10.1271/bbb.60.103] [PMID: 8824829]

[26] Sugiyama Y, Ohtani I, Isobe M, Takai A, Ubukata M, Isono K. Molecular shape analysis and activity of tautomycin, a protein phosphatase inhibitor. Bioorg Med Chem Lett 1996; 6: 3-8. [http://dx.doi.org/10.1016/0960-894X(95)00546-6]

[27] Mitsuhashi S, Matsuura N, Ubukata M, Oikawa H, Shima H, Kikuchi K. Tautomycetin is a novel and specific inhibitor of serine/threonine protein phosphatase type 1, PP1. Biochem Biophys Res Commun 2001; 287(2): 328-31. [http://dx.doi.org/10.1006/bbrc.2001.5596] [PMID: 11554729]

[28] Shenolikar S. Protein serine/threonine phosphatases--new avenues for cell regulation. Annu Rev Cell Biol 1994; 10: 55-86. [http://dx.doi.org/10.1146/annurev.cb.10.110194.000415] [PMID: 7888183]

[29] Ceulemans H, Stalmans W, Bollen M. Regulator-driven functional diversification of protein phosphatase-1 in eukaryotic evolution. BioEssays 2002; 24(4): 371-81. [http://dx.doi.org/10.1002/bies.10069] [PMID: 11948623]

[30] Sasaki K, Shima H, Kitagawa Y, Irino S, Sugimura T, Nagao M. Identification of members of the protein phosphatase 1 gene family in the rat and enhanced expression of protein phosphatase 1 alpha gene in rat hepatocellular carcinomas. Jpn J Cancer Res 1990; 81(12): 1272-80. [http://dx.doi.org/10.1111/j.1349-7006.1990.tb02690.x] [PMID: 2177460]

[31] Shima H, Hatano Y, Chun YS, *et al.* Identification of PP1 catalytic subunit isotypes PP1γ1, PP1δ and PP1α in various rat tissues. Biochem Biophys Res Commun 1993; 192: 1289-96. [http://dx.doi.org/10.1006/bbrc.1993.1556] [PMID: 7685164]

[32] Cohen PT. Protein phosphatase 1--targeted in many directions. J Cell Sci 2002; 115(Pt 2): 241-56. [PMID: 11839776]

[33] Deng X, Ito T, Carr B, Mumby M, May WS Jr. Reversible phosphorylation of Bcl2 following interleukin 3 or bryostatin 1 is mediated by direct interaction with protein phosphatase 2A. J Biol Chem 1998; 273(51): 34157-63. [http://dx.doi.org/10.1074/jbc.273.51.34157] [PMID: 9852076]

[34] Zhang W, Liu HT. MAPK signal pathways in the regulation of cell proliferation in mammalian cells. Cell Res 2002; 12(1): 9-18. [http://dx.doi.org/10.1038/sj.cr.7290105] [PMID: 11942415]

[35] Munshi A, Ramesh R. Mitogen-activated protein kinases and their role in radiation response. Genes

Cancer 2013; 4(9-10): 401-8.
[http://dx.doi.org/10.1177/1947601913485414] [PMID: 24349638]

[36] Keyse SM. Protein phosphatases and the regulation of mitogen-activated protein kinase signalling. Curr Opin Cell Biol 2000; 12(2): 186-92.
[http://dx.doi.org/10.1016/S0955-0674(99)00075-7] [PMID: 10712927]

[37] Eichhorn PJA, Creyghton MP, Bernards R. Protein phosphatase 2A regulatory subunits and cancer. Biochim Biophys Acta 2009; 1795: 1-15.

[38] Suganuma M, Fujiki H, Suguri H, *et al.* Okadaic acid: an additional non-phorbol-12-tetradecanote-13-acetate-type tumor promoter. Proc Natl Acad Sci USA 1988; 85(6): 1768-71.
[http://dx.doi.org/10.1073/pnas.85.6.1768] [PMID: 3126494]

[39] Fujiki H, Suganuma M. Tumor promotion by inhibitors of protein phosphatases 1 and 2A: the okadaic acid class of compounds. Adv Cancer Res 1993; 61: 143-94.
[http://dx.doi.org/10.1016/S0065-230X(08)60958-6] [PMID: 8394044]

[40] Mitsuhashi S, Shima H, Li Y, *et al.* Tautomycetin suppresses the TNFα/NF-kB pathway *via* inhibition of IKK activation. Int J Oncol 2008; 33: 1027-35.
[PMID: 18949366]

[41] Sontag E, Sontag JM, Garcia A. Protein phosphatase 2A is a critical regulator of protein kinase C zeta signaling targeted by SV40 small t to promote cell growth and NF-kappaB activation. EMBO J 1997; 16(18): 5662-71.
[http://dx.doi.org/10.1093/emboj/16.18.5662] [PMID: 9312025]

[42] Shanley TP, Vasi N, Denenberg A, Wong HR. The serine/threonine phosphatase, PP2A: endogenous regulator of inflammatory cell signaling. J Immunol 2001; 166(2): 966-72.
[http://dx.doi.org/10.4049/jimmunol.166.2.966] [PMID: 11145674]

[43] Kim HJ. Hawke, N.; Baldwin, A.S. NF-kB and IKK as therapeutic targets in cancer. Cell Death Differ 2006; 13: 738-47.
[http://dx.doi.org/10.1038/sj.cdd.4401877] [PMID: 16485028]

[44] Jackson-Bernitsas DG, Ichikawa H, Takada Y, *et al.* Evidence that TNF-TNFR1-TRADD-TRF2-RIP-TAK1-IKK pathway mediates constitutive NF-kappaB activation and proliferation in human head and neck squamous cell carcinoma. Oncogene 2007; 26(10): 1385-97.
[http://dx.doi.org/10.1038/sj.onc.1209945] [PMID: 16953224]

[45] Ahmad R, Raina D, Trivedi V, *et al.* MUC1 oncoprotein activates the IkB kinase β complex and constitutive NF-kB signaling. Nat Cell Biol 2007; 9: 1419-27.
[http://dx.doi.org/10.1038/ncb1661] [PMID: 18037881]

[46] Lee DF, Kuo HP, Chen CT, *et al.* IKK beta suppression of TSC1 links inflammation and tumor angiogenesis *via* the mTOR pathway. Cell 2007; 130(3): 440-55.
[http://dx.doi.org/10.1016/j.cell.2007.05.058] [PMID: 17693255]

[47] Derheimer FA, Kastan MB. Multiple roles of ATM in monitoring and maintaining DNA integrity. FEBS Lett 2010; 584(17): 3675-81.
[http://dx.doi.org/10.1016/j.febslet.2010.05.031] [PMID: 20580718]

[48] Goodarzi AA, Jonnalagadda JC, Douglas P, *et al.* Autophosphorylation of ataxia-telangiectasia mutated is regulated by protein phosphatase 2A. EMBO J 2004; 23(22): 4451-61.
[http://dx.doi.org/10.1038/sj.emboj.7600455] [PMID: 15510216]

[49] Peng A, Maller JL. Serine/threonine phosphatases in the DNA damage response and cancer. Oncogene 2010; 29(45): 5977-88.
[http://dx.doi.org/10.1038/onc.2010.371] [PMID: 20838380]

[50] Shreeram S, Demidov ON, Hee WK, *et al.* Wip1 phosphatase modulates ATM-dependent signaling pathways. Mol Cell 2006; 23(5): 757-64.
[http://dx.doi.org/10.1016/j.molcel.2006.07.010] [PMID: 16949371]

[51] Matsuoka S, Ballif BA, Smogorzewska A, *et al.* ATM and ATR substrate analysis reveals extensive protein networks responsive to DNA damage. Science 2007; 316(5828): 1160-6. [http://dx.doi.org/10.1126/science.1140321] [PMID: 17525332]

[52] Lu X, Nguyen TA, Moon SH, Darlington Y, Sommer M, Donehower LA. The type 2C phosphatase Wip1: an oncogenic regulator of tumor suppressor and DNA damage response pathways. Cancer Metastasis Rev 2008; 27(2): 123-35. [http://dx.doi.org/10.1007/s10555-008-9127-x] [PMID: 18265945]

[53] Fenton TR, Gout IT. Functions and regulation of the 70kDa ribosomal S6 kinases. Int J Biochem Cell Biol 2011; 43(1): 47-59. [http://dx.doi.org/10.1016/j.biocel.2010.09.018] [PMID: 20932932]

[54] Belandia B, Brautigan D, Martín-Pérez J. Attenuation of ribosomal protein S6 phosphatase activity in chicken embryo fibroblasts transformed by Rous sarcoma virus. Mol Cell Biol 1994; 14(1): 200-6. [http://dx.doi.org/10.1128/MCB.14.1.200] [PMID: 8264587]

[55] Hutchinson JA, Shanware NP, Chang H, Tibbetts RS. Regulation of ribosomal protein S6 phosphorylation by casein kinase 1 and protein phosphatase 1. J Biol Chem 2011; 286(10): 8688-96. [http://dx.doi.org/10.1074/jbc.M110.141754] [PMID: 21233202]

[56] Ruvinsky I, Meyuhas O. Ribosomal protein S6 phosphorylation: from protein synthesis to cell size. Trends Biochem Sci 2006; 31(6): 342-8. [http://dx.doi.org/10.1016/j.tibs.2006.04.003] [PMID: 16679021]

[57] Li Y, Mitsuhashi S, Ikejo M, *et al.* Relationship between ATM and ribosomal protein S6 revealed by the chemical inhibition of Ser/Thr protein phosphatase type 1. Biosci Biotechnol Biochem 2012; 76(3): 486-94. [http://dx.doi.org/10.1271/bbb.110774] [PMID: 22451389]

[58] Kelker MS, Page R, Peti W. Crystal structures of protein phosphatase-1 bound to nodularin-R and tautomycin: a novel scaffold for structure-based drug design of serine/threonine phosphatase inhibitors. J Mol Biol 2009; 385(1): 11-21. [http://dx.doi.org/10.1016/j.jmb.2008.10.053] [PMID: 18992256]

[59] Han DJ, Jeong YL, Wee YM, *et al.* Tautomycetin as a novel immunosuppressant in transplantation. Transplant Proc 2003; 35(1): 547. [http://dx.doi.org/10.1016/S0041-1345(02)03905-2] [PMID: 12591525]

[60] Liao H, Li Y, Brautigan DL, Gundersen GG. Protein phosphatase 1 is targeted to microtubules by the microtubule-associated protein Tau. J Biol Chem 1998; 273(34): 21901-8. [http://dx.doi.org/10.1074/jbc.273.34.21901] [PMID: 9705329]

[61] Ambach A, Saunus J, Konstandin M, Wesselborg S, Meuer SC, Samstag Y. The serine phosphatases PP1 and PP2A associate with and activate the actin-binding protein cofilin in human T lymphocytes. Eur J Immunol 2000; 30(12): 3422-31. [http://dx.doi.org/10.1002/1521-4141(2000012)30:12<3422::AID-IMMU3422>3.0.CO;2-J] [PMID: 11093160]

[62] Liu S, Yu Z, Yu X, *et al.* SHP2 is a target of the immunosuppressant tautomycetin. Chem Biol 2011; 18(1): 101-10. [http://dx.doi.org/10.1016/j.chembiol.2010.10.015] [PMID: 21276943]

[63] Niu M, Sun Y, Liu B, Tang L, Qiu R. Differential effects of tautomycetin and its derivatives on protein phosphatase inhibition, immunosuppressive function and antitumor activity. Korean J Physiol Pharmacol 2012; 16(2): 145-51. [http://dx.doi.org/10.4196/kjpp.2012.16.2.145] [PMID: 22563261]

[64] Tang H, Zhao ZJ, Huang XY, Landon EJ, Inagami T. Fyn kinase-directed activation of SH2 domain-containing protein-tyrosine phosphatase SHP-2 by Gi protein-coupled receptors in Madin-Darby canine kidney cells. J Biol Chem 1999; 274(18): 12401-7.

[http://dx.doi.org/10.1074/jbc.274.18.12401] [PMID: 10212213]

[65] Van Hoof C, Goris J. PP2A fulfills its promises as tumor suppressor: which subunits are important? Cancer Cell 2004; 5(2): 105-6.
[http://dx.doi.org/10.1016/S1535-6108(04)00027-3] [PMID: 14998482]

[66] Eichhorn PJA, Creyghton MP, Bernards R. Protein phosphatase 2A regulatory subunits and cancer. Biochim Biophys Acta 2009; 1795: 1-15.

[67] Janssens V, Longin S, Goris J. PP2A holoenzyme assembly: in cauda venenum (the sting is in the tail). Trends Biochem Sci 2008; 33(3): 113-21.
[http://dx.doi.org/10.1016/j.tibs.2007.12.004] [PMID: 18291659]

[68] Mumby M. PP2A: unveiling a reluctant tumor suppressor. Cell 2007; 130(1): 21-4.
[http://dx.doi.org/10.1016/j.cell.2007.06.034] [PMID: 17632053]

[69] Westermarck J, Hahn WC. Multiple pathways regulated by the tumor suppressor PP2A in transformation. Trends Mol Med 2008; 14(4): 152-60.
[http://dx.doi.org/10.1016/j.molmed.2008.02.001] [PMID: 18329957]

[70] Chen W, Possemato R, Campbell KT, Plattner CA, Pallas DC, Hahn WC. Identification of specific PP2A complexes involved in human cell transformation. Cancer Cell 2004; 5(2): 127-36.
[http://dx.doi.org/10.1016/S1535-6108(04)00026-1] [PMID: 14998489]

[71] Connor JH, Weiser DC, Li S, Hallenbeck JM, Shenolikar S. Growth arrest and DNA damage-inducible protein GADD34 assembles a novel signaling complex containing protein phosphatase 1 and inhibitor 1. Mol Cell Biol 2001; 21(20): 6841-50.
[http://dx.doi.org/10.1128/MCB.21.20.6841-6850.2001] [PMID: 11564868]

[72] Bollen M, Peti W, Ragusa MJ, Beullens M. The extended PP1 toolkit: designed to create specificity. Trends Biochem Sci 2010; 35(8): 450-8.
[http://dx.doi.org/10.1016/j.tibs.2010.03.002] [PMID: 20399103]

[73] O'Connell N, Nichols SR, Heroes E, *et al.* The molecular basis for substrate specificity of the nuclear NIPP1:PP1 holoenzyme. Structure 2012; 20(10): 1746-56.
[http://dx.doi.org/10.1016/j.str.2012.08.003] [PMID: 22940584]

[74] Valin A, Ouyang J, Gill G. Transcription factor Sp3 represses expression of p21CIP[1]*via* inhibition of productive elongation by RNA polymerase II. Mol Cell Biol 2013; 33(8): 1582-93.
[http://dx.doi.org/10.1128/MCB.00323-12] [PMID: 23401853]

[75] Choi WJ. The heterochoromatin-1 phosphorylation contributes to TPA-induced AP-1 expression. Biomol Ther (Seoul) 2014; 22(4): 308-13.
[http://dx.doi.org/10.4062/biomolther.2014.057] [PMID: 25143809]

[76] Sassone-Corsi P, Mizzen CA, Cheung P, *et al.* Requirement of Rsk-2 for epidermal growth factor-activated phosphorylation of histone H3. Science 1999; 285(5429): 886-91.
[http://dx.doi.org/10.1126/science.285.5429.886] [PMID: 10436156]

[77] Thomson S, Clayton AL, Hazzalin CA, Rose S, Barratt MJ, Mahadevan LC. The nucleosomal response associated with immediate-early gene induction is mediated *via* alternative MAP kinase cascades: MSK1 as a potential histone H3/HMG-14 kinase. EMBO J 1999; 18(17): 4779-93.
[http://dx.doi.org/10.1093/emboj/18.17.4779] [PMID: 10469656]

[78] Yamamoto Y, Verma UN, Prajapati S, Kwak YT, Gaynor RB. Histone H3 phosphorylation by IKK-alpha is critical for cytokine-induced gene expression. Nature 2003; 423(6940): 655-9.
[http://dx.doi.org/10.1038/nature01576] [PMID: 12789342]

[79] MacCallum DE, Losada A, Kobayashi R, Hirano T. ISWI remodeling complexes in Xenopus egg extracts: identification as major chromosomal components that are regulated by INCENP-aurora B. Mol Biol Cell 2002; 13(1): 25-39.
[http://dx.doi.org/10.1091/mbc.01-09-0441] [PMID: 11809820]

[80] Murnion ME, Adams RR, Callister DM, Allis CD, Earnshaw WC, Swedlow JR. Chromatin-associated protein phosphatase 1 regulates aurora-B and histone H3 phosphorylation. J Biol Chem 2001; 276(28): 26656-65.
[http://dx.doi.org/10.1074/jbc.M102288200] [PMID: 11350965]

[81] Prigent C, Dimitrov S. Phosphorylation of serine 10 in histone H3, what for? J Cell Sci 2003; 116(Pt 18): 3677-85.
[http://dx.doi.org/10.1242/jcs.00735] [PMID: 12917355]

[82] Sugiyama K, Sugiura K, Hara T, *et al.* Aurora-B associated protein phosphatases as negative regulators of kinase activation. Oncogene 2002; 21(20): 3103-11.
[http://dx.doi.org/10.1038/sj.onc.1205432] [PMID: 12082625]

[83] Mitsuhashi S, Shima H, Tanuma N, *et al.* Protein phosphatase type 2A, PP2A, is involved in degradation of gp130. Mol Cell Biochem 2005; 269(1-2): 183-7.
[http://dx.doi.org/10.1007/s11010-005-3089-x] [PMID: 15786731]

[84] Gu M, Ouyang C, Lin W, *et al.* Phosphatase holoenzyme PP1/GADD34 negatively regulates TLR response by inhibiting TAK1 serine 412 phosphorylation. J Immunol 2014; 192(6): 2846-56.
[http://dx.doi.org/10.4049/jimmunol.1302537] [PMID: 24534530]

[85] Wang L, Popko B, Tixier E, Roos RP. Guanabenz, which enhances the unfolded protein response, ameliorates mutant SOD1-induced amyotrophic lateral sclerosis. Neurobiol Dis 2014; 71: 317-24.
[http://dx.doi.org/10.1016/j.nbd.2014.08.010] [PMID: 25134731]

[86] Stone S, Lin W. The unfolded protein response in multiple sclerosis. Front Neurosci 2015; 9: 264.
[http://dx.doi.org/10.3389/fnins.2015.00264] [PMID: 26283904]

[87] Koyanagi S, Hamadan AM, Horiguchi M, *et al.* cAMP-response element (CRE)-mediated transcription by activating transcription by activating transcription factor-4 (ATF4) is essential for circadian expression of the Period2 gene. J Biol Chem 2011; 286: 32416-23.
[http://dx.doi.org/10.1074/jbc.M111.258970] [PMID: 21768648]

[88] Obrietan K, Impey S, Smith D, Athos J, Storm DR. Circadian regulation of cAMP response element-mediated gene expression in the suprachiasmatic nuclei. J Biol Chem 1999; 274(25): 17748-56.
[http://dx.doi.org/10.1074/jbc.274.25.17748] [PMID: 10364217]

[89] O'Neill JS, Maywood ES, Chesham JE, Takahashi JS, Hastings MH. cAMP-dependent signaling as a core component of the mammalian circadian pacemaker. Science 2008; 320(5878): 949-53.
[http://dx.doi.org/10.1126/science.1152506] [PMID: 18487196]

[90] Schmutz I, Wendt S, Schnell A, Kramer A, Mansuy IM, Albrecht U. Protein phosphatase 1 (PP1) is a post-translational regulator of the mammalian circadian clock. PLoS One 2011; 6(6): e21325.
[http://dx.doi.org/10.1371/journal.pone.0021325] [PMID: 21712997]

[91] Partch CL, Green CB, Takahashi JS. Molecular architecture of the mammalian circadian clock. Trends Cell Biol 2014; 24(2): 90-9.
[http://dx.doi.org/10.1016/j.tcb.2013.07.002] [PMID: 23916625]

[92] Lee HM, Chen R, Kim H, Etchegaray JP, Weaver DR, Lee C. The period of the circadian oscillator is primarily determined by the balance between casein kinase 1 and protein phosphatase 1. Proc Natl Acad Sci USA 2011; 108(39): 16451-6.
[http://dx.doi.org/10.1073/pnas.1107178108] [PMID: 21930935]

[93] Swingle M, Ni L, Honkanen RE. Small-molecule inhibitors of ser/thr protein phosphatases: specificity, use and common forms of abuse. Methods Mol Biol 2007; 365: 23-38.
[PMID: 17200551]

[94] Yamashita T, Chai HL, Teramoto S, *et al.* Rescue of amyotrophic lateral sclerosis phenotype in a mouse model by intravenous AAV9-ADAR2 delivery to motor neurons. EMBO Mol Med 2013; 5(11): 1710-9.
[http://dx.doi.org/10.1002/emmm.201302935] [PMID: 24115583]

[95] Terajima H, Yoshitane H, Ozaki H, *et al.* ADARB1 catalyzes circadian A-to-I editing and regulates RNA rhythm. Nat Genet 2017; 49(1): 146-51.
[http://dx.doi.org/10.1038/ng.3731] [PMID: 27893733]

[96] Masri S, Sassone-Corsi P. Plasticity and specificity of the circadian epigenome. Nat Neurosci 2010; 13(11): 1324-9.
[http://dx.doi.org/10.1038/nn.2668] [PMID: 20975756]

[97] Woldemichael BT, Jawaid A, Kremer EA, *et al.* The microRNA cluster miR-183/96/182 contributes to long-term memory in a protein phosphatase 1-dependent manner. Nat Commun 2016; 7: 12594.
[http://dx.doi.org/10.1038/ncomms12594] [PMID: 27558292]

[98] Tam GW, van de Lagemaat LN, Redon R, *et al.* Confirmed rare copy number variants implicate novel genes in schizophrenia. Biochem Soc Trans 2010; 38(2): 445-51.
[http://dx.doi.org/10.1042/BST0380445] [PMID: 20298200]

[99] Kohen R, Dobra A, Tracy JH, Haugen E. Transcriptome profiling of human hippocampus dentate gyrus granule cells in mental illness. Transl Psychiatry 2014; 4: e366.
[http://dx.doi.org/10.1038/tp.2014.9] [PMID: 24594777]

[100] Mansuy IM, Shenolikar S. Protein serine/threonine phosphatases in neuronal plasticity and disorders of learning and memory. Trends Neurosci 2006; 29(12): 679-86.
[http://dx.doi.org/10.1016/j.tins.2006.10.004] [PMID: 17084465]

[101] Genoux D, Haditsch U, Knobloch M, Michalon A, Storm D, Mansuy IM. Protein phosphatase 1 is a molecular constraint on learning and memory. Nature 2002; 418(6901): 970-5.
[http://dx.doi.org/10.1038/nature00928] [PMID: 12198546]

[102] Gräff J, Koshibu K, Jouvenceau A, Dutar P, Mansuy IM. Protein phosphatase 1-dependent transcriptional programs for long-term memory and plasticity. Learn Mem 2010; 17(7): 355-63.
[http://dx.doi.org/10.1101/lm.1766510] [PMID: 20592054]

[103] Koshibu K, Gräff J, Beullens M, *et al.* Protein phosphatase 1 regulates the histone code for long-term memory. J Neurosci 2009; 29(41): 13079-89.
[http://dx.doi.org/10.1523/JNEUROSCI.3610-09.2009] [PMID: 19828821]

[104] Gräff J, Woldemichael BT, Berchtold D, Dewarrat G, Mansuy IM. Dynamic histone marks in the hippocampus and cortex facilitate memory consolidation. Nat Commun 2012; 3: 991.
[http://dx.doi.org/10.1038/ncomms1997] [PMID: 22871810]

[105] Bennett D. Transcriptional control by chromosome-associated protein phosphatase-1. Biochem Soc Trans 2005; 33(Pt 6): 1444-6.
[http://dx.doi.org/10.1042/BST0331444] [PMID: 16246142]

[106] Breinbauer R, Hillisch A, Wakdmann H. Reverse chemical genetics revisited.Chemical Biology From Small Molecules to System Biology and Drug Design. Weinheim: WILEY-VCH Verlag GmbH & Co. KGaA 2007; Vol. 1: pp. 355-84.

[107] Ferrer I, Blanco-Aparicio C, Peregrina S, *et al.* Spinophilin acts as a tumor suppressor by regulating Rb phosphorylation. Cell Cycle 2011; 10(16): 2751-62.
[http://dx.doi.org/10.4161/cc.10.16.16422] [PMID: 21772120]

CHAPTER 7

Phosphodiesterases as Drug Targets in Selected Respiratory Diseases

Juraj Mokry[1,2,*] and **Daniela Mokra**[2,3]

[1] *Department of Pharmacology, Jessenius Faculty of Medicine in Martin, Comenius University in Bratislava, Slovakia*

[2] *Biomedical Center Martin, Jessenius Faculty of Medicine in Martin, Comenius University in Bratislava, Slovakia*

[3] *Department of Physiology, Jessenius Faculty of Medicine in Martin, Comenius University in Bratislava, Slovakia*

Abstract: Airway diseases associated with inflammation and airway obstruction are known for an accumulation of inflammatory blood cells in the lung tissue and airways and for a typical inflammatory symptomatology. An increased leak of inflammatory cells into the target tissue and their activation prolong the survival of pro-inflammatory cells. The complex cell interactions in the airways play an important role in the development of chronic inflammation, *e.g.* in bronchial asthma (BA). The inflammation in BA is directed by the activation of Th_2 lymphocytes that produce various inflammatory cytokines (interleukins IL-4, IL-13, and IL-5), eotaxin *etc.* In chronic obstructive pulmonary disease (COPD), an involvement of alveolar macrophages and neutrophils with the release of IL-8, neutrophil elastase, tumor necrosis factor (TNF)-α and myeloperoxidase has been confirmed.

Currently, 11 families of PDEs are known. In contrast to non-selective PDE inhibitors (*e.g.* theophylline), selective PDE inhibitors are able to act with a significantly reduced risk of adverse effects. Several groups of selective PDE inhibitors, *i.e.* PDE3, PDE4, PDE5, and PDE7, exert a possible relationship to immune responses in the airways. The activity of PDE3 in the airways has been confirmed in alveolar macrophages, endothelial cells, platelets and airway smooth muscle. Simultaneous administration of siguazodan (PDE3 inhibitor) and rolipram (PDE4 inhibitor) has led to enhanced relaxing response, suggesting a synergism and potential benefit in parallel inhibition of PDE3 and PDE4. The PDE4 enzyme family is expressed in various inflammatory cells; therefore, its inhibition has shown anti-inflammatory and immunomodulating effects. Rolipram suppresses neutrophil- and eosinophil-mediated inflammation, secretion of interleukins and activation of pro-inflammatory cells. Agents from the second generation of PDE4 inhibitors (*e.g.* roflumilast approved for clinical use in COPD patients) are able to suppress airway hyperresponsiveness, release of TNF-α, fibrotisation and remodeling of the lung parenchyma. An inhibition of PDE7 can

[*] **Corresponding author Juraj Mokrý:** Biomedical Center Martin and Department of Pharmacology, Jessenius Faculty of Medicine in Martin, Comenius University in Bratislava, Mala Hora 4C, 03601 Martin, Slovakia; Tel: +421 43 2633619; E-mail: mokry@jfmed.uniba.sk

Atta-ur-Rahman & M. Iqbal Choudhary (Eds.)

contribute to anti-inflammatory action as its expression has been confirmed in T-cells, epithelial cells, airway and vascular smooth muscle cells, lung fibroblasts, and eosinophils. While their effectiveness in individual administration is relatively weak, the simultaneous PDE4 and PDE7 inhibition seems to be promising. The anti-inflammatory activity of dual PDE inhibitors (*e.g.* PDE3/PDE4, or PDE4/PDE7) has been studied only in COPD conditions, and only in limited extent in allergic BA. An inhibition of PDE5 leads to decreased pulmonary vascular resistance, suppressed pulmonary hypertension and vascular remodeling. Sildenafil (PDE5 inhibitor) inhibits cell proliferation, has pro-apoptotic action, and suppresses the bronchial hyperresponsiveness, infiltration by white blood cells, and concentrations of exhaled NO. These findings suggest the possible benefits of combining the PDE4 and PDE5 inhibitors in the therapy of BA and COPD. There are several studies with roflumilast and cilomilast already conducted in patients with BA; however, there is a lack of relevant data from experimental and clinical studies on simultaneous effects of PDE4 and PDE5, PDE3 and PDE4, and PDE4 and PDE7 up to now.

Acute lung injury (ALI) represents another disorder often associated with the inflammatory changes. ALI arises from various reasons: a direct injury to the lungs by aspiration of the gastric content, by aspiration of neonatal meconium, by pneumonia, or by inhalation of toxic gases. In addition, ALI is generated indirectly as a consequence of serious systemic injury, *e.g.* in sepsis, severe trauma with shock, or pancreatitis. These situations may progress into a more severe form of ALI, acute respiratory distress syndrome (ARDS). Several PDE inhibitors, including non-selective PDE inhibitor aminophylline, have been used in various animal models of ALI/ARDS. In experiments, the selective PDE3, PDE4, and PDE5 inhibitors have demonstrated potent anti-inflammatory, anti-oxidative, anti-edematous and vasodilation effects associated with lower occurrence of side effects than observed with non-selective PDE inhibitors, as their influence is limited to one type of PDE only. However, their use in these disorders is still under research.

Keywords: Phosphodiesterase, Non-Selective PDE Inhibitors, Theophylline, Selective PDE Inhibitors, PDE3, PDE4, PDE5, Olprinone, Roflumilast, Tadalafil, Sildenafil, Bronchial Asthma, Chronic Obstructive Pulmonary Disease, Acute Lung Injury, Meconium Aspiration Syndrome, Therapy.

INTRODUCTION

There are several airway diseases associated with inflammation and airway obstruction, with bronchial asthma (BA) and chronic obstructive disease (COPD) as the main representatives. Both of these diseases are characterized by an accumulation of inflammatory blood cells in the lung tissue and airways and by a typical inflammatory symptomatology. An increased leak of inflammatory cells into the target tissue and their activation lead to a prolonged survival of pro-inflammatory cells. A crucial role in the development of chronic inflammation play various cell interactions in the airways. The inflammation in BA is directed by the activation of Th_2 lymphocytes producing various inflammatory cytokines (interleukins IL-4, IL-5, and IL-13), eotaxin *etc.* In contrast, COPD is characterized by an involvement of alveolar macrophages and neutrophils with the release of IL-8, neutrophil elastase, tumor necrosis factor (TNF)-α, and myeloperoxidase.

COPD and BA represent major health problems, especially in developing and western countries. Both of these diseases are treated based on well-defined guidelines [1, 2]. However, new therapies are still to be developed, as resistance to therapy may occur.

Recently, a selective phosphodiesterase (PDE) inhibition has been proposed, tested, and approved as one of the therapeutic approaches. There are tens of experimental and clinical studies with selective PDE3 and PDE4 inhibitors tested both in BA and COPD. It is widely known that maintaining the intracellular levels of 3` cyclic phosphate bonds in cyclic 3`,5` adenosine and/or guanosine monophosphate (cAMP and cGMP) stabilizes the cells and leads to a suppression of inflammatory responses and bronchodilation [3, 4]. Therefore, an inhibition of PDE may lead to better control and keeping the intracellular levels of cAMP and cGMP high. The use of PDE inhibitors either alone or in combination has undergone substantial re-evaluation. In this chapter, new trends in the use of non-selective and selective PDE inhibitors are discussed, including the presentation of some recent data on their efficacy in experimental and clinical settings. Furthermore, their potential in influencing the respiratory distress syndrome in neonates, children and adults is described, especially due to their anti-inflammatory and antioxidant action.

PHOSPHODIESTERASES

PDEs represent a super-family of phosphohydrolases – enzymes able to hydrolyze cAMP and cGMP. They are classified into 11 super-families, comprising different 21 genes. Furthermore, an existence of more than 100 mRNA products [3] is supposed due to alternative splicing and the use of different transcriptional start sites.

It is widely known that almost all cells are able to produce several different PDE subtypes. Thus, their localization and origin are considered to be the major regulatory mechanisms in regulating the intracellular levels of cyclic nucleotides (cAMP and cGMP) in these cells and tissues. The major role of PDEs remains in the control the cyclic nucleotides degradation. The regulation of degradation of second messengers leads to faster and more significant changes in their concentrations compared to influencing their production, therefore the use of PDE inhibitors has been appreciated and accepted as an effective strategy in the therapeutic management of several diseases. Caffeine was found to be a significant PDE inhibitor. Theophylline, another representative of methylxanthines, plays an important role in the therapy of various lung diseases. However, these PDE inhibitors possess only narrow therapeutic indices, especially due to their pharmacokinetic properties and their ability to inhibit several PDE subtypes in different tissues and organs, *i.e.* their effect is considered as non-selective regarding the PDE [3].

Functions of respective enzymes depend on the function of original tissue or organ. The reason for that is based on a specific localization of the PDE isoforms. Their selective inhibition can lead to more selective influence on respective tissue/organ in physiological, and especially in pathological condition. Fig. (**1**) summarizes functions of PDE isoforms and their tissue expression.

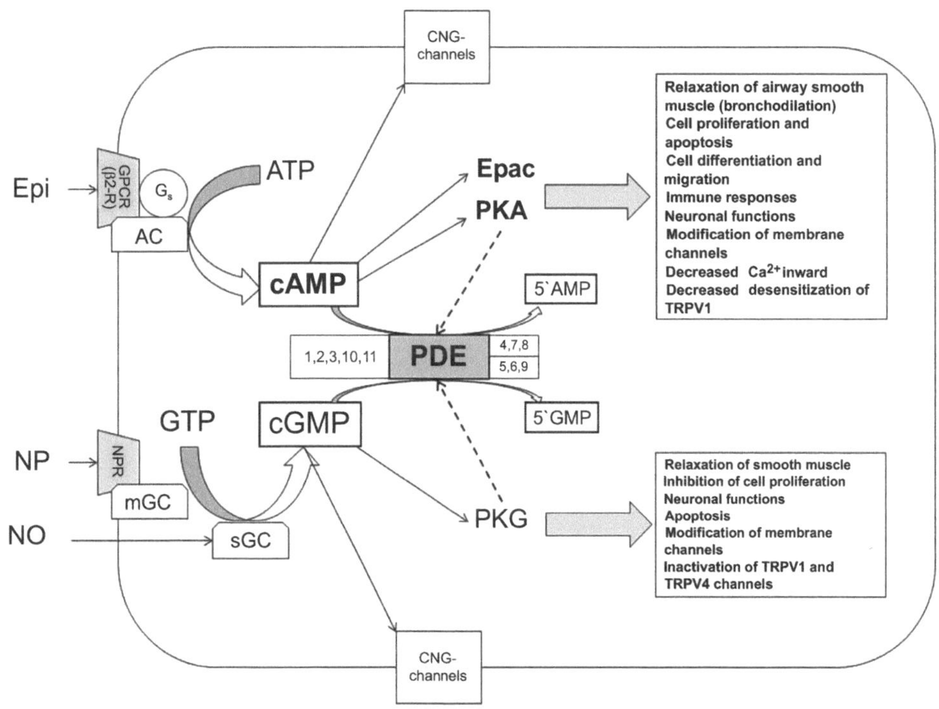

Fig. (1). Changes in cAMP levels and cGMP levels in the cells – involvement of PDE inhibition.
Abbreviations: AC – adenylate cyclase, β2-R – β_2-receptor, CNG-channels – cyclic nucleotide gated channels, Epac - exchange protein directly activated by cAMP, Epi – epinephrine, GPCR – G-protein coupled receptor, Gs – stimulatory G protein, sGC – soluble guanylate cyclase, mGC – membrane-bound guanylate cyclase, NO – nitric oxide, NP – natriuretic peptide, NPR – receptor for natriuretic peptide, PDE – phosphodiesterase, PKA – proteinkinase A, PKG – proteinkinase G, TRPV – transient potential receptor vanilloid [7]. The numbers on the left from PDE represent isoforms hydrolyzing both cAMP and cGMP (1,2,3,10,11). The numbers on the right from PDE represent isoforms hydrolyzing selectively cAMP (top, 4,7,8), or cGMP (bottom, 5,6,9).

PDE INHIBITORS

Non-selective PDE Inhibitors

Methylxanthines (xanthine derivatives) are naturally found as purine alkaloids in plants, *e.g.* fruits of cocoa and coffee as well as tea plant leaves. The most famous xanthine derivative, caffeine (1,3,7-trimethylxanthine), was isolated from bean of green coffee in Germany in 1820 for the first time and completely synthetized in 1885 [9]. Central effects (*e.g.* psychostimulation) prevail among its effects, being in contrast to theophylline and theobromine, where bronchodilating and cardiostimulating (peripheral) effects dominate.

Theophylline has been used in clinical practice for around 80 years. It is the mostly used methylxanthine and is known for its narrow therapeutic range. Its doses required for reaching plasmatic concentrations (from 5 to 15 mg/l) are considered to be safe [10]. However, only at higher plasma concentrations (>10 mg/l) a significant bronchodilating action can be observed, while lower concentrations (5-10 mg/l) are associated with anti-inflammatory effects. Currently, theophylline is used predominantly in the treatment of COPD and BA. Nevertheless, a development of newer anti-asthma drugs moved theophylline into a position of additional bronchodilating drug [10]. In addition to bronchodilating and anti-inflammatory actions, both theobromine and theophylline possess also significant antitussive activity [11, 12] as confirmed in naïve as well as ovalbumin sensitized guinea pigs [13], and healthy cats [14].

The effects of aminophylline (a mixture of theophylline and ethylenediamine in a ratio of 2:1) are generally similar to those of theophylline. Ethylenediamine is added for increasing the solubility and absorption of theophylline. Furthermore, it has demonstrated potent antioxidant effects, used in scavenging hypochlorous acid (HOCl) and hydroxyls (OH.) in *in vivo* conditions, and thus preventing an oxidative burst in tissues [15]. Aminophylline is used for its bronchodilating and anti-inflammatory effects, and for central stimulation of breathing [10, 16].

Pentoxifylline (another representative of dimethylmethyxanthines) possesses outstanding hemorheologic properties, suppressive effects on aggregation of platelets, generation of thrombus and it improves the peripheral blood flow. Therefore, it is used in the therapy of *cludicatio intermittens* and dementias of vascular origin. Furthermore, after incubation with meconium in *in vitro* conditions, the administration of pentoxifylline inhibited polymorphonuclear leukocytes and suppressed the production of TNFα [17]. In an *in vivo* model of piglets with meconium aspiration syndrome (MAS), the administration of pentoxifylline ameliorated local ventilation/perfusion ratio and led to prevention of the increase of TNFα, proteins and alveolar macrophages in bronchoalveolar lavage (BAL) fluid. However, it did not act on the neutrophils accumulation in lungs [18]. Therapy with aminophylline in our experiments significantly ameliorated respiratory parameters compared to non-treated group of rabbits with experimentally induced MAS. Aminophylline led to decreased pulmonary vascular resistance expressed as a percentage of right-to-left pulmonary shunts and central venous pressure, and it increased the oxygenation. Furthermore, aminophylline led to the suppressed inflammation and edema, to reduced *in vitro* reactivity of tracheal and lung smooth muscle strips to histamine and it reduced the oxidation impairment and the number of neutrophils in lungs compared to non-treated rabbits with MAS. The significant improvement of these pulmonary functions was observable already 30 minutes after administration of aminophylline [19, 20].

Mechanism of Action

The exact mechanisms of methylxanthines action (theophylline is used as a reference in the text below) is still intensively discussed. Theophylline added to the strips of human airway smooth muscle in *in vitro* conditions leads to their direct relaxation and has similar final part of action to β_2-agonists, known as so-called functional antagonist, *i.e.* it blocks and prevents the effects of all agents associated with bronchoconstricting [21].

It is known that non-selective inhibition of PDEs leads to increased intracellular levels of cAMP and cGMP, and to bronchodilation and vasodilation. However, these changes are observed only at higher plasmatic concentrations of theophylline. The increase of cAMP induced by theophylline leads to opening the ion maxiK^+ channels, thus leading to the bronchodilation. Furthermore, theophylline leads to decrease of intracellular levels of Ca^{2+}, suppresses release of acetylcholine and monoamines, and suppresses production, release and effects of various inflammatory and bronchoconstricting mediators. Similarities in the structure of theophylline and other methylxanthines are important in their antagonistic effects at purinergic (adenosine) receptors. Theophylline suppresses the activity of basophils and mast cells, stimulates production of surfactant and mucocilliar transportation, and increases the capacity to scavenge reactive oxygen species [10, 21, 22]. Pentoxifylline, similarly to theophylline, is known as a non-selective inhibitor of PDEs and as an antagonist of receptors for adenosine; however, it inhibits TNFα and synthesis of leukotrienes, and suppresses inflammation [23, 24].

The methylxanthine properties differ according to their ability to influence different PDE isoenzymes or at the level of their interactions with adenosine receptors. Pentoxifylline and aminophylline have demonstrated antioxidant activity; aminophylline has reduced neutrophil accumulation in lungs, suppressed lipid peroxidation, levels of TNFα and IL-8, and led to oxygenation improvement in patients during replacement of heart valve [25]. Pentoxifylline suppressed production of oxygen species by neutrophils after their exposure to TNFα in *in vitro* conditions [26].

Mechanism of Action at Higher Doses

Most of confirmed molecular mechanisms of methylxanthine action are clinically relevant only at higher concentrations in plasma (mostly > 10 mg/l) [27]. There is only a limited evidence of clinical state amelioration and anti-inflammatory effects of xanthine derivatives at lower concentrations (5-10 mg/l). Major mechanisms involved at higher concentrations are associated with non-selective inhibition of PDE and antagonism of adenosine receptors [27].

Theophylline is only weak and non-selective inhibitor of PDEs. PDEs are involved in degradation of cyclic nucleotides, leading to increased intracellular concentrations of cAMP and cGMP. In *in vitro* tests, theophylline demonstrated a relaxing effect on airway smooth muscle by inhibition of PDE3 (as well as PDE4 and PDE5); however, high concentrations of theophylline were needed to induce the maximum relaxing effects [28]. A more recent work by Polson *et al.* [29] demonstrated that only 5-10% inhibition of PDEs could be observed at therapeutic concentrations of theophylline in plasma. The "non-selectivity" of theophylline effects on PDEs is responsible for common unwanted effects, *e.g.* headache, nausea and vomiting, mostly due to increased plasma levels.

Methylxanthines have been described also as potent antagonists of receptors for adenosine. Adenosine is a signal nucleoside with several physiological functions; in airways, its exposure leads to bronchoconstricting and inflammatory actions. Adenosine significantly influences the activity of eosinophils, neutrophils, macrophages and lymphocytes [30]. As the adenosine receptors stimulation by ATP leads to the bronchoconstriction, their antagonism could be considered as the major mechanism of theophylline action. However, the antagonism of adenosine receptors (similarly to the inhibition of PDEs) is responsible for serious adverse effects, especially in cardiovascular system (*e.g.* arrhythmias caused by antagonism of A_1 receptors).

Theophylline has further actions, including increased release of anti-inflammatory acting interleukin (IL)-10; the release of IL-10 is typically suppressed in inflammatory conditions. However, the increase in IL-10 expression is probably associated with an inhibition of PDE [31].

Theophylline modulates also the transcription process - it inhibits the penetration of pro-inflammatory transcription factor – nuclear factor kappa B (NF-κB) into nucleus, thus preventing the expression of genes responsible for inflammation. PDE inhibition is involved in these effects, so it is observed only at higher plasma concentrations of theophylline [32].

Other mechanisms of theophylline manifested only at higher plasma concentrations include inhibition of fosfoinositid-3-kinase (PI3K) – an enzyme involved in the response to oxidative stress), stimulation of apoptotic processes in eosinophils and neutrophils (probably through antagonism at receptors for adenosine – A_{2A}) [27]. The PDE inhibition induces also the apoptosis of T-lymphocytes [10]. Higher therapeutic and toxic concentrations of theophylline in plasma (> 20 mg/l) can induce other changes, such as increase in concentrations of circulating catecholamines, suppression of Ca^{2+} entry into inflammatory cells, inhibition of prostaglandin function and antagonizing the action of TNFα [10].

Mechanism of Action at Lower Doses

In lower therapeutical concentrations, other mechanisms of action of theophylline can be observed, including histone-deacetylase (HDAC) activation [33]. Glucocorticosteroids are characteristic by their ability to suppress transcription and translation of inflammatory genes by binding to and activation of specific glucocorticosteroid receptors (trans-repression). The receptors after activation are able to include HDAC2 into a transcription complex of genes responsible for inflammation. This leads to activation of the complex and reverses the histone acetylation, thus suppresses genes activated by inflammatory stimuli [34]. Theophylline is able to increase the HDAC activity and to suppress the expression of inflammatory genes [35]. This effect is, however, observed only at lower concentrations of theophylline (around 5 mg/l) and tapers at increased levels. This observation has been recently confirmed by our study in guinea pigs, where stronger anti-inflammatory action (suppression of IL-4 and IL-5 release and TNF-α production suppression) was observed at theophylline doses of 5 and 10 mg/kg for seven consecutive days, leading to plasma concentrations below therapeutic range, *i.e.* less than 8 mg/l [36]. The mechanism of action of theophylline on HDAC is different from that of glucocorticoids. Latest data have suggested that HDAC was directly activated, without any involvement of HDAC2 in the transcription complex. The mechanism of this HDAC activation has not been completely elucidated yet; however, neither PDE inhibition nor adenosine receptor antagonism is involved. Theophylline is able to combat the resistance to glucocorticosteroids present in some patients with COPD or bronchial asthma [21]. We should take into account also mild increase in intracellular concentrations of cAMP due to PDE inhibition. The cAMP after activation of protein kinase A may directly stimulate the cAMP-response element binding protein (CREB) or activating transcription factor 1 (ATF-1) and stimulate cAMP response elements (CRE), that are located in genes (either up-regulated by glucocorticosteroids or genes that are not responsive to glucocorticosteroids). Therefore, the final effect of combining a glucocorticosteroid with a cAMP/elevating drug (including theophylline) can lead to an increased expression of anti-inflammatory genes [4]. Involvement of other mechanisms in pharmacological action of theophylline, different from PDE inhibition, has been suggested by Dubuis *et al.* [37]. They found that theophylline inhibited capsaicin-induced cough under both control and disease conditions by decreasing the excitability of sensory nerves through activation of small- and intermediate-conductance calcium-activated potassium channels [37].

It is important to mention that not all xanthine derivatives confer the same effect as described above. While there have not been any published studies about its effects on cough, doxophylline (also known as Novophylline), a theophylline derivative with a dioxolane group, has been tested recently for its mechanism of action. However, the precise mechanism of action has not been revealed, but lack of adenosine antagonism, PDE inhibition and activation of HDAC enzymes at therapeutic doses suggest an action different from theophylline and selective PDE4 inhibitors like roflumilast [38].

Non-selective inhibition of PDEs has been proposed recently by other authors when elucidating the mechanisms of bronchodilator action of various plants. Dual inhibition of calcium influx and PDE was demonstrated by Arshad *et al.* [39] in *Chrozophora prostrata* and by Gilani *et al.* [40] a potassium channel activation with PDE inhibition in *Salvia officinalis*. However, neither the level of inhibition or participation of PDE isoforms nor their antitussive potential was tested in those studies.

Adverse Effects

One of the major limitations of wider theophylline use is relatively high occurrence of unwanted effects, especially when surpassing the upper border of therapeutic plasma concentrations 15 – 20 mg/l [41]. Gradually increasing the doses to the concentration required may prevent these effects partially. Nevertheless, monitoring of theophylline plasmatic concentration is required in some patients.

The adverse effects can be magnified due to pharmacokinetic interactions. Theophylline is metabolized by a P-450 cytochrome (CYP1A2 isoenzyme). This isoenzyme can be inhibited by several drugs, including cimetidine, macrolides, chinolons, or fluvoxamine. Their co-administration can lead to slower metabolism of theophylline, and thus increase in its plasmatic concentrations and adverse effects occurrence.

Headache, abdominal pain, nausea, vomiting, and restlessness belong to the most frequent adverse effects. Furthermore, theophylline may lead to an increased secretion of gastric acid, gastroesophageal reflux, and increase in diuresis [10]. Higher plasmatic concentrations of theophylline can induce convulsions, changes in cardiovascular system (hypertension, tachycardia, arrhythmias), and in most severe cases can lead to death [42].

There are several mechanisms causing the unwanted side effects of theophylline. Adenosine receptor A_1 antagonism participates in the stimulation of central nervous system, increased gastric secretion as well as in arrhythmias. The central PDE inhibition (especially PDE4) is responsible for the development of the most often observed unwanted effects, *e.g.* gastrointestinal symptomatology, nausea, and abdominal pain. These effects can limit the use of the first generation of selective PDE4 inhibitors (the vomiting center in brain is influenced by PDE4 inhibition).

The unwanted effects observed due to the antagonism of adenosine receptors and inhibition of PDEs are usually present at higher plasmatic concentrations of theophylline (10-15 mg/l) and may they often lead to withdrawal of theophylline from the therapy. By using the lower concentrations (5-10 mg/l) these adverse effects and interactions can be significantly prevented [10].

In our experiments in experimentally induced MAS we noticed relatively severe cardiovascular adverse effects associated with intravenous administration of aminophylline. During the first five minutes of administration, increased blood pressure and heart rate were observed, followed by an increase of several heart rate variability parameters characterizing a parasympathetic activity [43].

There are several ways to prevent adverse effects of aminophylline, or theophylline. The first is to reduce the drug dosage. In our previous experiments we compared a therapeutic effect of high dose (2 mg/kg) and low dose (1 mg/kg) of aminophylline on pulmonary functions – the results showed more pronounced improvement of some respiratory parameters at high doses, while a decrease of several markers of oxidative stress as well as *in vitro* reactivity of pulmonary strips to bronchoconstricting mediators were more significant in the group of rabbits with lower aminophylline dose [44].

Another way to prevent the adverse effects of aminophylline is to co-administer it with other suitable drugs, *e.g.* with glucocorticosteroids. An additive effect of both agents enables administration of lower doses, thus reducing their adverse effects. In our experiments we demonstrated a superior effectiveness of a combination of aminophylline and intratracheally administered budesonide compared to aminophylline alone [44].

Recent therapeutic approaches (introduced to prevent cardiovascular adverse effects) include the administration of selective PDE inhibitors; targeted inhibition of specific PDE isoforms should lead to minimization of adverse effects induced by nonselective PDE inhibition and by previously mentioned mechanisms of methylxanthines independent on PDE inhibition (*e.g.* antagonism of adenosine receptors) [4].

Selective PDE Inhibitors

There are several PDE isoforms in mammals; they possess structurally-distinct active sites. Many PDE isoforms are involved in different body functions and associated with various diseases. Therefore, it is generally suggested that selective PDE inhibitors may influence different diseased systems/organs without affecting other ones and without an occurrence of severe adverse effects.

In contrast to theophylline (non-selective PDE inhibitor), selective PDE inhibitors can act with significantly lower risk of unwanted effects. Several groups of selective PDE inhibitors, *i.e.* PDE3, PDE4, PDE5, and PDE7, exert a possible relationship to immune responses in the airways. The activity of PDE3 in the airways has been confirmed in alveolar macrophages, endothelial cells, platelets, and airway smooth muscle. Simultaneous administration of siguazodan (PDE3 inhibitor) and rolipram (PDE4 inhibitor) has led to enhanced relaxing response, suggesting a synergism and potential benefit in parallel inhibition of PDE3 and PDE4 [27]. The PDE4 enzyme family is expressed in various inflammatory cells; therefore, its inhibition has shown anti-inflammatory and immunomodulating effects. Rolipram suppresses neutrophil- and eosinophil-mediated inflammation, secretion of interleukins and activation of pro-inflammatory cells. Second generation PDE4 inhibitors (roflumilast – approved for clinical use in COPD patients) suppress airway hyperresponsiveness, release of TNF-α, fibrotisation and remodeling of the lung parenchyma [45]. PDE7 inhibition can contribute to anti-inflammatory action as its expression has been confirmed in T-cells, epithelial cells, airway and vascular smooth muscle cells, lung fibroblasts, and eosinophils [46]. While their effectiveness in individual administration is relatively weak, simultaneous PDE4 and PDE7 inhibition seems to be promising [47]. The anti-inflammatory activity of dual PDE inhibitors (*e.g.* PDE3/PDE4, or PDE4/PDE7) has been studied mostly in COPD conditions, and only in limited extent in allergic BA. PDE5 inhibition leads to decreased pulmonary vascular resistance, suppressed pulmonary hypertension and vascular remodeling. Sildenafil (PDE5 inhibitor) inhibits cell proliferation, has pro-apoptotic action, and suppresses the bronchial hyperresponsiveness, infiltration by white blood cells and concentrations of exhaled NO [6]. These findings suggest possible benefits of combining the PDE4 and PDE5 inhibitors in the therapy of BA and COPD. There are several studies with roflumilast and cilomilast already conducted in patients with BA; however, there is a lack of relevant data from experimental and clinical studies on simultaneous effects of PDE4 and PDE5, PDE3 and PDE4, and PDE4 and PDE7 up to now.

While methylxanthines are characteristic for their non-selective PDE inhibition, there are several selective PDE inhibitors with successful experimental and clinical "experience". The mostly studied selective PDE inhibitors related to immune responses in respiratory system are isoforms PDE 3, 4, 5, and 7 [27].

PDE3 Inhibitors

PDE3 is an enzyme hydrolyzing cAMP (and to certain extent also cGMP) and occurring in two major isoforms – PDE3A, and PDE3B [48]. PDE3 inhibitors are currently used as non-glycoside cardiotonic agents. Their effect is related to a selective inhibition of PDE3A, which is considered as a cardiovascular type. The major representatives include milrinon, amrinon, and vesnarinon.

In the respiratory system, the PDE3 activity is observed predominantly in airway smooth muscle cells, endothelial cells, alveolar macrophages, and platelets. Both PDE3 and PDE4 are considered as the most important PDE isoforms responsible for hydrolyzis of cAMP in the airways [49]. Besides a direct bronchodilating effect, which is more significant after administration of selective inhibitors of PDE3 compared to PDE4, the PDE3 inhibition leads to a potentiation of smooth muscle relaxation evoked by agonists of β_2-receptors. A simultaneous administration of PDE3 inhibitor siguazodan and PDE4 inhibitor rolipram led to enhancement of the relaxation. This suggests a synergistic action and possible benefits of simultaneous use of PDE3 and PDE4 inhibitors. PDE3 inhibitors have only mild action on activation of T-cell [50]; however, they can enhance these effects characteristic for PDE4 inhibitors.

In our previous experiments with guinea pigs, a cough-suppressing effect of a PDE3 inhibitor cilostazol has been confirmed [51], especially in ovalbumin-sensitized animals. Furthermore, another study by Koike *et al.* [52] demonstrated anti-inflammatory action of a PDE3 inhibitor olprinon in an acute lung injury model in rabbits. These data suggest that PDE3 inhibitors could be used in cough therapy [53] as well as in some neonatal diseases, like MAS. Milrinone (PDE3 inhibitor) improved oxygenation index and survival rate in 4 patients with MAS [54]. Olprinone in rabbits with MAS ameliorated lung functions, prevented development of pulmonary edema, led to decrease in neutrophil count in lungs and the percentage of right-to-left shunts [55, 56]. After further and detailed analyzing the markers of oxidative stress in lungs, a significant suppression of protein and lipid oxidation both in isolated mitochondria of alveolar cells and in lung homogenate was observed in the olprinone treated animals with MAS [55]. These results demonstrated a benefit of selective PDE inhibitors; however, further studies are essential, including other selective PDE inhibitors (*e.g.* PDE4 inhibitors).

PDE4 Inhibitors

The family of PDE4 enzymes is encoded in 4 different genes (PDE4A, PDE4B, PDE4C, PDE4D). Products of all these genes hydrolyze specifically the cAMP. Except for PDE4C, all of them are present in inflammatory cells, *e.g.* B-cells, T-cells, neutrophils, eosinophils, airway epithelial and endothelial cells [57]. Therefore, the selective PDE4 inhibitors have predominantly anti-inflammatory and immunomodulating actions [58].

The most famous representative of the first generation PDE4 inhibitors is rolipram. It is often used in experimental medicine, leading to suppression of both eosinophil and neutrophil inflammation, secretion of interleukins from activated T-helper cells (Th-1 and Th-2), airway epithelial cells, monocytes, macrophages, and basophils [58]. The PDE4 inhibition suppresses neutrophil, eosinophil, monocytes and basophil functions. However, the maximum decline in macrophage activity and airway smooth muscle relaxation can be observed only after so called dual inhibition, *i.e.* after simultaneous use of PDE3 and PDE4 inhibitors.

Unfortunately, rolipram often leads to severe nausea and vomiting, disqualifying it from the clinical use. Therefore newer group of selective PDE4 inhibitors have been developed and introduced. As was described previously [49], nausea and vomiting resulted from activation of a high-affinity binding site of PDE for rolipram, whereas the suppression of inflammation was associated with binding to a low-affinity binding site. Second generation agents have similar affinity to both of these binding sites, leading to a stronger bronchodilation and activity against inflammation with fewer unwanted effects [49]. Roflumilast, cilomilast and piclamilast belong to this newer (second generation) group of PDE4 inhibitors. These agents decrease the airway hyperresponsiveness, suppress release of TNF-α, fibrotization, and remodeling the pulmonary parenchyma (probably due to an inhibition of special matrix metalloproteinases) and inhibition of other anti-inflammatory mechanisms characteristic for chronic inflammatory changes in bronchial asthma and COPD [49]. Based on positive results in phase III of clinical drug testing, they have been recently introduced into clinical practice for treatment of moderate to severe COPD [59, 60].

Despite a natural affinity between special subtypes of PDE4, the current tendency is to search for even more selective PDE4 inhibitors, acting only in a specific subtype of PDE4. Currently known inhibitors of PDE4 have different affinity to respective PDE4 subtypes, thus leading to various incidence of unwanted effects. In our previous experiments in guinea pigs we demonstrated the effects of PDE4 inhibitors in ovalbumin-induced allergic inflammation; both rolipram and citalopram led to suppression of *in vivo* and *in vitro* airway reactivity [51].

Our recent study on 2nd generation PDE4 inhibitor roflumilast effects in ovalbumin-sensitized guinea pigs revealed suppression of cough and pro-apoptotic action on inflammatory cells (suppressed viability of white blood cells in blood and bronchoalveolar lavage fluid) already at a dose of 1.0 mg/kg in ovalbumin-sensitized guina pigs [61]. Furthermore, roflumilast has demonstrated inhibitory effects on leukotriene production and bronchoconstriction in ovalbumin-sensitized and challenged guinea pigs [62]. In humans, roflumilast was shown to be effective in cough suppression when added to the daily therapy of severe/very severe COPD with inhaled corticosteroids and long-acting β_2-agonists or long-acting muscarinic antagonists over 52 weeks period [63]. Therefore, roflumilast has been confirmed as a promising therapeutic tool not only in COPD, but also in patients with bronchial asthma [64 - 66], and asthma-COPD overlap [67].

This is in accordance with our previous experiments with another PDE4 inhibitor YM976 at the dose of 1.0 mg/kg intraperitoneally, where potent antitussive effects were observed both after single dose and seven daily doses, associated with decrease in *in vivo* and *in vitro* airway reactivity [68]. Similar results were demonstrated using citalopram after a single dose of 1.0 mg/kg b.w [51], but not with rolipram at the same dose, where no significant antitussive effects were observed [47]. The discrepancy can be partly explained by different affinity to the PDE4B isoform, responsible for anti-inflammatory and bronchodilating effects. Aoki *et al.* [69, 70] have demonstrated only a little emetogenic activity of YM976 (no emesis observed at the oral dose of 10.0 mg/kg b.w. in ferrets), suggesting relatively safe profile of this PDE4 inhibitor compared to rolipram and still limiting the use of roflumilast in asthmatic patients [71]. The antitussive effect of YM976 was associated with a decrease in specific airway resistance, where the involvement of relaxation of smooth muscle, decreased mucosal edema, and secretion is presumed to occur. However, the suppression of tracheal strips contractility to cumulative doses of both acetylcholine and histamine were observed, thus suggesting a direct link between relaxation of airway smooth muscle and antitussive effect [68]. In contrast, previously tested citalopram led to a significant suppression of airway reactivity only in *in vitro* conditions and not *in vivo* [51]. Moriuchi *et al.* [72] have demonstrated a potent relaxing effect of YM976 and rolipram on bovine tracheal smooth muscle contracted by histamine, but not metacholine. Both PDE4 inhibitors used in their study were more potent compared to theophylline, and they led to a significant increase in cAMP levels measured in tracheal smooth muscle cells. However, YM976 inhibited inflammatory cell functions significantly more than rolipram [69]. This suggests that YM976 may suppress antigen-induced bronchoconstriction *via* inhibition of mediator release from the leukocytes, and not by direct relaxation effect [72].

One should note that bronchodilating and anti-inflammatory effects of PDE4 inhibitors can be amplified by simultaneous administration of PDE3 inhibitors, suggesting potential benefit of agents with dual action, *i.e.* inhibiting both PDE3 and PDE4 isoforms. Based on these results, an administration of dual PDE3/4 inhibitors (*e.g.* RPL554) should be considered. Previous findings have demonstrated more significant airway relaxing effects of siguazodan (PDE3 inhibitor) compared to rolipram [73, 74]. Meanwhile, their simultaneous administration led to an additive relaxation, suggesting an interaction or even synergism between inhibition of PDE3 and PDE4 [4, 75]. Furthermore, zardaverine (dual inhibitor of PDE3/4) after its inhalational administration led to a significant bronchodilation [49].

PDE5 Inhibitors

In contrast to PDE3 and PDE4 inhibitors, PDE5 inhibitors selectively hydrolyze cGMP. PDE1 and PDE5 are considered as the major cGMP-degrading enzymes in vascular smooth muscle. PDE5 is expressed in smooth muscle of vascular vessels (especially pulmonary arteries and veins as well as bronchial vessels) and in airway smooth muscle [76].

Thus, PDE5 inhibition leads to reduction of pulmonary vascular resistance [77 - 79], decrease of pulmonary hypertension, and suppression of vascular remodeling [76]. These effects were demonstrated after increased production of cGMP as well as after inhibition of PDE5 with sildenafil. This led to an inhibition of cell proliferation and pro-apoptotic activity [80]. Therefore, the use of PDE5 inhibition has been tested also in COPD. Furthermore, anti-inflammatory effects similar to PDE4 inhibitors were demonstrated with PDE5 inhibitors [81].

In our previous experiments we showed an antitussive effect of zaprinast (known as selective PDE5 inhibitor at higher doses), which was more significant in guinea pigs sensitized by ovalbumin. This effect was accompanied only with a mild bronchodilating action in *in vivo* conditions and non-significant decrease of *in vitro* airway responsiveness, suggesting participation of other mechanisms in its antitussive effect besides smooth muscle relaxation (anti-inflammatory action or direct modulation of airways cough receptors) [82]. However, these results are in concordance with the study of Sebkhi *et al.* [76], who did not demonstrate any effect of zaprinast on eosinophil accumulation and bronchial reactivity. In contrast, sildenafil as a more selective and potent PDE5 inhibitor, significantly inhibited bronchial reactivity (hyperresponsiveness), infiltration by leukocytes and levels of exhaled NO [81], suggesting benefits of dual PDE4/PDE5 inhibition in the therapy of COPD and bronchial asthma.

Both PDE1 and PDE5 are responsible for degradation of cGMP, suggesting involvement of nitric oxide-soluble guanylate cyclase pathway in obstructive airway diseases associated with cough [83]. Only limited data about the effects of PDE5 inhibitors on the cough have been described. In our recent study, we have demonstrated potent and dose-dependent suppressive effects on *in vivo* and *in vitro* airway reactivity as well as on markers of inflammation (*e.g.* IL-4, IL-5, TNF-α) after roflumilast and tadalafil administration in ovalbumin-sensitized guinea pigs [84]. These changes were associated with cough suppressive effects [7] of tadalafil and roflumilast only at higher doses (1 mg/kg b.w.). Furthermore, their co-administration at lower doses (both tadalafil and roflumilast at the dose of 0.5 mg/kg b.w.) led to similar results as observed by a single therapy with higher doses (1 mg/kg b.w.), suggesting a benefit of dual PDE4/PDE5 inhibition [85]. Similar results were observed by Takeda *et al.* [86] in mice challenged with ovalbumin, where an additive effect of roflumilast and tadalafil was found at substantially lower doses, suggesting clinical use of the combination of PDE4 and PDE5 inhibitors in asthma and in the therapy of asthma- and COPD-related cough.

There have been relatively few selective PDE inhibitors tested in experimentally induced MAS. Besides previously mentioned PDE3 inhibitor olprinone, an administration of sildenafil reversed a blood pressure increase in *arteria pulmonalis* within 1 hour after its administration in piglets with MAS. Furthermore, there was no observable influence on systemic circulation in these animals [87].

PDE7 Inhibitors

PDE7 isoform is an enzyme specifically hydrolyzing cAMP, but insensitive to rolipram. Its expression was confirmed in brain, skeletal muscle cells and especially in T-cells [88, 89]. PDE7 enzymes are encoded by two genes PDE7A and PDE7B with significant expression of PDE7A1 protein in T-cell lines, peripheral T-lymphocytes, epithelial cell lines, airway and vascular smooth muscle cells, lung fibroblasts, and eosinophils [90]. Thus, the inhibition of this isoenzyme could be an effective tool in suppressing the airway hyperresponsiveness and inflammation. There are several selective PDE7 inhibitors, *e.g.* VP1.15 and S14 synthesized in the Instituto de Quimica Medica [91], pyrimidine inhibitors [92], purine inhibitors [93], ASB16165 synthesized in Asubio Pharma Co Ltd in Japan [94], thiadiazol derivative PF0332040 synthesized by Pfizer [95]. Another selective PDE4 inhibitor synthesized by the Discovery Research Laboratory in Japan has some activity against PDE7 isoenzyme – T-2585 [96]. BRL50481 is the only commercially available selective PDE7 inhibitor. However, it was developed against PDE7A and has a published IC50 against PDE7B of only 12.1 uM [97]. It was previously demonstrated to be effective in suppressing some inflammatory cells in *in vitro* conditions, *e.g.* lung macrophages, monocytes, and CD8+ T-lymphocytes [97]. Recently, several other effects were confirmed in animal models of Parkinson disease, *e.g.* protection of dopaminergic neurons [98]. In our experiments, the intraperitoneal administration of BRL50481 significantly decreased *in vivo* and *in vitro* airway reactivity. The most significant suppression of contractile responses to histamine was present in ovalbumin-sensitized animals [46], suggesting its benefits in clinical states associated with allergic inflammation, which was simulated by sensitization with ovalbumin. However, the anti-inflammatory action of BRL 50481 administered as a single treatment is only mild. In contrast, the combination of inhibitors of PDE7 (BRL 50481) and PDE4 (rolipram) at half doses increased the effects observed after the single use of respective inhibitors at double doses [47, 99]. These observations are relevant also for a possible reduction of the dose of PDE4 inhibitor and thus preventing its possible adverse effects [21]. The concept of the dual inhibition was described in literature and new dual PDE inhibitors have been studied intensively (*e.g.* PDE3/4 and PDE4/7) for their anti-inflammatory action [100].

Examples of available selective PDE inhibitors as well as their potential use can be found in Table **1**. The effects of selective and non-selective inhibition of PDE isoforms and subsequent increase of cAMP in the cells are shown on Fig.(**1**).

Dual/multi-isoform Selective PDE Inhibitors

Besides selective PDE inhibitors described above, there are numerous multi-isoform inhibitors (inhibitors of multiple PDEs), *i.e.* their selectivity is "imperfect" and they are able to inhibit several PDE isoforms [4, 75]. Thus, drugs with suitable combinations of inhibiting more isoforms have been developed – *e.g.* PDE3 and PDE4, PDE4 and PDE1, PDE4 and PDE5, and PDE4 and PDE7. They could positively potentiate their action in some pathological conditions and minimize their adverse effects through decreased doses – so called multicomponent approach [3, 100, 101].

Table 1. Functions of PDE isoforms with examples of their selective inhibitors. Involvement of PDEs in airway diseases is marked in bold [3 - 8].

PDE family	Preference to cAMP or cGMP	Selective inhibitor	Tissue expression	Function
PDE1	Both	Vinpocetin 8MM-IBMX	Brain, **smooth muscle**, heart, testis	PDE1A – **regulation of (vascular) smooth muscle**, sperm function PDE1B – involved in dopaminergic transmission, **influence on surviving and activation of immune cells** PDE1C – **essential for proliferation of smooth muscle**, regulates sperm and neuronal functions
PDE2	Both	Bay 607550	Adrenal cortex, brain, heart	PDE2 – mediates connection between cGMP and cAMP metabolism, regulates secretion of aldosterone from suprarenal glad, phosphorylation of Ca^{2+} channels in heart *via* cAMP and protein kinase A (PKA), and *via* cGMP in neurons, long-term memory, **barrier function of endothelial cells in inflammation**
PDE3	Both	Cilostamide Cilostazol Olprinon	Heart, **smooth muscle**, adipose tissue, platelets	PDE3A – regulates myocardium contractility, platelets aggregation, vascular and **airway smooth muscle contraction**, maturation of oocyte and rennin release PDE3B – mediates response to insulin (especially its antilipolytic action), regulates cell cycle/proliferation and mediates inhibitory effects of leptin and other influences on insulin and rennin secretion
PDE4	cAMP	Rolipram Roflumilast Cilomilast Apremilast GSK256006 CHF6001 YM976	**Ubiquitous**	PDE4 – plays an important role in wide spectrum of processes, neural functions, **activation of monocytes and macrophages, infiltration of neutrophils, proliferation of smooth muscle cells**, fertility, vasodilation and myocardium contractility
PDE5	cGMP	Zaprinast Sildenafil Tadalafil	**Smooth muscle**, platelets, cerebellum	PDE5 – **regulates (vascular) smooth muscle** in penis and **lungs**, is involved in mediation of NO-cGMP in platelets (clotting control) and plays role in regulation of cGMP mediation in brain
PDE6	cGMP	-	Retina	PDE6 – is involved in signal transmission and photoreaction in eye, regulates melatonin release from epiphysis

(Table 1) contd.....

PDE family	Preference to cAMP or cGMP	Selective inhibitor	Tissue expression	Function
PDE7	cAMP	ASB16165 BRL50481 (more selective to PDE7A)	Skeletal muscle, **immune cells**, brain	PDE7 – plays a probable role in **activation of T-cells as well as other inflammatory cells**
PDE8	cAMP	PF04957325	Immune cells, liver, kidney, testis, thyroid	PDE8 – plays a role in activation of T-cells as well as in function of sperm and Leydig`s cells
PDE9	cGMP	BAY736691	Brain, kidney	PDE9 – its role is not completely elucidated, influences signal pathway NO-cGMP in brain
PDE10	Both	Papaverine	Brain, testis	PDE10A – regulates cGMP in brain and can play a role in learning and memory processes
PDE11	Both	BC11-38	Prostate, testis, skeletal muscle	PDE11 – plays a potential role in development of sperm function and in adrenal gland (Cushing syndrome)

An example of dual inhibitor of PDE4 and PDE1 was described two decades ago – KF19514 demonstrated a suppressive effect on inflammation and airway remodeling in a murine model of chronic bronchial asthma [102]. This indicates possible benefit of PDE1 inhibition in regulation of smooth muscle proliferation. However, there have been no further studies discussing this combination in influencing airway diseases [4].

Zardaverine (inhibiting PDE3 and PDE4), or trequinsin (inhibiting PDE4 and PDE2) also belong to this group of agents and their potential use in bronchial asthma and COPD needs to be evaluated.

The administration of drugs with dual inhibitory effect on PDE4 and PDE7, *e.g.* YM-393059, can be considered as a very promising mode of therapy [103]. These agents inhibit several functions of T-cells, *e.g.* they are able to decrease the cytokine production and release, as well as activation and proliferation of various inflammatory mediators. Stepwise development and tests of dual PDE4/PDE7 inhibitors led to the classification of a new independent group of PDE inhibitors, where especially blocking the T-cells functions due to PDE7 inhibition and anti-inflammatory effect due to PDE4 inhibition are expected [104, 105]. In our previous animal study on guinea pigs sensitized with ovalbumin we demonstrated beneficial suppressive effects on *in vivo* and *in vitro* airway reactivity and several inflammatory mediators after 7-days lasting administration of a combination of rolipram and BRL50481 in half-reduced doses compared to single therapy [47]. Furthermore, these drugs have lower risk of adverse effect (like nausea and vomiting) compared to selective PDE4 inhibitors administered alone, especially due to decreased doses required to reach the same effect. Unfortunately, there has not been confirmed superior anti-inflammatory activity of this combination over PDE4 inhibitors [4]. Another multiple PDE inhibitor is dipyridamole, which inhibits PDE5, PDE6, PDE7, PDE8, PDE10 and PDE11.

It is important to mention that some multi-isoform inhibitors may lead in various unwanted effects, which should not be observed with "pure" selectivity. As examples, sildenafil (inhibiting PDE1, PDE5, and PDE6), tadalafil (inhibiting PDE5, PDE6, and PDE11) or vardenafil (inhibiting PDE5 and PDE6) could be mentioned.

ROLE OF PDE AND POTENTIAL USE OF PDE INHIBITORS IN CHRONIC OBSTRUCTIVE PULMONARY DISEASE

COPD is known to be a major global health problem worldwide. In many patients, COPD is leading to other extra-pulmonary pathologies. Usually it is associated with progressive and irreversible decline in lung functions (especially expiratory airflow), oedema and inflammation, mucus hypersecretion, airway collapse, fibrosis, pulmonary hypertension, heart failure and venous thromboembolism. Tobacco smoking is one of the most important risk factors [4]. Despite a preventability of this disease (*e.g.* smoking cessation), COPD is expected to be the third leading cause of death in 2030 [106].

In the therapy of obstructive airway diseases associated with inflammation, several groups of drugs are currently used. They are generally divided to the group of controller and relievers. Short-acting β_2-agonists and anticholinergics (muscarinic antagonists) belong to the group of relievers. The most important controllers used for suppression of inflammation are represented by glucocorticosteroids, as well as long-acting β_2-agonists and methylxanthines (*e.g.* theophylline) with sustained release [107, 108]. Except of theophylline, inhalation routes of administration are preferred. This way of administration leads to significantly reduced incidence of unwanted effects. The adverse effects associated with only limited effects due to necessity to use lower (safer) doses lead to further development and searching for new, more effective and safer drugs suitable for therapy of bronchial asthma and COPD.

In COPD, several groups of drugs influencing intracellular levels of cAMP are used: sympatomimetic bronchodilators – selective β_2-agonists (short-acting – SABA, *e.g.* salbutamol, long-acting – LABA, *e.g.* formoterol), combinations of inhaled glucocorticosteorids with long-acting β_2-agonist (or even triple combination with muscarinic antagonists, *e.g.* beclomethasone+formoterol+ glycopyrrhonium), non-selective PDE inhibitors (*e.g.* theophylline) with several other mechanisms involved in their pharmacological effects as described above

[109, 110]. Another group of drugs approved for COPD therapy less than 10 years ago is known as selective PDE4 inhibitors (*e.g.* roflumilast) added on to the existing therapy in patients with severe disease associated with chronic bronchitis and a history of frequent exacerbations [4, 27].

There are several 2nd generation PDE4 inhibitors already in clinical practice or in clinical phase of drug testing (roflumilast, cilomilast, piclamilast), with roflumilast being registered for severe COPD as an add-on therapy with glucocorticosteroids in 2010 [45, 111, 112]. Their effectiveness after the oral administration is similar to theophylline, but with relatively lower incidence of adverse effects [113, 114]. These are predominantly in gastrointestinal tract (*e.g.* nausea, vomiting, diarrhea), as well as headache. Contrary to theophylline, they usually do not require therapeutic drug monitoring. Nevertheless, occurrence of these side effects limits increasing their doses (as well as the anti-inflammatory potential). An inhalation administration has reduced adverse effects; however, it was associated with decrease of their effectiveness under required level [22]. GSK256066, an inhaled PDE4 inhibitor, demonstrated suitable results in the control of moderate COPD after inhalatory administration with limited adverse effects and increased safety and tolerability profile [115].

Post-marketing surveillance will determine in larger clinical use, whether PDE4 inhibitors will be used only in combination with other anti-inflammatory agents (glucocorticosteroids), or also as monotherapy [116, 117]. Some care should be taken, as there were recently published negative results of a study with a new inhaled PDE4 inhibitor UK-500,001 [118]. In contrast, studies with oglemilast and Zl-n-91 brought positive results in experimental phase of testing [119, 120]. In addition, dual inhibition of PDE3 and PDE4 is of interest in COPD patients, as RPL554 was demonstrated as an effective and well-tolerated bronchodilator, bronchoprotector, and anti-inflammatory drug with potential in the therapy of COPD and bronchial asthma [121 - 123].

There are several experimental studies suggesting positive effects of PDE5 inhibitors in cigarette smoke-induced changes (*e.g.* emphysema and pulmonary hypertension). Seimetz *et al.* [124] used preventive treatment with tadalafil (PDE5 inhibitor), piclamilast (PDE4 inhibitor) or their combination and demonstrated almost complete prevention of emphysema development, increase in lung compliance, tidal volume, structural remodeling of lung vasculature right ventricular systolic pressure, and right ventricular hypertrophy induced by cigarette smoke exposure in mice. In another experimental study in rats, sildenafil attenuated airway inflammation and mucus production after acrolein exposure [125], possibly through the nitric oxide/cGMP pathway. This suggests therapeutic potential of selective PDE5 inhibitors for chronic inflammatory airway diseases

[83, 126]. In humans, however, sildenafil did not demonstrate beneficial effects on exercise capacity in patients with COPD and emphysema without pulmonary hypertension. Furthermore, sildenafil significantly worsened gas exchange at rest and quality of life. Therefore, routine use of PDE5 inhibitors in the therapy of COPD is questionable [127].

ROLE OF PDE AND POTENTIAL USE OF PDE INHIBITORS IN BRONCHIAL ASTHMA

Bronchial asthma is a chronic inflammatory disease usually associated with airway responsiveness, bronchoconstriction, cough and eosinophil accumulation in lungs and airways. As cAMP may reduce the cytokine release from inflammatory cells and stabilize them, the selective inhibition of PDEs involved in its degradation is of interest. Selective inhibitors of PDE (especially PDE3 and PDE4) have been extensively studied for their ability to influence the airway inflammation and hyperresponsiveness of the airway smooth muscle [49, 128]. Several clinical studies testing the second generation of PDE4 inhibitors (piclamilast, cilomilast, and roflumilast) confirmed the bronchodilating, anti-inflammatory, and antitussive effects, which were demonstrated also in the experimental conditions. Nevertheless, these promising results emphasize the necessity of testing the inhibitors of other PDE isoforms in the respiratory diseases with an inflammatory component [45, 75, 129].

The clinical experience in children is limited to theophylline; however, its administration is recommended only in failure of therapy with first-line anti-inflammatory drugs. Generally, in most developed countries, theophylline is considered as a third-line therapy used only in severe bronchial asthma and COPD [130]. Its advantage is an oral administration and a relatively low price, as well as strong anti-inflammatory action observable in lower, less toxic doses [131, 132]. In children, there was demonstrated a benefit of aminophylline (theophylline with ethylenediamine) compared to salbutamol in young asthmatics: after an intravenous administration in severe asthma, similar effectiveness was observed with shorter stay in hospital in theophylline-treated patients [133]. Despite this observation, the most recent edition of guidelines for bronchial asthma therapy in children characterized effects of regularly administered theophylline as weak and non-significant, nevertheless with amelioration of clinical condition of patients [134]. As the effectiveness of theophylline is lower compared to low doses of inhaled glucocorticosteroids, the concomitant occurrence of adverse effects led to its classification into level IV according to evidence-based medicine. Furthermore, several studies have demonstrated beneficial effects of theophylline, aminophylline and theobromine on cough, which represents one of the major symptoms in bronchial asthma [82, 130, 135]. However, a direct relationship to

PDE inhibition has not been established [37].

In our previous experiments, selective PDE4 inhibitor roflumilast significantly decreased the specific airway resistance (sR_{aw}, volumic airway resistance), a marker of *in vivo* airway reactivity, and reduced *in vitro* airway reactivity in both tracheal and lung tissue strips in ovalbumin-sensitized guinea pigs, suggesting a direct bronchodilating action (61). Another tested drug citalopram (known as a selective serotine reuptake inhibitor used for the therapy of depression) increases the cAMP levels *via* selective PDE4 inhibition [136, 137]. It has led to a significant suppression of the airway reactivity only in *in vitro* conditions but not *in vivo* [51], whereas rolipram has demonstrated potent *in vivo* and *in vitro* activity in the ovalbumin-sensitized animals [68].

PDE5 inhibition by tadalafil alone and in combination with roflumilast led to the decrease of both *in vivo* and *in vitro* airway reactivity, as well [85]. However, a significant suppression of contractile responses to the cumulative doses of acetylcholine and histamine was found in ovalbumin-sensitized animals only in the lung tissue. Similar results were observed also in the guinea pigs exposed to ovalbumin after administration of sildenafil [81]. Different response of the tracheal and lung tissue strips is probably caused by a different distribution of PDEs in various types of cells. Thus, more obvious effects on the *in vitro* airway reactivity in the lung tissue strips could be explained by a higher expression of PDE5 in the vascular smooth muscle, which is present in a higher proportion in the lung tissue compared to the tracheal tissue.

Inhibitors of PDE4 (rolipram, cilomilast, roflumilast, or piclamilast) have demonstrated a potent anti-inflammatory and immunomodulation effects with reduced adverse effects (for roflumilast represented predominantly by gastrointestinal side effects, nausea and vomiting, and unexplained weight loss) compared to non-selective PDE inhibitors [138]. Potent anti-inflammatory, antitussive and bronchodilating action have been recently demonstrated in our study with the selective PDE4 inhibitor YM976 [68]. Roflumilast decreased the relative counts of both neutrophils and eosinophils in the blood and of eosinophils in the BAL fluid [85]. These results are in concordance with the study published by Kim *et al.* [139], who found a significant decline in the eosinophils and total white blood cells counts in mice sensitized by ovalbumin after the administration of roflumilast at the dose of 1.0 mg/kg b.w. In another *in vivo* study, the changes in the mobilization of neutrophils and eosinophils and/or their activation, probably due to an increase in intracellular cAMP levels after roflumilast therapy, were observed. This prevented the cells to release their content, and ameliorated the antigen-evoked contractions of the tracheal rings [140].

Another novel PDE4 inhibitor, MEM 1414, demonstrated a significant reduction of late responses following inhaled allergen challenge after oral pretreatment in steroid naïve atopic asthmatics [141]. These changes were associated with suppressed cytokine production from inflammatory cells. However, MEM 1414 was associated with a typical adverse event profile of PDE4 inhibitors (*e.g.* nausea and vomiting), although these were relatively mild [141].

Selective PDE5 inhibition by the long-acting PDE5 inhibitor tadalafil confirmed its effectiveness in the model of allergic inflammation [84]. Tadalafil reduced the total leukocyte count and the relative count of eosinophils in the blood; this finding is in accordance with several previous studies [81, 142, 143].

The co-administration of two selective PDE inhibitors provided significant anti-inflammatory action, *i.e.* decreased total leukocyte count and percentage of lymphocytes in the blood and decreased the percentages of neutrophils and eosinophils in the BAL fluid compared to the sensitized non-treated group [85]. Despite the fact that roflumilast and tadalafil were co-administered in reduced doses, their combination reached even more significant decline in eosinophil count in the BAL fluid than it was observed by dexamethasone, suggesting a significant anti-inflammatory potential in this dual PDE inhibition. Similar advantages of the concomitant administration of two PDE inhibitors (rolipram and BRL50481) have been recently demonstrated by our research group [47]. On the other hand, in a model of airway hyperresponsiveness in mice, the pretreatment with selective PDE4 inhibitor RO-1724 in a co-administration with sildenafil did not show any additional benefit [142]. Similarly, an earlier study published by Banner and Page [144] failed to show any significant effects of acute or chronic administration of PDE5 inhibitor in the ovalbumin-induced eosinophil infiltration in guinea pigs; they suggested that PDE4 isoenzyme played a role in the control of the allergen-induced eosinophil infiltration into the airways, and indicated that a period of low dose chronic treatment with a PDE4 inhibitor or dual PDE3/4 inhibitor was necessary for eosinophil accumulation in the airways to be reduced [144]. Furthermore, there are several studies suggesting benefits of selective PDE4 inhibitors and dual of PDE3/PDE4 inhibitors such as RPL 554 and zardaverine in the therapy of bronchial asthma [5, 120, 145 - 147].

The changes in the white blood cell counts in blood and BAL fluid are tightly associated with a secretion of several cytokines. Immunohistochemical analyses of bronchial biopsies from bronchial asthma patients have indicated the mast cells as a main source of cytokines, including IL-3, IL-4, IL-5, IL-6, IL-8, IL-10, IL-13, and TNF-α [148]. The secretion of IL-4 and IL-13 stimulates B-lymphocytes to an increased production of allergen-specific IgE. This is bound to receptors on mast cells and granulocytes and upon contact with an allergen, the rapid degranulation and release of various mediators occurs. Furthermore, IL-4 has been described as a crucial mediator responsible for a differentiation of T-cells to Th2 subtype [149]. IL-5 acts as an eosinophil-protecting and maturation factor and plays an important role in the stimulation of eosinophils in the allergic inflammation. TNF-α released from the mast cells enhances an expression of adhesion molecules (E-selectin, ICAM-1) on the endothelial cells. This enables the infiltration of leukocytes into the site of inflammation and offers an important signal for maturation and migration of dendritic cells, thus supporting the immune response. TNF-α activates alveolar macrophages and neutrophils, and together with TGF-β released from the mast cells, participates in the fibrotic changes of bronchi in chronic bronchial asthma [150]. Roflumilast and tadalafil significantly reduced the release of IL-5, and roflumilast additionally decreased the secretion of IL-4 in the lung tissue [84]. Roflumilast showed the tendency to decrease the concentration of TNF-α; however, without reaching the level of statistical significance. Page and Spina [151] suggested that the inhibition of PDE4 suppressed the expression of inflammatory mediators *via* increased intracellular concentrations of cAMP, with simultaneous suppression of TNF-α and granulocytes/monocytes colony stimulating factor (GM-CSF) secretion. The inflammation in the airways leads to tissue remodeling due to an over-expression of matrix metalloproteases (MMP), which is induced also by TNF-α. PDE4 inhibitors are potent blockers of pro-MMP secretion induced by TNF-α, and therefore may be of benefit in the diseases associated with abnormal tissue destruction and remodeling [152]. In a murine model of bronchial asthma, the selective inhibition of PDE4 with roflumilast led to a significant decrease of IL-13 and IL-4 [139], which is in concordance with our data obtained from a guinea pig lung tissue.

In other murine model of bronchial asthma, an inhaled administration of roflumilast was effective in the suppression of IL-4, IL-5, TNF-α, IgE specific for ovalbumin, and white blood cell count in the BAL fluid. However, its combination with a corticosteroid fluticasone propionate was even more potent, suggesting the utility of the combined use of PDE4 inhibitors with other anti-inflammatory or bronchodilating drugs [153]. Nevertheless, information on the effectiveness of selective PDE4 inhibitors co-administered with other drugs (corticosteroids, or bronchodilating drugs) in bronchial asthma regarding a potential dose reduction in maintaining, or even improving their effects is still lacking [153]. Roflumilast combined with tadalafil decreased concentrations of three inflammatory mediators (IL-4, IL-5, TNF-α) in ovalbumin-sensitized guinea pigs, suggesting a potential of the dual PDE4 and PDE5 inhibition [85].

Patel *et al.* [154] have recently demonstrated that selective PDE4 inhibition (cilomilast, piclamilast, and rolipram), but not PDE3 inhibition (cilostamid, cilostazol, and milrinon), enhanced the anti-inflammatory action of long-acting β_2-agonists (formoterol). This effect is mainly related to the decreased production of TNF-α and subsequently of IL-8 (CXCL-8), due to increased intracellular levels of cAMP in the smooth muscle cells after activation of β_2-receptors and induction of mitogen-activated protein (MAP) kinase phosphatase (MKP-1) [154].

The induction of inflammatory genes in bronchial asthma and COPD is stimulated also by oxidative stress. Redox-sensitive transcription factor NF-κB, an important element in a wide spectrum of inflammatory pathways regulating the activity of cytokines, chemokines, and adhesion molecules (*e.g.* TNF-α, GM-CSF, IL-8, IL-2 and IL-6, adhesion protein, E-selectin, ICAM-1 and VCAM-1) [155], is regulated by a reduction/oxidation potential, *i.e.* it is influenced by an imbalance between oxidants (reactive oxygen species, ROS) and antioxidants [156]. In our recent experiments, NF-κB present in the lung homogenate was suppressed only in the guinea pigs treated with roflumilast alone or in the combination with tadalafil, demonstrating the superior role of PDE4 on its expression [85]. Furthermore, we measured concentrations of thiobarbituric acid reactive substances (TBARS), representing one of the markers of lipid peroxidation [85]. The reason for this analysis was based on the hypothesis that ovalbumin-sensitization could increase the oxidation stress [157]. We found that the activation of eosinophils and neutrophils in the ovalbumin-sensitized guinea pigs elevated concentrations of TBARS in the lung homogenate, suggesting an overproduction of ROS. These data are in concordance with other studies showing increased concentrations of markers of oxidation stress in bronchial asthma patients [158, 159]. Oxidation stress enhances pro-inflammatory responses *via* activation of redox-susceptible transcription factors, *e.g.* NF-κB and activation protein (AP)-1 [55]. ROS stimulate PDEs with subsequent decrease of cAMP and increase in Ca^{2+} signaling in the inflammatory cells. Therefore, antioxidant action of PDE inhibitors may prevent the increase of Ca^{2+} levels in macrophage and suppress the production of TNF-α [160]. The administration of roflumilast and tadalafil caused a drop in the measured concentrations of TBARS. However, an expected decrease in TBARS was not observed in the group treated by a combination of these two drugs in reduced doses, indicating a dose-dependent antioxidant potential of selective PDE4 and PDE5 inhibitors [85]. The doses of roflumilast and tadalafil used in our study were lower than in previous experiments [161] for better comparison and examination of the efficacy with decreased risk of adverse effects. Despite the dose reduction, all of the agents conferred significant bronchodilating and anti-inflammatory effects, including action on circulating leukocytes, and lung tissue levels of cytokines and NF-κB. The difference between the action of roflumilast and tadalafil can be explained by different doses necessary to reach the same effect, as confirmed by a previous study by Takeda *et al.* [86]. They found that both roflumilast and tadalafil were (after a single dose) potent modulators of allergen-induced alterations in tracheal smooth muscle responsiveness in mice, with higher dose of tadalafil necessary. Furthermore, they confirmed a potential of co-administration at lower doses (0.1 mg/kg of roflumilast and 2 mg/kg of tadalafil). In contrary to that study, we administered these drugs for seven days, with potentially stronger action both on airway reactivity and inflammatory mediators. The doses of roflumilast and tadalafil used in our experiments were chosen in order to compare the results of roflumilast and tadalafil at the same doses in guinea pigs [84]. Higher doses of roflumilast (5 mg/kg) have been used in the study by Forkuo *et al.* [162], who observed attenuation of the asthma-like phenotype promoted by β_2-agonists by PDE4 inhibition induced by roflumilast or rolipram in phenylethanolamine N-methyltransferase-knockout mice. Finally, the effects of roflumilast and tadalafil on TNF-α and NF-κB were more obvious than those of dexamethasone, whereas the antioxidant activity of roflumilast and tadalafil was comparable to dexamethasone at the same doses [85]. Both of selective PDE inhibitors (roflumilast and tadalafil) can be used in diseases associated with allergic inflammation, which was in this study mimicked by the ovalbumin sensitization. Furthermore, patients may benefit from their synergistic effect, as only a half dose was used for their simultaneous administration, and thus the potentially observed adverse effects of PDE4 inhibitors could be minimized. The concept of dual PDE inhibition has been previously described in the literature and new dual PDE inhibitors (*e.g.* PDE3/PDE4, PDE4/PDE5, and PDE4/PDE7) are intensively studied for their anti-inflammatory action [68, 75].

There are several subtypes of bronchial asthma, potentially resistant to glucocorticosteroids and often associated with neutrophil inflammation [163]. PDE4 inhibitors are considered in their management as one of the therapeutic options due to their neutrophil-suppressive effects. Asthma-COPD overlap (ACO, previously known as ACOS, *i.e.* asthma-COPD overlap syndrome) could represent another indication of selective PDE4 inhibitors [85, 164 - 166].

ROLE OF PDE AND POTENTIAL USE OF PDE INHIBITORS IN ACUTE LUNG INJURY

Definitions of Acute Lung Injury

Acute lung damage is another respiratory indication in which PDE inhibitors can be of benefit. An older definition of this situation by the American-European Consensus Conference (AECC) [167] included: acute onset, bilateral infiltrates on chest radiography, pulmonary artery wedge pressure $\leq$18 mm Hg or the absence of clinical evidence of left atrial hypertension, and acute severe hypoxemia. If values of ratio between arterial partial pressure of oxygen and fraction of inspired oxygen (PaO_2/FiO_2) were >200 mm Hg (or 26.7 kPa) and <300 mm Hg (or 40 kPa), the situation was called "acute lung injury" (ALI). In $PaO_2/FiO_2 \leq 200$ mm Hg the situation was called "acute respiratory distress syndrome" (ARDS).

A newer definition of ARDS (also known as a Berlin Definition) states 3 categories of ARDS based on the degree of hypoxemia: mild (PaO_2/FiO_2 200-300 mm Hg), moderate (PaO_2/FiO_2 100-200 mm Hg), and severe ($PaO_2/FiO_2 \leq 100$ mm Hg) [168, 169]. For experimental studies where respiratory insufficiency is induced artificially and other clinically relevant signs besides hypoxemia cannot be determined, the term "acute lung injury" is still preferred as a general term to describe acute damage of the lung.

ALI/ARDS may result from direct (pulmonary) reasons (*e.g.,* pneumonia, aspiration, near-drowning or inhalation of toxic gases) or from indirect (extrapulmonary) reasons (*e.g.,* sepsis, trauma, blood transfusion or pancreatitis) [170]. Acute lung damage can occur in all age groups, whereas the triggering insults and clinical courses can differ a little in individual forms of ALI/ARDS. From this reason, ALI/ARDS in adults and older children is discussed in this chapter separately from special forms of ALI in the newborns.

Acute Lung Injury in Adults and Older Children

Pathophysiological Features

Although the mechanisms that initiate lung injury are not completely understood, the insult (pulmonary or extrapulmonary) increases the permeability of microvascular barriers to water, proteins and other solutes that leads to a rapid formation of perivascular and intraalveolar edema. Acute lung damage is associated with an increased coagulation and a depressed fibrinolysis that results in a diffuse alveolar deposition of fibrin, which amplifies lung inflammation and injury [171]. The loss of epithelial integrity enables alveolar flooding and alters removal of fluids from the alveolar spaces. Damage to epithelial cells type II alters transport of ions and production of surfactant. Increased permeability of the alveolar-capillary barriers is also linked to a massive transfer of leukocytes and erythrocytes into the alveolar space [172, 173]. Damaged cells trigger the release of a broad spectrum of mediators including reactive oxygen species (ROS), pro-inflammatory cytokines such as interleukins (IL)-1β, IL-6, IL-8 or tumor necrosis factor (TNF)α, which activate polymorphonuclears (PMN), particularly neutrophils, as well as macrophages and other cells, and stimulate neutrophil chemotaxis into the site of injury. Pro-inflammatory substances increase apoptosis of lung epithelial cells, causing further damage and dysfunction of the alveolar lining. Delayed apoptosis of neutrophils, providing a prolonged persistence of activated neutrophils in the lung, aggravates potential damage of the tissue by their products [174]. A schematic displaying the complex pathophysiology of

ALI/ARDS is provided in Fig. (**2**) [173].

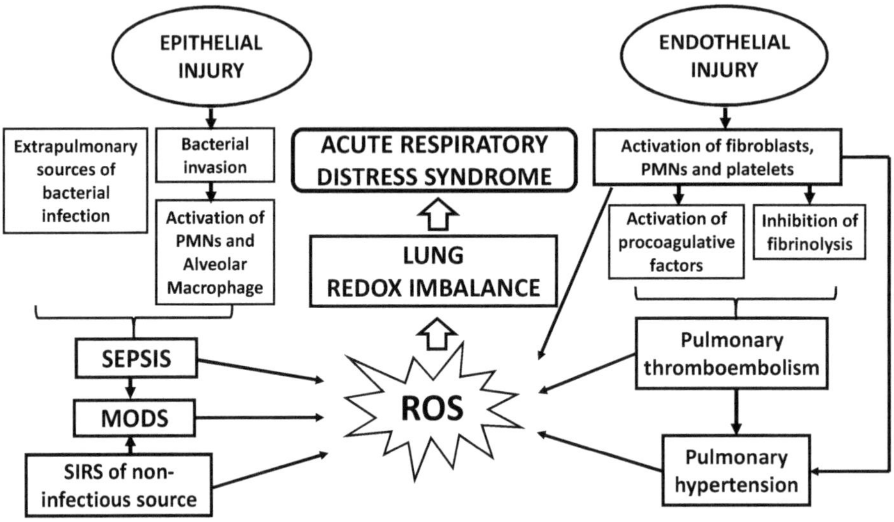

Fig. 2. Pathophysiology events related to the development of ARDS. MODS: multiple organ dysfunction syndrome, PMNs: polymorphonuclear leukocytes, PMNn: polymorphonuclear neutrophils, SIRS: systemic inflammatory response syndrome, ROS: reactive oxygen species (modified according to [173]).

The substances released in high concentrations from the injured lung cells or produced by the activated cells spill over into the circulation and thereby can serve as valuable biomarkers of ARDS, reflecting dynamics of the process and the future prognosis [175, 176]. In addition, some of these mediators can evoke a systemic inflammatory response (or systemic inflammatory response syndrome, SIRS) for which the magnitude of SIRS determines the severity of ARDS. SIRS can originate from infective causes (such as sepsis, defined as SIRS with identifiable focus of infection) or non-infective causes (haemorrhagic shock, acute pancreatitis, burns *etc.*). Excessive systemic cytokine-mediated leukocyte activation can lead to injury of organ microcirculation, local hypoperfusion and hypoxia, and finally to multiple organ dysfunction syndrome (MODS) and multiple organ failure. *Vice versa*, in severe extrapulmonary injury, the activated leukocytes participating in SIRS migrate into the pulmonary circulation where they increase endothelial permeability and cause lung edema [177].

The differences between the ARDS pathomechanisms of pulmonary and extrapulmonary origin provoked debates on a clinical impact of these changes on the therapeutic strategies [178], as some therapies can have a different efficiency in the primary epithelial injury (ARDS of pulmonary origin) and the primary endothelial injury (ARDS of extrapulmonary origin) [179]. For instance, methylprednisolone more effectively attenuated the inflammatory changes, lung mechanics and morphometrics in direct as compared to indirect experimental ALI [180]. On the other hand, threapeutic response to recruitment maneuvers and positive end-expiratory pressure (PEEP) on oxygenation and lung mechanics was more significant in animals with indirect ALI than in direct ALI [181]. However, the number of experimental and clinical studies comparing a response to various pharmacological agents in the pulmonary and extrapulmonary forms of ARDS is insufficient to bring any conclusions in this topic.

A resolution of ALI/ARDS within several days-weeks after the triggering insult is realized by synchronous reabsorption of alveolar edema, repair of the epithelial and endothelial barriers, and removal of inflammatory cells and exudate from the distal airspaces [172, 182]. The situation can finish with a complete healing or can progress into interstitial lung fibrosis.

Treatment Approaches in ALI/ARDS

Despite improvement in understanding the pathophysiology of ARDS, the only recommended therapy of this syndrome at the moment is a lung-protective mechanical ventilation. There are two generally accepted rules for an appropriate ventilation of the injured lung: 1) use of low tidal volumes (< 6 ml/kg of the body weight) combined with limited inspiratory plateau pressures (< 30 cm H_2O) to prevent a lung overdistension („barotrauma/volutrauma“); 2) application of high levels of PEEP to prevent a repetitive opening and closing of the terminal lung units („atelectrauma“) [183].

Several groups of medications have been also used [184, 185]. However, none has proven to be commonly effective, although some may be beneficial in subgroups of patients with specific causes of the lung injury that might make them more responsive than others [172, 186]. One of these perspective approaches is the use of PDE inhibitors, as these agents exert a wide spectrum of actions potentially beneficial in ALI/ARDS.

Use of PDE Inhibitors in ALI/ARDS

Both non-selective and selective PDE inhibitors have been tested in *in vitro* conditions and in *in vivo* models of ALI or in patients with ARDS.

In *in vitro* studies, changes similar to ARDS can be evoked by exposure of isolated cells to *e.g.* lipopolysaccharide (LPS). In peripheral blood mononuclear cells stimulated with LPS or recombinant human IL-1β, non-selective PDE inhibitors theophylline and pentoxifylline suppressed production of TNFα and IL-1β [187, 188]. Similarly, pentoxifylline, caffeine, and theophylline derivative aminophylline attenuated TNF-induced PMN-mediated pulmonary artery endothelial cell cytotoxicity [26]. In pulmonary artery endothelial cells incubated with endotoxin, with or without the presence of neutrophils, pentoxifylline attenuated an increase in albumin clearance, and lessened cell lysis and morphological changes [189].

Furthermore, in the blood of COPD-patients and healthy controls *ex vivo* pre-incubated with a derivative of caffeine, 1,7-dimethylxanthine, LPS-induced production of IL-6 and TNFα was significantly suppressed [190]. Another xanthine derivative, 3-isobutyl-1-methyl-xanthine (IBMX), increased both cAMP and cGMP within 30 min of incubation with bovine pulmonary endothelial cells stimulated by LPS and thereby protected the cells from LPS effects [191]. IBMX also suppresses LPS-induced synthesis of TNFα in mononuclear cells [187].

Anti-inflammatory action was observed in *in vitro* studies for several PDE4 inhibitors. Roflumilast reduced the LPS-induced release of TNFα and chemokines involved in the recruitment of monocytes and T-cells, but not of neutrophils [192]. Similarly, treatment of monocytes with other agents inhibiting PDE4 enzymes abolished LPS-induced TNFα generation [193]. In addition, roflumilast and other PDE4 inhibitors reduced fibroblast chemotaxis, fibroblast-driven collagen gel contraction [194], their transition to myofibroblasts [195], release of matrix metalloproteases [152], and suppressed markers of proliferation, myofibroblast transition, extracellular matrix formation and inflammation in the lung fibroblasts [196] suggesting that PDE4 inhibitors might be useful also in mitigation of the fibrotic changes in ALI/ARDS. Recent studies have shown that roflumilast reduces LPS-induced stress fibers, an essential requirement for a direct migration of PMN into the alveolar space [197], whereas cilomilast treatment of deletion of the PDE4B gene inhibited the LPS-induced NF-κB activation and inflammatory response in multiple cell types, including lung epithelial cells, pulmonary microvascular endothelial cells and vascular smooth muscle cells [198]. Similarly, the PDE3 inhibitor olprinone was able to inhibit TNFα and IL-6 release, and to augment IL-10 release in rat alveolar macrophages exposed to LPS [52].

In vivo models of ALI can be distinguished according to the primary target: 1) models with a primary dysfunction of the pulmonary epithelium, such as intratracheal LPS-induced model, saline lavage-induced model of surfactant depletion, hyperoxia-induced model, acid-aspiration model, intratracheal bleomycin induced model of ALI *etc*.; 2) models with a primary dysfunction of the endothelium, such as intravenous LPS-induced model, oleic acid-induced model of ALI *etc*.; and 3) models with a dysfunction of both epithelium and endothelium, such as models of sepsis or ischemia-reperfusion [199, 200].

Regarding the findings that ARDS from pulmonary and extrapulmonary origin can exert not only differences in the pathophysiology, but also in the response to the treatment [178, 201], we have reviewed the effects of PDE inhibitors in *in vivo* studies and in clinical studies separately for ARDS of pulmonary and extrapulmonary origins.

ARDS of Pulmonary Origin

The most frequent reason for pulmonary ARDS is pneumonia, which is often invoked by intratracheal LPS instillation in *in vivo* models.

As listed below, administration of non-selective PDE inhibitors theophylline, aminophylline, pentoxifylline, lisofylline or caffeine effectively suppressed lung inflammation and vascular permeability in animal models, while some appeared to be of benefit also in clinical studies. For instance, theophylline inhibited neutrophil recruitment, TNFα release and cellular activation in the bronchoalveolar lavage (BAL) fluid in LPS-induced ALI in rats [202]. Pretreatment with aminophylline, a derivative of theophylline, prevented an increase in pulmonary artery pressure and lung weight, and prevented an increase in pulmonary vascular permeability in two models of ALI [203, 204]. In a rabbit lung exposed to phosgene, posttreatment with aminophylline lowered LTC4/D4/E4-mediated pulmonary capillary permeability, pulmonary artery pressure, tracheal pressure and lung weight gain, and reduced lipid peroxidation expressed by thiobarbituric acid-reactive substances (TBARS) [205]. Pretreatment with pentoxifylline attenuated an increase in pulmonary artery pressure and lung weight gain in a protamine-induced ALI [204], and prevented increases in lung water and albumin lung-to-plasma ratio as well as increases in cell numbers in the BAL fluid in IL2-induced ALI in guinea pigs [206]. Lisofylline prevented both neutrophil-dependent and neutrophil-independent oxidant-induced capillary leak in isolated rat lungs [207]. In rats given IL-1 intratracheally, lisofylline decreased lung leak and lung lipid alterations without decreasing neutrophil accumulation or lung lavage cytokine-induced neutrophil chemoattractant (CINC) levels [208].

Chronic caffeine administration for 2 weeks before ALI, as well as acute administration at a high dose 30 min before oleic-acid induced ALI, attenuated lung edema, hemorrhage, neutrophil recruitment and expression of TNFα and IL-1β in both wild type mice and A_{2A} receptor knockout mice. This was accompanied by increased cAMP levels and up-regulation of A_{2B} receptor mRNA in the lungs. In contrast, acute caffeine administration at a low dose 30 min before induction of ALI enhanced the inflammation and lung damage in wild type mice with decreasing cAMP levels, but not in A_{2A} receptor knockout mice. These results suggested that caffeine either enhances lung damage by antagonizing A_{2A} receptor or protects against lung damage *via* A_{2A} receptor-independent mechanisms, depending on the time of exposure (chronic *vs.* acute) and dose of administration (high *vs.* low) [209]. A derivative of caffeine, 1,7-dimethylxanthine (or paraxanthine), has been reported to inhibit the nuclear enzyme poly(ADP-ribose) polymerase-1 (PARP-1), which plays an important role in LPS-induced lung inflammation. Paraxanthine attenuated lung myeloperoxidase, transcription of IL-6, TNFα, MIP1α and MIP2 genes, and PAR-polymer formation in mice with LPS-induced ALI. Serum amyloid P component and plasma IL-6 also decreased after the treatment, indicating a reduced systemic inflammatory response [190].

Several selective PDE inhibitors have been used successfully in models of ALI/ARDS. Considering a distribution of individual PDEs in the epithelial lung cells and immune cells, PDE3, PDE4 and PDE5 inhibitors are of particular attention in the treatment of the acutely damaged lung, whereas the anti-inflammatory potential of PDE4 inhibitors seems to be superior. In rats with LPS-induced ALI, the PDE4 inhibitor rolipram inhibited neutrophil recruitment, TNFα release and cellular activation in the BAL fluid [2012], while inhaled PDE4 inhibitor GSK256066 inhibited LPS-induced pulmonary neutrophilia and increase in exhaled nitric oxide [210]. In addition, Turner *et al.* [211] found that pretreatment with rolipram in LPS-injured rats improved mortality, decreased plasma TNFα, reduced lung edema formation and alveolar hemorrhage, reduced hemoconcentration and thrombocytopenia, and inhibited LPS-induced alterations in renal and hepatic function, but did not influence LPS-induced pulmonary neutrophilia. In fact, LPS-induced changes in circulating leukocyte populations were actually exacerbated by rolipram. Pretreatment with rolipram in a murine model of ALI induced by LPS and zymosan reduced extravascular albumin accumulation and neutrophil sequestration in the lung, attenuated ALI-induced increase in serum TNFα, however, dissociation between the TNFα production and induction of injury in this model suggests that the anti-inflammatory effect of rolipram is independent of TNFα inhibition [212]. These findings were confirmed by another study, where pretreatment with rolipram or SB 207499 (another PDE4 inhibitor) in LPS-injured rats conferred a dose-dependent attenuation of the lung inflammation. Nevertheless, the lung production of TNFα and IL-1β was unaffected by either SB 207499 or rolipram, indicating that the inhibitory actions of rolipram or SB207499 are not linked to the production of TNFα or the inhibition of IL-1β [213].

In mice with LPS-induced ALI, roflumilast and rolipram decreased transendothelial and transepithelial migration, even when administered after the inflammation was evoked, with superior effect of roflumilast on the epithelium. Both agents decreased TNFα, IL-6, and CXCL2/3. CXCL1, a strong PMN chemoattractant secreted by the epithelium, was more reduced by roflumilast

[197]. A recent *in vivo* study by Ma *et al.* [198] on knockout mice has demonstrated that a PDE4B deletion can reduce LPS-induced vascular permeability, attenuate the lung water content, histological signs of lung injury, and elevate the PaO_2/FiO_2 ratio [198]. In addition, PDE4B showed a predominant role in controlling proliferation of lung fibroblasts [214]. This finding was confirmed by *in vivo* studies where selective PDE4 inhibitors suppressed early inflammatory stage and attenuated the late stage of bleomycin-induced lung fibrosis [215, 216]. With regard to the importance of fibroproliferative changes in the late phase of ARDS, PDE4 inhibitors may be beneficial not only for mitigation of inflammation but also to prevent fibrotic changes in the injured lung.

Although the use of PDE4 inhibitors in ALI/ARDS is well described, information on the use of PDE3 inhibitors in ALI/ARDS is limited. In one study, pretreatment with PDE3 inhibitor olprinone in rats with LPS-induced ALI inhibited neutrophil influx into the lung, suppressed pro-inflammatory cytokines TNFα and IL-6 and increased anti-inflammatory cytokine IL-10 [52].

In contrast to the many pre-clinical studies, there is a lack of information on the use of PDE inhibitors in patients with ARDS. Two decades ago, a non-selective PDE inhibitor lisofylline appeared to show potential for the treatment of ARDS, as it inhibits a lysophosphatidic acyltransferase and decreases a release of cell membrane-derived free fatty acids, which are elevated in patients with ARDS. In a small group of septic patients, lisofylline reduced an ARDS-induced increase in the ratio of serum free fatty acids (*i.e.* the ratio of C18 unsaturated fatty acids linoleate and oleate to fully saturated palmitate) [217]. However, although the randomized, double-blind, placebo-controlled, multicenter study showed a decrease in circulating free fatty acids, it found no additional benefit in patients with ARDS and was stopped early [218]. For this reason, lisofylline is actually not recommended for the treatment of ARDS [186, 219].

ARDS of Extrapulmonary Origin (Sepsis, Shock and Multiple Organ Failure)

Similarly to ARDS of pulmonary origin, PDE inhibitors have been widely used also in sepsis. In septic guinea pigs, the non-selective PDE inhibitor aminophylline decreased albumin leak, but with a decrease in mean arterial pressure, while pentoxifylline decreased albumin leak without a decrease in the mean arterial pressure [220]. In contrast, in rats with *E.coli*-induced sepsis, pentoxifylline reduced serum TNFα, but aggravated a systemic hypotension, did not modify neutropenia, thrombocytopenia or microvascular permeability assessed by organ wet/dry weight ratios and did not improve 24 h survival [221]. In a porcine model of gram-negative sepsis, both pretreatment and posttreatment with pentoxifylline attenuated lung injury and improved oxygenation, but plasma

TNFα, blood PMN counts, and CD18 expression were unaffected. The cardiac index was improved by administration of pentoxifylline if given prior to the onset of sepsis or given 1 h after the onset of bacterial infusion. Administration of pentoxifylline 2 h after the onset of *Pseudomonas aeruginosa* infusion (in established septic shock) caused uncontrolled, fatal systemic hypotension in two of the four animals from the group [222]. Similarly, in spontaneously breathing rats with endotoxemia, aminophylline at a dose of 1 mg/kg failed to prevent the endotoxemia-induced respiratory and hemodynamic manifestations of sepsis [42]. In contrast, in a rat model of infrarenal aortic cross-clamping induced ALI, pentoxifylline attenuated the secondary ALI as it reduced histological scores and wet lung contents, improved blood gases, decreased cell counts and protein content in the BAL fluid, inhibited myeloperoxidase (MPO) activity and ICAM-1 expression in the lung and lowered plasma levels of TNFα, IL-6, IL-1β and soluble ICAM-1, and stimulated a production of anti-inflammatory IL-10 [223]. In agreement with the study by Li *et al.* [223], pentoxifylline given for 3 days in rats with a nitrogen mustard-induced ALI reduced the histological signs of lung injury, suppressed inflammation, reduced oxidative stress, and upregulated the markers of wound repair including pro-surfactant protein-C and proliferating nuclear cell antigen by type II cells [224]. The anti-inflammatory action of pentoxifylline has been recently demonstrated also in mice with cecal ligature and puncture (CLP)-induced ALI model, where pretreatment with pentoxifyllin attenuated lung injury and reduced mortality, decreased secretions of IL-2, IL-6, IL-10 and IL-17. Furthermore, the raising cAMP levels partly restored the ratio of T-regulatory cells (Treg) and interleukin (IL)-17-producing T-helper lymphocytes (Th17) by modulating the transcription of Foxp3 and RORγt through the STAT3 pathway [225].

Lisofylline, a pentoxifylline derivative, given 30 min prior to sepsis or 1 h after injection of bacteria attenuated sepsis-induced pulmonary hypertension, neutropenia, and hypoxemia, and reduced wet-to-dry weight ratio, tissue albumin index, content of TBARS, and myeloperoxidase (MPO) activity in sepsis-injured minipigs. However, its efficacy was blunted in the group treated 2 h after bacteria injection [226]. In the later work of this research group, lisofylline given as a pretreatment or a treatment 1 h after sepsis significantly attenuated sepsis-induced neutrophil attachment to fibronectin. Sepsis-enhanced phagocytic activity was reduced in the lisofylline pretreatment sepsis group, but not in the post-treatment groups. Sepsis caused neutropenia, pretreatment produced neutrophilia, and posttreatment 1 h after injection of bacteria caused the neutropenia to recover to control levels [227].

In addition to non-selective PDE inhibitors, selective PDE4 and PDE3 inhibitors are effective in the treatment of extrapulmonary forms of ARDS. In sepsis and systemic inflammation, increased microvascular permeability and consecutive breakdown of microcirculatory flow contribute to organ failure and death. Therefore, Schick *et al.* [228] have studied the role of cAMP levels in the endothelial cells of rats with systemic LPS-induced systemic inflammation to maintain capillary endothelial barrier properties in an acute inflammation and evaluated the potential of PDE4 inhibitors to modulate these changes. It was found that rolipram blocked a reduction in cAMP levels in mesenteric microvessels. Increased endothelial cAMP by administration of either rolipram or roflumilast led to stabilization of endothelial barrier properties, improved microcirculation, and improved metabolism as indicated by blood gas analyses [228]. The therapeutic potential of PDE4 inhibitors in stabilizing endothelial cAMP levels followed by activation of Rho-GTPase Rac1 has been recently confirmed also in other studies [229]. Microcirculatory dysfunction is correlated with increased mortality among septic patients and is believed to be a major contributor to the development of acute kidney injury, as well. In mice with ALI model, rolipram acutely restored capillary perfusion, increased renal blood flow despite transiently decreasing mean arterial pressure, reduced increased renal microvascular permeability. Delayed treatment with rolipram at 6 h after induction of sepsis restored the renal microcirculation, reduced blood urea nitrogen and serum creatinine, and increased glomerular filtration rate at 18 h. However, delayed treatment with rolipram did not reduce serum nitrate/nitrite levels, a marker of nitric oxide production, nor reactive nitrogen species generation in renal tubules [230]. In rats with polymicrobial sepsis due to peritonitis, systemic administration of low-dose rolipram improved microcirculatory flow, whereas high-dose therapy had adverse effects. Furthermore, rolipram decreased IL-1α, IL-1β, IL-12, and TNFα levels, which appeared not to correlate with barrier-stabilizing effects of rolipram, and alleviated sepsis-induced acute kidney injury and lung edema [231].

In vivo studies showed that PDE3 inhibitors can exert valuable effects in sepsis, as well. For instance, in a polymicrobial sepsis induced by cecal ligation and puncture (CLP) in BALB/c mice, olprinone given following CLP prevented the ALI development in CLP mice, as indicated by alleviated hypoxemia, pulmonary vascular hyperpermeability, and histological lung damage. Furthermore, administration of olprinone suppressed apoptosis induction in septic lungs and improved the survival of CLP mice [232]. Olprinone appeared to have beneficial therapeutic effects in restoring not only systemic circulation, but also improved hepatic circulation and mitochondrial oxidation in the liver in a porcine model of endotoxic shock [233]. In zymosan-induced multiple organ failure in mice, treatment with olprinone attenuated the peritoneal exudation and the migration of polymorphonuclear cells, alleviated the lung, liver and pancreatic injury, renal dysfunction and mitigated lung and intestine myeloperoxidase (MPO) activity increased by zymosan. Immunohistochemical analysis showed a reduced degree of staining for nitrotyrosine, iNOS, PAR, TNFα and IL-1β in tissue sections obtained from zymosan-injected mice, treated with olprinone. In addition, olprinone reduced the development of systemic toxicity, loss in body weight and mortality, caused by zymosan [234]. As recently demonstrated in mice with peritoneal sepsis, PDE inhibition can attenuate also an impaired mitochondrial cAMP-PKA signaling in the septic myocardium [235].

Despite the fact that several non-selective and selective PDE inhibitors appear to be beneficial in the animal studies, the information on their use in patients with sepsis is rather rare. Bacher *et al.* [236] published that administration of pentoxifylline in septic patients significantly improved hemodynamic performance (increased heart rate and cardiac index and decreased systemic vascular resistance index and peripheral vascular resistance), which was accompanied by an increase in oxygen transport and oxygen uptake with unchanged oxygen extraction ratio. Recent analysis of data from six small studies performed by Cochrane Database System showed that pentoxifylline decreased mortality and length of hospital stay in patients with neonatal sepsis without any adverse effects [237].

Pulmonary Hypertension Related to ALI/ARDS

Pulmonary hypertension associated with ALI/ARDS is related to alterations in the pulmonary blood flow and injury to lung microcirculation. The situation is complicated by procoagulant and thrombotic events in the pulmonary arteries. These changes further impair gas exchange, where hypoxia and hypercapnia progressively increase pulmonary vascular resistance and worsen pulmonary hypertension [238].

Pulmonary hypertension negatively influences the performance of the right ventricle. Acute *cor pulmonale*, described as right ventricle dilation with septal dyskinesia, impairment of left ventricle diastolic filling and compensatory tachycardia to preserve cardiac output occurs in 25% of patients with ARDS [239]. In this situation, administration of systemic vasodilators can further aggravate the ventilation/perfusion mismatch due to increases in intrapulmonary shunt and hypoxemia induced by the simultaneous dilation of systemic and pulmonary vessels, and might favor systemic hypotension [240]. To avoid systemic effects, selective pulmonary vasodilators have been introduced in the treatment of ARDS. However, although inhaled NO and aerosolized prostacyclin reduced pulmonary hypertension, improved ventilation/perfusion matching and ameliorated hypoxemia, no survival advantage was found [241, 242].

PDE inhibitors, particularly selective PDE5 inhibitors (*e.g.*, sildenafil) are the other drugs potentially beneficial for this purpose. In older studies, pretreatment with the non-selective PDE inhibitor aminophylline in an isolated rabbit lung reduced pulmonary hypertension and lung edema in HCl-induced ALI [243], and treatment with aminophylline dose-dependently attenuated both pulmonary hypertension and bronchoconstriction in mongrel dogs with serotonin-induced model of pulmonary hypertension and bronchoconstriction, whereas these two effects were equipotent [244]. Acute hypoxic pulmonary vasoconstriction was markedly blunted by the PDE5 inhibitor sildenafil in the isolated perfused lung of wild-type and endothelial NO synthase (eNOS)-deficient mice, as well. Wild-type mice dosed orally with sildenafil throughout 3 weeks of exposure to hypoxia exhibited a significant reduction in right ventricular systolic pressure coupled with a small reduction in right ventricular hypertrophy and inhibition of pulmonary vascular remodeling. In eNOS mutant mice, sildenafil attenuated the increase in right ventricular systolic pressure, but without a significant effect on right ventricular hypertrophy or vascular remodeling. The results indicated that the response to sildenafil is partially mediated through the eNOS-NO-cGMP pathway, but other biochemical sources of cGMP also play a role [245].

Positive effects on pulmonary hypertension were expected also for PDE3 inhibitors, as they enhance myocardial contraction, produce vasodilation, and suppress platelet aggregation. The PDE3 inhibitor olprinone dose-dependently attenuated both pulmonary hypertension and bronchoconstriction, whereas the vasodilation effect was more potent to bronchodilation [244], while other PDE3 inhibitor milrinone seemed to have a greater impact on the airway smooth muscle than in the pulmonary vascular smooth muscle [246]. In isolated and perfused rabbit lungs, both olprinone and milrinone inhibited hypoxic pulmonary vasoconstriction [247]. In a canine model of hypoxic pulmonary hypertension, intravenous olprinone at a high dose increased heart rate and decreased mean aortic pressure, mean pulmonary arterial pressure, pulmonary vascular resistance, systemic vascular resistance and right ventricular stroke work index, while at lower doses the changes were less obvious [248].

Compared to experimental studies, results from clinical studies on the use of PDE inhibitors are rather contradictory. In a randomized double-blind study, sildenafil at a dose of 100 mg given orally to healthy volunteers 1 h before breathing 11% O_2 for 30 min almost abolished the acute increase in mean pulmonary artery pressure produced by hypoxia, with no significant effect on systemic blood pressure [245]. Sildenafil at a dose of 50 mg given *via* nasogastric tube decreased mean pulmonary arterial pressure and pulmonary artery occlusion pressure, but decreased systemic mean arterial pressure in patients with pulmonary hypertension. Sildenafil did not improve pulmonary arterial oxygen tension and arterial oxygenation, even resulted in a further increase in the shunt fraction [249].

The effects of another PDE5 inhibitor, tadalafil, were evaluated from a small number of studies available on Medline and International Pharmaceutical Abstracts. Tadalafil was effective and well tolerated in patients with pulmonary arterial hypertension and was associated with significant improvements in exercise capacity and reduced time to clinical worsening [250].

Acute Lung Injury in the Neonates

Respiratory Distress Syndrome (RDS)

RDS (or also called hyaline membrane disease) is a life-threatening condition that occurs particularly in preterm neonates. Signs of respiratory distress develop soon after birth, resulting from immaturity of the lung and insufficient synthesis of pulmonary surfactant [251, 252]. To prevent RDS, antenatal steroids are given to mothers to enhance maturation of the fetal lung and to promote surfactant production. Mild forms of RDS respond to supplemental oxygen and application of continuous positive airway pressure (CPAP), while more severe cases require endotracheal intubation and administration of exogenous surfactant with or without a need of mechanical ventilation [252, 253]. Recent studies have shown that administration of PDE inhibitors can be a valuable part of the therapeutical protocol as they can additionally improve the status of the neonates who suffer from respiratory distress.

In an *in vitro* study by Schüller *et al.* [254], pentoxifylline downregulated expression of CD14, CD11b, CD64, CD71 and CD80 in blood monocytes in a concentration-dependent manner, with the greatest effect on CD14 and CD11b observed for the samples of preterm infants. Pentoxifylline suppressed production of TNFα, IL-1β and IL-6 in samples from all age groups (adults, term and preterm neonates), reduced an early production of IL-10 in monocytes of the preterm and term neonates and downregulated TLR4 expression, decreased signaling and suppressed phagocytosis, confirming strong anti-inflammatory properties of pentoxifylline [254].

Similar favorable effects were observed in animal models of RDS, in which the surfactant depletion resembling the changes in RDS is evoked by a repetitive saline lung lavage. In our recent experiments carried out in saline-lavaged rabbits, the non-selective PDE inhibitor aminophylline reduced transmigration of leukocytes into the lung, decreased lung edema formation, and improved gas exchange and right-to-left pulmonary shunts compared to the non-treated group with ALI. Furthermore, aminophylline decreased oxidative stress and markers of inflammation and tissue injury (RAGE, TNFα, and IL-6) in the lung, and decreased lung epithelial cell apoptosis (unpublished observations).

Beneficial effects have been published also for selective PDE inhibitors. For instance, in a lung lavage model of ALI in rats, intratracheally delivered PDE4 inhibitor roflumilast prevented hyaline membrane formation and lung infiltration of neutrophils, but did not influence oxygenation compared with untreated control animals. However, additional improvement of the therapy on oxygenation and HM formation was observed when roflumilast was added to rSP-C surfactant [255]. In our recent experiments performed on saline-lavaged rabbits, PDE4 inhibitor roflumilast, PDE3 inhibitor olprinone and PDE5 inhibitor sildenafil reduced the leak of cells, particularly of neutrophils, into the lung, which was associated with a decrease of circulating neutrophils in the blood. Roflumilast and olprinone decreased lung edema formation expressed as a wet/dry lung weight ratio. All tested agents improved oxygenation and reduced right-to-left pulmonary shunts, decreased pro-inflammatory cytokines and markers of tissue injury, diminished markers of oxidative stress and prevented an ALI-induced decrease in total antioxidant status. Finally, treatment with the PDE inhibitors decreased the number of apoptotic lung epithelial cells [256].

Bronchopulmonary Dysplasia (BPD) or Chronic Lung Disease (CLD)

Impaired alveolar and vascular development in prematurity may result in bronchopulmonary dysplasia (BPD) or chronic lung disease (CLD), which is defined as a supplemental oxygen dependency for at least 28 days from the birth, and at 36 weeks corrected gestational age [251, 252]. When RDS in extreme prematurity progresses into BPD, besides supportive treatment and supplemental oxygen, corticosteroids are used to reduce the lung inflammation [252, 253]. However, PDE inhibitors have appeared to be of benefit also.

For instance, the non-selective PDE inhibitor caffeine was shown to Smad signalling and TGF-β1 regulated genes involved in the airway remodelling, suggesting its potential in BPD [257]. Caffeine was effective also in animal models of hyperoxia-induced ALI, which serve as models of BPD. In hyperoxia-induced lung injury, early caffeine treatment elevated cAMP levels in the lungs, increased levels of phosphorylated endothelial nitric oxide synthase (eNOS) at serine1177, total and serine51 phosphorylated GTP cyclohydrolase 1 (GCH1), and tetrahydrobiopterin (BH4) levels, with improved alveolar structure and angiogenesis [258]. Teng *et al.* [259] found that caffeine treatment attenuated hyperoxia-impaired alveolar formation and reduced endoplasmic reticulum stress. Caffeine also suppressed the hyperoxia-induced activation of cyclooxygenase-2 and markers of apoptosis [259]. In another study, caffeine attenuated chemokine and cytokine upregulation, as well as the influx of CD11b(+), ED-1(+) and myeloperoxidase(+) leukocytes, suggesting that the protective effects of caffeine in the neonatal lung are mediated, at least in part, by reducing pulmonary inflammation [260]. Caffeine is used also to treat apnea of prematurity [261, 262]. Due to complex anti-oxidant, anti-inflammatory, and anti-apoptotic properties, and potential neuroprotective effects on the developing brain [263], caffeine has become an important part of the therapeutic protocol in the preterm infants [264 - 266].

Pentoxifylline is another non-selective PDE inhibitor potentially beneficial in CLD [267], displaying both potent anti-inflammatory properties and unique hemorrheologic effects, *i.e.* lowers blood viscosity, and improves microcirculation and tissue perfusion, with very rare adverse effects [268].

In the treatment of CLD in very preterm infants, selective PDE4 inhibitors can be also valuable. In preterm rat pups exposed to hyperoxia, treatment with PDE4 inhibitors rolipram and piclamilast prolonged a median survival by up to 7 days and reduced alveolar fibrin deposition, lung inflammation and vascular leakage by decreasing an influx of monocytes and macrophages and protein efflux in the BAL. Analysis of mRNA expression demonstrated a significant reduction of expression in key genes involved in inflammation, fibrin deposition and alveolarization suggesting the utility of these agents in CLD [269].

Meconium Aspiration Syndrome (MAS)

MAS is a serious respiratory disorder that is usually observed in term neonates. Aspiration of meconium (the first stools of the newborn) into the lungs causes airway obstruction, surfactant dysfunction, neutrophil-mediated lung inflammation, lung edema formation, and pulmonary vasocontriction resulting in persistent pulmonary hypertension of the newborn (PPHN) [20, 270]. Treatment approaches have been historically based on oro- and nasopharyngeal suctioning after delivery of the head but before delivery of the shoulders. Routine suctioning on the perineum is no longer recommended as it does not reduce incidence of MAS and endotracheal suctioning is limited to seriously depressed infants [271]. Severe cases of MAS require oxygen supplemention, *via* CPAP or mechanical ventilation. Further improvement can be seen with the use of exogenous surfactant and inhaled NO, or anti-inflammatory drugs, *e.g.* corticosteroids [20, 272]. However, PDE inhibitors can be also beneficial in this indication.

The non-selective PDE inhibitor pentoxifylline in an *in vitro* study decreased a meconium-induced degranulation of elastase and lactoferrin from azurophilic and specific granules of PMNs [17]. In piglets with meconium-induced lung injury, pentoxifylline prevented an increase in the concentrations of TNFα and protein as well as counts of alveolar macrophages in the BAL fluid, but had no significant effect on pulmonary neutrophil accumulation [18]. In rats with meconium-induced ALI, pentoxifylline pretreatment showed a trend to prevent an increase in

TNFα and protein content in the BAL fluid and to improve oxygenation index compared to non-treated animals [273].

In our study performed on meconium-instilled rabbits, intravenous aminophylline improved gas exchange, decreased right-to-left pulmonary shunts, and ventilatory pressures, reduced lung edema formation, airway hyperreactivity to histamine, neutrophil counts in the BAL fluid, and diminished oxidative modifications of proteins and lipids in the lung [19]. In the following experiments, comparison of two doses showed that aminophylline delivered at a high dose (2 mg/kg) was superior to aminophylline at a low dose (1 mg/kg) with regard to improvement of lung function, and reduction of lung edema, lipid oxidation, tracheal reactivity to histamine and migration of neutrophils into the lung. On the other hand, low-dose aminophylline suppressed protein oxidation and meconium-induced lung tissue hyperreactivity to histamine proving that low-dose methylxanthines exert some anti-inflammatory properties [274].

As mentioned above, treatment with PDE inhibitors is often associated with serious side effects. To evaluate acute cardiovascular effects of PDE inhibitors associated with their delivery, we analyzed changes in the blood pressure, heart rate (HR) and heart rate variability (HRV) during and within 5 h after intravenous aminophylline and PDE3 inhibitor olprinone administration in a rabbit model of MAS. In meconium-instilled animals, both agents acutely increased blood pressure and heart rate, whereas HRV remained higher until the end of the experiment and correlated well with markers of cardiovascular injury [43, 56]. This finding of acute cardiovascular changes associated with both intravenous aminophylline and olprinone indicate that if these drugs are used for treatment of MAS, their possible cardiovascular effects should be considered and thoroughly monitored, particularly in patients with cardiovascular instability. On the other hand, olprinone in meconium-instilled rabbits effectively reduced the numbers of PMNs in the BAL fluid, decreased the formation of oxidation markers in the lungs, reduced lung edema and prevented a decrease in total antioxidant status (TAS) in the lung and plasma [55].

PDE inhibitors can be easily combined with other anti-inflammatory drugs, *e.g.* with corticosteroids, providing better therapeutic effect in combination. In our study, combination of budesonide and aminophylline more effectively reduced intrapulmonary shunts, lung edema and the number of lung neutrophils, and improved gas exchange in meconium-instilled rabbits when compared to aminophylline-alone treatment, without significant cardiovascular effects, suggesting possible benefits of such a combination also in patients with MAS [44].

Persistent Pulmonary Hypertension of the Newborn (PPHN)

PPHN is characterized by a failure of the pulmonary vasculature to adapt to the *ex-utero* environment after birth. This results in a persistence of high pulmonary vascular resistance (PVR), right-to-left shunting including shunts *via* patent ductus arteriosus and foramen ovale, leading to pulmonary hypoperfusion, hypoxia and acidosis. PPHN can be primary or secondary to an associated lung condition, *e.g.* lung parenchymal disease, infection, asphyxia or MAS. In an effort to reduce PVR, oxygen and pulmonary vasodilators, *e.g.* inhaled NO, are used in the treatment of PPHN [252, 253]. However, in PPHN or in abnormally constricted pulmonary vasculature in sepsis, RDS, MAS *etc.*, selective PDE3 and PDE4 inhibitors, but particularly PDE5 inhibitors seem to be of an interest [275, 276].

PDE5 inhibitors cause larger reductions in the pulmonary than in the systemic vascular resistance, therefore they appear to be ideal for the treatment of PPHN [277]. In a hypoxia-induced piglet model of neonatal pulmonary hypertension, orally given tadalafil decreased pulmonary arterial pressure on average by 54% and increased cardiac output by 88% following tadalafil administration. An increase in PaO_2 after tadalafil resulted from a reduction in the alveolar-arterial oxygen gradient expressing reduced lung shunt fraction [278]. Intravenous delivery of the PDE5 inhibitor sildenafil in a piglet model of neonatal pulmonary hypertension induced by intratracheal instillation of meconium completely reversed an increase in the pulmonary vascular resistance induced by meconium within 1 h after commencing the infusion, whereas no effect on the systemic hemodynamics was observed. Sildenafil also improved cardiac output, which was not accompanied by a deterioration in oxygenation (due to changes in intrapulmonary shunts) [87]. In another study, effects of increasing doses of intravenous sildenafil on hemodynamics and oxygenation were evaluated in piglets with meconium-induced pulmonary hypertension. Sildenafil reduced mean pulmonary artery pressure and PVR by 30%. This effect was achieved at the lowest dose, without subsequent changes at higher doses. However, sildenafil produced a dose-related increase in oxygenation index [279].

Recent analysis of data published in the Cochrane Database System showed that sildenafil has a potential for reducing mortality and improving oxygenation also in neonates with PPHN, especially in resource-limited settings where iNO is not available. However, large-scale randomized trials comparing sildenafil to active controls (other pulmonary vasodilators) and providing follow-up for survivors are needed to assess the comparative effectiveness and long-term safety of sildenafil *versus* other pulmonary vasodilators [280]. In pediatric patients with pulmonary hypertension, dipyridamole (a non-selective PDE5, PDE6, PDE7, PDE10 and

PDE11 inhibitor) reduced an indexed pulmonary vascular resistance (PVRI), which was primarily mediated through an increase in cardiac index, as is was not associated with decrease in pulmonary artery pressure, but was associated with a decrease in mean aortic pressure and indexed systemic vascular resistance (SVRI). Pretreatment with dipyridamole blunted the severity of acute hypoxic pulmonary vasoconstriction in children with exaggerated hypoxic pressor responses [281]. Although there are no randomized controlled studies assessing the efficacy and safety of milrinone (*e.g.* a risk of cardiomyocyte apoptosis), this drug appears to show potential in the treatment of PPHN [275, 276, 282, 283].

Congenital Pneumonia and Neonatal Sepsis

Congenital pneumonia is usually associated with bacterial (mainly group B *Streptococcus*), viral or fungal transplacental infection. The clinical picture includes signs of respiratory distress and temperature instability within 48 h of age [252, 253]. Infants with neonatal pneumonia as a part of a generalized sepsis illness require broad-spectrum antibiotic therapy and supportive care, often with a need for supplemental oxygen, *via* CPAP or mechanical ventilation [284].

However, recent studies have shown that some improvement can be found also after treatment with PDE inhibitors. Promising results of the non-selective PDE inhibitor pentoxifylline in preterm neonates with sepsis [285] have been supported by *in vitro* findings that pentoxifyllin suppresses LPS-induced and TLR4-mediated cytokine production in neonatal cord blood cells in a concentration-dependent manner [286]. Similarly, administration of the PDE4 inhibitor rolipram in a model of pediatric sepsis associated with development of both acute cardiac dysfunction and acute kidney injury using cecal ligation and puncture in rat pups improved 4-days survival, restored cardiac function, restored renal microcirculation and reduced microvascular leakage, suggesting that rolipram could be of value as an adjunctive therapy for septic infants with cardiorenal syndrome [287].

CONCLUSIONS

Development of selective PDE inhibitors is by some authors described as “only” increasing the selectivity of theophylline. For several decades, theophylline has been used as a bronchodilator for the treatment of COPD and bronchial asthma. However, it acts only as a mild and non-selective inhibitor of PDEs and part of its mechanism of action leading to clinical effects is PDE-independent.

Therefore, selective inhibitors of PDE4 should be never categorized as theophylline successors. Their broader clinical use requires further (particularly clinical) testing, focusing both on higher effectiveness and minimizing the incidence of adverse effects in comparison to non-selective PDE inhibitors. Nevertheless, this pharmacological group shows great potential in influencing several respiratory system diseases, especially those associated with chronic inflammation.

CONSENT FOR PUBLICATION

Not applicable.

CONFLICT OF INTEREST

The authors confirm that they have no conflicts of interest to declare for this publication.

ACKNOWLEDGEMENTS

This study was supported by VEGA grants 1/0255/18, and 1/0356/18, Slovak Research and Development Agency under the contract number APVV-0305-12 and APVV-17-0052, and by the project "Biomedical Center Martin", ITMS code: 26220220187, the project is co-financed from EU sources

LIST OF SELECTIVE PDE INHIBITORS

vinpocetin	PDE1 inhibitor;
olprinone	*cilostazol, siguazodane* – PDE3 inhibitors;
rolipram	*roflumilast, cilomilast, YM976* – PDE4 inhibitors;
zaprinast	*sildenafil, tadalafil* – PDE5 inhibitors;
BRL50481	PDE7 inhibitor;
zardaverine	*RPL554* – dual PDE3/PDE4 inhibitors

REFERENCES

[1] GINA 2017. 2017 GINA Report, Global Strategy for Asthma Management and Prevention [online] 2017. Available from: http://www.ginasthma.org

[2] GOLD 2017. Global Strategy for Diagnosis, Management, and Prevention of COPD – 2017 2017. Available from: http://goldcopd.org

[3] Bender AT, Beavo JA. Cyclic nucleotide phosphodiesterases: Molecular regulation to clinical use. Pharmacol Rev 2006; 58: 488-520. [http://dx.doi.org/10.1124/pr.58.3.5]

[4] Giembycz MA, Maurice DH. Cyclic nucleotide-based therapeutics for chronic obstructive pulmonary disease. Curr Opin Pharmacol 2014; 16: 89-107. [http://dx.doi.org/10.1016/j.coph.2014.04.001]

[5] Matera MG, Page C, Cazzola M. PDE inhibitors currently in early clinical trials for the treatment of asthma. Expert Opin Investig Drugs 2014; 23(9): 1267-75. [http://dx.doi.org/10.1517/13543784.2014.921157]

[6] Mokry J, Mokra D. Immunological aspects of phosphodiesterase inhibition in the respiratory system. Respir Physiol Neurobiol 2013; 187: 11-7. [http://dx.doi.org/10.1016/j.resp.2013.02.004]

[7] Mokry J, Urbanova A, Kertys M, Mokra D. Inhibitors of phosphodiesterases in the treatment of cough. Respir Physiol Neurobiol in press

[8] Omori K, Kotera J. Overview of PDEs and their regulation. Circ Res 2007; 100: 309-27. [http://dx.doi.org/10.1161/01.RES.0000256354.95791.f1]

[9] Nosalova G. New opinions on the mechanism of action of xanthine derivatives. Slovakofarma Revue 2000; 10: 51-5.

[10] Barnes PJ. Theophylline in chronic obstructive pulmonary disease. New horizons. Proc Am Thorac Soc 2005; 2: 334-9. [http://dx.doi.org/10.1513/pats.200504-024SR]

[11] Matera MG, Rogliani P, Zanasi A, Cazzola M. Bronchodilator therapy for chronic cough. Pulm Pharmacol Ther 2017; pii: S1094-5539(17)30029-9. [http://dx.doi.org/10.1016/j.pupt.2017.05.011]

[12] Morice AH, McGarvey L, Pavord ID, Higgins B, Chung KF, Birring SS. Theobromine for the treatment of persistent cough: a randomised, multicentre, double-blind, placebo-controlled clinical trial. J Thorac Dis 2017; 9(7): 1864-72. [http://dx.doi.org/10.21037/jtd.2017.06.18]

[13] Mokry J, Nosalova G, Mokra D. Influence of xanthine derivatives on cough and airway reactivity in guinea pigs. J Physiol Pharmacol 2009; 60: 87-91.

[14] Nosalova G, Mokry J. The mechanism of action of xanthine derivatives and suppression of cough. Acta Med Mart 2001; 1: 14-8.

[15] Lapenna D, De Gioia S, Mezzetti A, Ciofani G, Festi D, Cuccurullo F. Aminophylline: could it act as an antioxidant *in vivo*? Eur J Clin Invest 1995; 25: 464-70. [http://dx.doi.org/10.1111/j.1365-2362.1995.tb01731.x]

[16] Salari P, Mojtahedzadeh M, Najafi A, *et al.* Comparison of the effect of aminophylline and low PEEP *vs.* high PEEP on EGF concentration in critically ill patients with ALI/ARDS. J Clin Pharm Ther 2005; 30: 139-44. [http://dx.doi.org/10.1111/j.1365-2710.2004.00621.x]

[17] Tegtmeyer FK, Heilemann A, Reiss I, Fischer T. Inhibition of meconium induced activation of granulocytes from neonates and adults by pentoxyphylline. Klin Padiatr 2002; 214: 347-52. [http://dx.doi.org/10.1055/s-2002-35367]

[18] Korhonen K, Kiuru A, Svedström E, Kääpä P. Pentoxifylline reduces regional inflammatory and ventilatory disturbances in meconium-exposed piglet lungs. Pediatr Res 2004; 56: 901-6. [http://dx.doi.org/10.1203/01.PDR.0000145256.19073.E4]

[19] Mokra D, Mokry J, Tatarkova Z, Redfors B, Petraskova M, Calkovska A. Aminophylline treatment in meconium-induced acute lung injury in a rabbit model. J Physiol Pharmacol 2007; 58 (Suppl. 5): 399-07.

[20] Mokra D, Mokry J. Meconium aspiration syndrome: from pathomechanisms to treatment. New York: Nova Biomedical Books 2010.

[21] Barnes PJ. Frontrunners in novel pharmacotherapy of COPD. Curr Opin Pharmacol 2008; 8: 300-7. [http://dx.doi.org/10.1016/j.coph.2008.03.001]

[22] Barnes PJ. Future treatments for chronic obstructive pulmonary disease and its comorbities. Proc Am Thorac Soc 2008; 5: 857-64. [http://dx.doi.org/10.1513/pats.200807-069TH]

[23] Deree J, Martins JO, Melbostad H, Loomis WH, Coimbra R. Insights into the Regulation of TNF-α Production in Human Mononuclear Cells: The Effects of Non-Specific Phosphodiesterase Inhibition. Clinics (São Paulo) 2008; 63: 321-8.
[http://dx.doi.org/10.1590/S1807-59322008000300006]

[24] Peters-Golden M, Canetti C, Mancuso P, Coffey MJ. Leukotrienes: underappreciated mediators of innate immune responses. J Immunol 2005; 174: 589-94.
[http://dx.doi.org/10.4049/jimmunol.174.2.589]

[25] Luo WJ, Ling X, Huang RM. Effects of aminophylline on cytokines and pulmonary functions undergoing valve replacement. Eur J Cardiothorac Surg 2004; 25: 766-71.
[http://dx.doi.org/10.1016/j.ejcts.2004.02.024]

[26] Zheng H, Crowley JJ, Chan JC, *et al.* Attenuation of tumor necrosis factor-induced endothelial cell cytotoxicity and neutrophil chemiluminescence. Am Rev Respir Dis 1990; 142: 1073-8.
[http://dx.doi.org/10.1164/ajrccm/142.5.1073]

[27] Spina D, Page CP. Xanthines and Phosphodiesterase Inhibitors. Handb Exp Pharmacol 2017; 237: 63-91.
[http://dx.doi.org/10.1007/164_2016_71]

[28] Rabe KF, Magnussen H, Dent G. Theophylline and selective PDE inhibitors as bronchodilators and smooth muscle relaxants. Eur Respir J 1995; 8: 637-42.

[29] Polson JB, Kazanowski JJ, Goildman AL, Szentivanyi A. Inhibition of human pulmonary phosphodiesterase activity by therapeutic levels of theophylline. Clin Exp Pharmacol Physiol 1978; 5: 535-9.
[http://dx.doi.org/10.1111/j.1440-1681.1978.tb00707.x]

[30] Spicuzza L, Di Maria G, Polosa R. Adenosine in the airways: Implications and applications. Eur J Pharmacol 2006; 533: 77-88.
[http://dx.doi.org/10.1016/j.ejphar.2005.12.056]

[31] Mascali JJ, Cvietusa P, Negri J, Borish L. Anti-inflammatory effects of theophylline: modulation of cytokine production. Ann Allergy Asthma Immunol 1996; 77: 34-8.
[http://dx.doi.org/10.1016/S1081-1206(10)63476-X]

[32] Ischiyama T, Hasegawa S, Matsubara T, Hayashi T, Furukawa S. Theophylline inhibits NF-κB activation and IκBα degradation in human pulmonary epithelial cells. Naunyn Schmiedebergs Arch Pharmacol 2001; 364: 558-61.
[http://dx.doi.org/10.1007/s00210-001-0494-x]

[33] Hansel TT, Tennant RC, Tan AJ, *et al.* Theophylline: mechanism of action and use in asthma and chronic obstructive pulmonary disease. Drugs Today (Barc) 2004; 40: 55-69.
[http://dx.doi.org/10.1358/dot.2004.40.1.799438]

[34] Barnes PJ. Reduced histone deacetylase in COPD: clinical implications. Chest 2006; 129: 151-5.
[http://dx.doi.org/10.1378/chest.129.1.151]

[35] Ito K, Lim S, Caramori G, *et al.* A molecular mechanism of action of theophylline: induction of histone deacetylase activity to decrease inflammatory gene expression. Proc Natl Acad Sci USA 2002; 99: 8921-6.
[http://dx.doi.org/10.1073/pnas.132556899]

[36] Urbanova A, Kertys M, Simekova M, *et al.* Bronchodilator and Anti-Inflammatory Action of Theophylline in a Model of Ovalbumin-Induced Allergic Inflammation. Adv Exp Med Biol 2016; 935: 53-62.
[http://dx.doi.org/10.1007/5584_2016_31]

[37] Dubuis E, Wortley MA, Grace MS, *et al.* Theophylline inhibits the cough reflex through a novel mechanism of action. J Allergy Clin Immunol 2014; 133: 1588-98.
[http://dx.doi.org/10.1016/j.jaci.2013.11.017]

[38] van Mastbergen J, Jolas T, Allegra L, Page CP. The mechanism of action of doxofylline is unrelated to HDAC inhibition, PDE inhibition or adenosine receptor antagonism. Pulm Pharmacol Ther 2012; 25: 55-61.
[http://dx.doi.org/10.1016/j.pupt.2011.10.007]

[39] Arshad U, Bashir S, Ur-Rehman N, Yaqub T, Gilani AH. Dual Inhibition of Ca(+2) Influx and Phosphodiesterase Enzyme Provides Scientific Base for the Medicinal Use of Chrozophora prostrata Dalz. in Respiratory Disorders. Phytother Res 2016; 30: 1010-5.
[http://dx.doi.org/10.1002/ptr.5610]

[40] Gilani AH, Rehman NU, Khan A, Alkharfy KM. Studies on Bronchodilator Activity of Salvia officinalis (Sage): Possible Involvement of K+ Channel Activation and Phosphodiesterase Inhibition. Phytother Res 2015; 29(9): 1323-9.
[http://dx.doi.org/10.1002/ptr.5384]

[41] Minton NA, Henry JA. Acute and chronic human toxicity of theophylline. Hum Exp Toxicol 1996; 15: 471-81.
[http://dx.doi.org/10.1177/096032719601500603]

[42] Fakioglu H, Gelvez J, Torbati D, *et al.* Aminophylline therapy during endotoxemia in anesthetized spontaneously breathing rats. Pharmacol Res 2004; 49: 45-50.
[http://dx.doi.org/10.1016/j.phrs.2003.03.001]

[43] Mokra D, Tonhajzerova I, Mokry J, Petraskova M, Hutko M, Calkovska A. Cardiovascular side effects of aminophylline in meconium-induced acute lung injury. Adv Exp Med Biol 2013; 756: 341-7. a
[http://dx.doi.org/10.1007/978-94-007-4549-0_41]

[44] Mokra D, Drgova A, Mokry J, *et al.* Combination of budesonide and aminophylline diminished acute lung injury in animal model of meconium aspiration syndrome. J Physiol Pharmacol 2008; 59 (Suppl. 6): 461-71. b

[45] Rabe KF. Update on roflumilast, a phosphodiesterase 4 inhibitor for the treatment of chronic obstructive pulmonary disease. Br J Pharmacol 2011; 163(1): 53-67.
[http://dx.doi.org/10.1111/j.1476-5381.2011.01218.x]

[46] Christensen I, Miskovicova H, Porvaznik I, Joskova M, Mokra D, Mokry J. Selective inhibition of phosphodiesterase 7 (PDE7) by BRL50481 in healthy and ovalbumin-sensitized guinea pigs. Acta Med Martiniana 2012; 12: 16-23.

[47] Mokry J, Joskova M, Mokra D, Christensen I, Nosalova G. Effects of selective inhibition of PDE4 and PDE7 on airway reactivity and cough in healthy and ovalbumin-sensitized guinea pigs. Adv Exp Med Biol 2013; 756: 57-64.
[http://dx.doi.org/10.1007/978-94-007-4549-0_8]

[48] Shakur Y, Holst LS, Landstrom TR, Movesian M, Degerman E, Manganiello V. Regulation and function of the cyclic nucleotide phosphodiesterase (PDE3) gene family. Prog Nucleic Acid Res Mol Biol 2001; 66: 241-77.
[http://dx.doi.org/10.1016/S0079-6603(00)66031-2]

[49] Chung KF. Phosphodiesterase inhibitors in airways disease. Eur J Pharmacol 2006; 533: 110-7.
[http://dx.doi.org/10.1016/j.ejphar.2005.12.059]

[50] Giembycz MA, Corrigan CJ, Seybold J, Newton R, Barnes PJ. Identification of cyclic AMP phosphodiesterases 3, 4 and 7 in human $CD4^+$ and $CD8^+$ T-lymphocytes: role in regulating proliferation and the biosynthesis of interleukin-2. Br J Pharmacol 1996; 118: 1945-58.
[http://dx.doi.org/10.1111/j.1476-5381.1996.tb15629.x]

[51] Mokry J, Mokra D, Nosalova G, Beharkova M, Feherova Z. Influence of selective inhibitors of phosphodiesterase 3 and 4 on cough and airway reactivity. J Physiol Pharmacol 2008; 59 (Suppl. 6): 473-82.

[52] Koike T, Nadeen Qutab M, Tsuchida M, Takekubo M, Saito M, Hayashi J. Pretreatment with

olprinone hydrochloride, a phosphodiesterase III inhibitor, attenuates lipopolysaccharide-induced lung injury *via* an anti-inflammatory effect. Pulm Pharmacol Ther 2008; 21: 166-71. [http://dx.doi.org/10.1016/j.pupt.2007.01.007]

[53] Fujimura M, Liu Q. Selective inhibitors for phosphodiesterase 3 and 4 in antigen-induced increase of cough reflex sensitivity in guinea pigs. Pulm Pharmacol Ther 2007; 20: 543-8. [http://dx.doi.org/10.1016/j.pupt.2006.05.004]

[54] Bassler D, Choong K, McNamara P, Kirpalani H. Neonatal persistent pulmonary hypertension treated with milrinone: four case report. Biol Neonate 2006; 89: 1-5. [http://dx.doi.org/10.1159/000088192]

[55] Mokra D, Drgova A, Pullmann R Sr, Calkovska A. Selective phosphodiesterase 3 inhibitor olprinone attenuates meconium-induced oxidative lung injury. Pulm Pharmacol Ther 2012; 25(3): 216-22. [http://dx.doi.org/10.1016/j.pupt.2012.02.007]

[56] Mokra D, Tonhajzerova I, Pistekova H, *et al.* Short-term cardiovascular effects of selective phosphodiesterase 3 inhibitor olprinone *versus* non-selective phosphodiesterase inhibitor aminophylline in a meconium-induced acute lung injury. J Physiol Pharmacol 2013; 64: 751-9. b

[57] Houslay MD, Adams DR. PDE4 cAMP phosphodiesterase: modular enzymes that orchestrate signalling cross-talk, desensitization and compartmentalization. Biochem J 2003; 370: 1-18. [http://dx.doi.org/10.1042/bj20021698]

[58] Sanz MJ, Cortijo J, Morcillo EJ. PDE4 inhibitors as new antiinflammatory drugs: effects on cell trafficking and cell adhesion molecules expression. Pharmacol Ther 2005; 106: 269-97. [http://dx.doi.org/10.1016/j.pharmthera.2004.12.001]

[59] Rabe KF, Bateman ED, O'Donnell D, Witte S, Bredenbroker D, Bethke TD. Roflumilast – an oral anti-inflammatory treatment for chronic obstructive pulmonary disease: a randomised controlled trial. Lancet 2005; 366: 563-71. [http://dx.doi.org/10.1016/S0140-6736(05)67100-0]

[60] Matera MG, Page CP, Cazzola M. Novel bronchodilators for the treatment of chronic obstructive pulmonary disease. Trends Pharmacol Sci 2011; 32: 495-506. [http://dx.doi.org/10.1016/j.tips.2011.04.003]

[61] Medvedova I, Prso M, Eichlerova A, Mokra D, Mikolka P, Mokry J. Influence of roflumilast on airway reactivity and apoptosis in ovalbumin-sensitized Guinea pigs. Adv Exp Med Biol 2015; 838: 11-8. [http://dx.doi.org/10.1007/5584_2014_77]

[62] Wollin L, Bundschuh DS, Wohlsen A, Marx D, Beume R. Inhibition of airway hyperresponsiveness and pulmonary inflammation by roflumilast and other PDE4 inhibitors. Pulm Pharmacol Ther 2006; 19: 343-52. [http://dx.doi.org/10.1016/j.pupt.2005.09.002]

[63] Sethi S, Martinez FJ, Rabe KF, *et al.* Effect of Roflumilast on Cough and Sputum in Patients with Severe or Very Severe Chronic Obstructive Pulmonary Disease (COPD) Receiving Inhaled Combination Therapy: Evaluation of the EXAcerbation of Chronic Pulmonary Disease Tool-Patient Reported Outcomes (EXACT-PRO) Subdomain Scores. Am J Respir Crit Care Med 2017; 195: A1335.

[64] Kawamatawong T. Roles of roflumilast, a selective phosphodiesterase 4 inhibitor, in airway diseases. J Thorac Dis 2017; 9: 1144-54. [http://dx.doi.org/10.21037/jtd.2017.03.116]

[65] Bardin P, Kanniess F, Gauvreau G, Bredenbröker D, Rabe KF. Roflumilast for asthma: Efficacy findings in mechanism of action studies. Pulm Pharmacol Ther 2015; 35: S4-S10. [http://dx.doi.org/10.1016/j.pupt.2015.08.006]

[66] Bateman ED, O'Byrne PM, Buhl R, Rabe KF. Roflumilast for asthma: Weighing the evidence. Pulm

Pharmacol Ther 2015; 35: S1-3.
[http://dx.doi.org/10.1016/j.pupt.2015.08.001]

[67] Louie S, Zeki AA, Schivo M, *et al.* The asthma-chronic obstructive pulmonary disease overlap syndrome: pharmacotherapeutic considerations. Expert Rev Clin Pharmacol 2013; 6: 197-219.
[http://dx.doi.org/10.1586/ecp.13.2]

[68] Mokrý J, Urbanová A, Medvedová I, *et al.* Effects of Selective Inhibition of PDE4 by YM976 on Airway Reactivity and Cough in Ovalbumin-Sensitized Guinea Pigs. Adv Exp Med Biol 2016; 921: 61-70.
[http://dx.doi.org/10.1007/5584_2016_237]

[69] Aoki M, Fukunaga M, Kitagawa M, *et al.* Effect of a novel anti-inflammatory compound, YM976, on antigen-induced eosinophil infiltration into the lungs in rats, mice, and ferrets. J Pharmacol Exp Ther 2000; 295: 1149-55.

[70] Aoki M, Fukunaga M, Sugimoto T, *et al.* Studies on mechanisms of low emetogenicity of YM976, a novel phosphodiesterase type 4 inhibitor. J Pharmacol Exp Ther 2001; 298: 1142-9.

[71] Gupta S. Side-effects of roflumilast. Lancet 2012; 379: 710-1.
[http://dx.doi.org/10.1016/S0140-6736(12)60304-3]

[72] Moriuchi H, Nakahara T, Maruko T, Sakamoto K, Ishii K. Relaxant effect of YM976, a novel phosphodiesterase 4 inhibitor, on bovine tracheal smooth muscle. Eur J Pharmacol 2003; 470: 57-64.
[http://dx.doi.org/10.1016/S0014-2999(03)01754-0]

[73] Page CP. Phosphodiesterase inhibitors for the treatment of asthma and chronic obstructive pulmonary disease. Int Arch Allergy Immunol 2014; 165: 152-64.
[http://dx.doi.org/10.1159/000368800]

[74] Abbott-Banner KH, Page CP. Dual PDE3/4 and PDE4 inhibitors: novel treatments for COPD and other inflammatory airway diseases. Basic Clin Pharmacol Toxicol 2014; 114: 365-76.
[http://dx.doi.org/10.1111/bcpt.12209]

[75] Giembycz MA. Phosphodiesterase-4: Selective and Dual-Specificity Inhibitors for the Therapy of Chronic Obstructive Pulmonary Disease. Proc Am Thorac Soc 2005; 2: 326-33. a
[http://dx.doi.org/10.1513/pats.200504-041SR]

[76] Sebkhi A, Strange JW, Phillips SC, Wharton J, Wilkins MR. Phosphodiesterase type 5 as a target for the treatment of hypoxia-induced pulmonary hypertension. Circulation 2003; 107: 3230-5.
[http://dx.doi.org/10.1161/01.CIR.0000074226.20466.B1]

[77] 76. aratza AA, Bush A, Magee AG. Safety and efficacy of sildenafil therapy in children with pulmonary hypertension. Int J Cardiol 2005; 100: 267-73.
[http://dx.doi.org/10.1016/j.ijcard.2004.09.002]

[78] Sastry BK, Narasimhan C, Reddy NK, Raju BS. Clinical efficacy of sildenafil in primary pulmonary hypertension: a randomized, placebo-controlled, double-blind, crossover study. J Am Coll Cardiol 2004; 43: 1149-53.
[http://dx.doi.org/10.1016/j.jacc.2003.10.056]

[79] Sawamura F, Kato M, Fujita K, Nakazawa T, Beardsworth A. Tadalafil, a long-acting inhibitor of PDE5, improves pulmonary hemodynamics and survival rate of monocrotaline-induced pulmonary artery hypertension in rats. J Pharmacol Sci 2009; 111: 235-43.
[http://dx.doi.org/10.1254/jphs.09110FP]

[80] Wharton J, Strange JW, Moller GM, *et al.* Antiproliferative effects of phosphodiesterase type 5 inhibition in human pulmonary artery cells. Am J Respir Crit Care Med 2005; 172: 105-13.
[http://dx.doi.org/10.1164/rccm.200411-1587OC]

[81] Toward TJ, Smith N, Broadley KJ. Effect of phosphodiesterase-5 inhibitor, sildenafil (Viagra), in animal models of irways disease. Am J Respir Crit Care Med 2004; 169: 227-34.
[http://dx.doi.org/10.1164/rccm.200211-1372OC]

[82] Mokry J, Nosalova G. The influence of the PDE inhibitors on cough reflex in guinea pigs. Bratisl Lek Listy 2011; 112: 131-5.

[83] Dupont LL, Glynos C, Bracke KR, Brouckaert P, Brusselle GG. Role of the nitric oxide-soluble guanylyl cyclase pathway in obstructive airway diseases. Pulm Pharmacol Ther 2014; 29: 1-6. [http://dx.doi.org/10.1016/j.pupt.2014.07.004]

[84] Urbanova A, Medvedova I, Kertys M, *et al.* Dose dependent effects of tadalafil and roflumilast on ovalbumin-induced airway hyperresponsiveness in guinea pigs. Exp Lung Res 2017. [http://dx.doi.org/10.1080/01902148.2017.1386735]

[85] Mokry J, Urbanova A, Medvedova I, *et al.* Effects of Tadalafil (PDE5 Inhibitor) and Roflumilast (PDE4 Inhibitor) on Airway Reactivity and Markers of Inflammation in Ovalbumin-Induced Airway Hyperresponsiveness in Guinea Pigs. J Physiol Pharmacol in press

[86] Takeda K, Shiraishi Y, Domenico J, *et al.* Combined effects of the phosphodiesterase type 4 (PDE4) inhibitor, roflumilast, and the PDE5 inhibitor, tadalafil, in allergen-induced airway hyperresponsiveness (AHR). Am J Crit Care Med 2009; 179: A5717.

[87] Shekerdemian LS, Ravn HB, Penny DJ. Intravenous sildenafil lowers pulmonary vascular resistance in a model of neonatal pulmonary hypertension. Am J Respir Crit Care Med 2002; 165: 1098-02. [http://dx.doi.org/10.1164/ajrccm.165.8.2107097]

[88] Castro A, Jerez MJ, Gil C, Martinez A. Cyclic nucleotide phosphodiesterases and their role in immunomodulatory responses: advances in the development of specific phosphodiesterase inhibitors. Med Res Rev 2005; 25: 229-44. [http://dx.doi.org/10.1002/med.20020]

[89] Smith SJ, Brookes-Fazakerley S, Donnelly LE, Barnes PJ, Barnette MS, Giembycz MA. Ubiquitous expression of phosphodiesterase 7A in human proinflammatory and immune cells. Am J Physiol Lung Cell Mol Physiol 2003; 284: L279-89. [http://dx.doi.org/10.1152/ajplung.00170.2002]

[90] Ghosh R, Sawant O, Ganpathy P, Pitre S, Kadam VJ. Phosphodiesterase inhibitors: their role and implications. Int J Pharm Tech Res 2009; 1: 1148-60.

[91] Paterniti I, Mazzon E, Gil C, *et al.* PDE7 inhibitors: new potential drugs for the therapy of spinal cord injury. PLoS One 2011; 6: e15937. [http://dx.doi.org/10.1371/journal.pone.0015937]

[92] Guo J, Watson A, Kempson J, *et al.* Identification of potent pyrimidine inhibitors of phosphodiesterase 7 (PDE7) and their ability to inhibit T cell proliferation. Bioorg Med Chem Lett 2009; 19: 1935-8. [http://dx.doi.org/10.1016/j.bmcl.2009.02.060]

[93] Pitts WJ, Vaccaro W, Huynh T, *et al.* Identification of purine inhibitors of phosphodiesterase 7 (PDE7). Bioorg Med Chem Lett 2004; 14: 2955-8. [http://dx.doi.org/10.1016/j.bmcl.2004.03.021]

[94] Kadoshima-Yamaoka K, Murakawa M, Goto M, *et al.* ASB16165, a novel inhibitor for phosphodiesterase 7A (PDE7A), supresses IL-12-induced IFN- production by mouse activated T lymphocytes. Immunol Lett 2009; 122: 193-7. [http://dx.doi.org/10.1016/j.imlet.2009.01.004]

[95] Jones NA, Leport M, Holand T, *et al.* Phosphodiesterase (PDE) 7 in inflammatory cells from patients with asthma and COPD. Pulm Pharmacol Ther 2007; 20(1): 60-8. [http://dx.doi.org/10.1016/j.pupt.2005.11.010]

[96] Nakata A, Ogawa K, Sasaki T, *et al.* Potential role of phosphodiesterase 7 in human T cell function: comparative effects of two phosphodiesterase inhibitors. Clin Exp Immunol 2002; 128: 460-6. [http://dx.doi.org/10.1046/j.1365-2249.2002.01856.x]

[97] Smith SJ, Cieslinski LB, Newton R, *et al.* Discovery of BRL 50481 [3-(N,N-dimethylsulfonamido-

-4-methyl-nitrobenzene], a Selective Inhibitor of Phosphodiesterase 7: In Vitro Studies in Human Monocytes, Lung Macrophages, and CD8+ TLymphocytes. Mol Pharmacol 2004; 66: 1679-89. [http://dx.doi.org/10.1124/mol.104.002246]

[98] Morales-Garcia JA, Redondo M, Alonso-Gil S, *et al.* Phosphodiesterase 7 Inhibition Preserves Dopaminergic Neurons in Cellular and Rodent Models of Parkinson Disease. PLoS One 2011; 6: e17240. [http://dx.doi.org/10.1371/journal.pone.0017240]

[99] Fortin M, D'Anjou H, Higgins ME, *et al.* A multi-target antisense approach against PDE4 and PDE7 reduces smoke-induced lung inflammation in mice. Respir Res 2009; 10: 39. [http://dx.doi.org/10.1186/1465-9921-10-39]

[100] Giembycz MA. Life after PDE4: overcoming adverse events with dual-specificity phosphodiesterase inhibitors. Curr Opin Pharmacol 2005; 5: 238-44. b [http://dx.doi.org/10.1016/j.coph.2005.04.001]

[101] Keith C, Borisy A, Stockwell B. Multicomponent therapeutics for networked systems. Nat Rev Drug Discov 2005; 4: 71-8. [http://dx.doi.org/10.1038/nrd1609]

[102] Manabe H, Akuta K, Sejimo H, *et al.* Antiinflammatory and bronchodilator properties of KF19514, a phosphodiesterase 4 and 1 inhibitor. Eur J Pharmacol 1997; 332: 97-107. [http://dx.doi.org/10.1016/S0014-2999(97)01063-7]

[103] Yamamoto S, Sugahara S, Ikeda K, Shimizu Y. Pharmacological profile of a novel phosphodiesterase 7A and -4 dual inhibitor, YM-393059, on acute and chronic inflammation models. Eur J Pharmacol 2006; 550: 166-72. [http://dx.doi.org/10.1016/j.ejphar.2006.08.023]

[104] Fan C. Phosphodiesterase inhibitors in airway disease. Eur J Pharmacol 2006; 533: 110-7. [http://dx.doi.org/10.1016/j.ejphar.2005.12.059]

[105] Giembycz MA, Newton R. Harnessing the clinical efficacy of phosphodiesterase 4 inhibitors in inflammatory lung diseases: dual-selective phosphodiesterase inhibitors and novel combination therapies. Handb Exp Pharmacol 2011; 204: 415-46. [http://dx.doi.org/10.1007/978-3-642-17969-3_18]

[106] Lopez A, Shibuya K, Rao C, *et al.* Chronic obstructive pulmonary disease: current burden and future projections. Eur Respir J 2006; 27: 397-412. [http://dx.doi.org/10.1183/09031936.06.00025805]

[107] Spina D, Landells LJ, Page CP. The role of theophylline and phosphodiesterase-4 isoenzyme inhibitors as antiinflammatory drugs. Clin Exp Allergy 1998; 28: 24-34.

[108] Spina D. Current and novel bronchodilators in respiratory disease. Curr Opin Pulm Med 2014; 20(1): 73-86. [http://dx.doi.org/10.1097/MCP.0000000000000012]

[109] Barnes PJ. Theophylline. Am J Respir Crit Care Med 2013; 188: 901-6. [http://dx.doi.org/10.1164/rccm.201302-0388PP]

[110] Voduc N, Alvarez GG, Amjadi K, Tessier C, Sabri E, Aaron SD. Effect of theophylline on exercise capacity in COPD patients treated with combination long-acting bronchodilator therapy: a pilot study. Int J Chron Obstruct Pulmon Dis 2012; 7: 245-52. [http://dx.doi.org/10.2147/COPD.S29990]

[111] European Medicines Agency. European Medicines Agency. European Public Assessment Report (EPAR) for Daxas. 2010. Available from: http://www.ema.europa.eu/ema/index.jsp?curl =pages/medicines/human/medicines/001179/human_med_001363.jsp

[112] Chong J, Leung B, Poole P. Phosphodiesterase 4 inhibitors for chronic obstructive pulmonary disease. Cochrane Database Syst Rev 2017; 9: CD002309.

[113] Rennard S, Knobil K, Rabe KF, *et al.* The efficacy and safety of cilomilast in COPD. Drugs 2008; 68: 3-57.
[http://dx.doi.org/10.2165/0003495-200868002-00002]

[114] Antoniu SA. Roflumilast for the treatment of chronic obstructive pulmonary disease. Curr Opin Investig Drugs 2006; 7: 412-7.

[115] Watz H, Mistry SJ, Lazaar AL. IPC101939 investigators. Safety and tolerability of the inhaled phosphodiesterase 4 inhibitor GSK256066 in moderate COPD. Pulm Pharmacol Ther 2013; 26(5): 588-95.
[http://dx.doi.org/10.1016/j.pupt.2013.05.004]

[116] Currie GP, Butler CA, Anderson WJ, Skinner C. Phosphodiesterase 4 inhibitors in chronic obstructive pulmonary disease: a new approach to oral treatment. Br J Clin Pharmacol 2008; 65: 803-10.
[http://dx.doi.org/10.1111/j.1365-2125.2008.03155.x]

[117] Luo J, Wang K, Liu D, Liang BM, Liu CT. Can roflumilast, a phosphodiesterase-4 inhibitor, improve clinical outcomes in patients with moderate-to-severe chronic obstructive pulmonary disease? A meta-analysis. Respir Res 2016; 17: 18.
[http://dx.doi.org/10.1186/s12931-016-0330-y]

[118] Vestbo J, Tan L, Atkinson G. Ward J on behalf of the UK-500,001 Global Study Team. A controlled trial of 6-weeks' treatment with a novel inhaled phosphodiesterase type-4 inhibitor in COPD. Eur Respir J 2009; 33: 1039-44.
[http://dx.doi.org/10.1183/09031936.00068908]

[119] Spina D. PDE4 inhibitors: current sttus. Br J Pharmacol 2008; 155: 308-15.
[http://dx.doi.org/10.1038/bjp.2008.307]

[120] Wang YJ, Jiang YL, Tang HF, Zhao CZ, Chen JQ. Zl-n-91, a selective phosphodiesterase 4 inhibitor, supresses inflammatory response in a COPD-like rat model. Int Immunopharmacol 2010; 10: 252-8.
[http://dx.doi.org/10.1016/j.intimp.2009.11.008]

[121] Franciosi LG, Diamant Z, Banner KH, *et al.* Efficacy and safety of RPL554, a dual PDE3 and PDE4 inhibitor, in healthy volunteers and in patients with asthma or chronic obstructive pulmonary disease: findings from four clinical trials. Lancet Respir Med 2013; 1(9): 714-27.
[http://dx.doi.org/10.1016/S2213-2600(13)70187-5]

[122] Calzetta L, Cazzola M, Page CP, Rogliani P, Facciolo F, Matera MG. Pharmacological characterization of the interaction between the dual phosphodiesterase (PDE) 3/4 inhibitor RPL554 and glycopyrronium on human isolated bronchi and small airways. Pulm Pharmacol Ther 2015; 32: 15-23.
[http://dx.doi.org/10.1016/j.pupt.2015.03.007]

[123] Calzetta L, Page CP, Spina D, *et al.* Effect of the mixed phosphodiesterase 3/4 inhibitor RPL554 on human isolated bronchial smooth muscle tone. J Pharmacol Exp Ther 2013; 346(3): 414-23.
[http://dx.doi.org/10.1124/jpet.113.204644]

[124] Seimetz M, Parajuli N, Pichl A, *et al.* Cigarette Smoke-Induced Emphysema and Pulmonary Hypertension Can Be Prevented by Phosphodiesterase 4 and 5 Inhibition in Mice. PLoS One 2015; 10(6): e0129327.
[http://dx.doi.org/10.1371/journal.pone.0129327]

[125] Wang T, Liu Y, Chen L, *et al.* Effect of sildenafil on acrolein-induced airway inflammation and mucus production in rats. Eur Respir J 2009; 33(5): 1122-32.
[http://dx.doi.org/10.1183/09031936.00055908]

[126] Bull T, Badesch DB. Sildenafil for COPD: a randomized crossover trial. COPD 2012; 9(3): 211-2.
[http://dx.doi.org/10.3109/15412555.2012.683966]

[127] Lederer DJ, Bartels MN, Schluger NW, *et al.* Sildenafil for chronic obstructive pulmonary disease: a randomized crossover trial. COPD 2012; 9(3): 268-75.

[http://dx.doi.org/10.3109/15412555.2011.651180]

[128] Holgate ST. Novel targets of therapy in asthma. Curr Opin Pulm Med 2009; 15: 63-71. [http://dx.doi.org/10.1097/MCP.0b013e32831da867]

[129] Hatzelmann A, Morcillo EJ, Lungarella G, *et al.* The preclinical pharmacology of roflumilast--a selective, oral phosphodiesterase 4 inhibitor in development for chronic obstructive pulmonary disease. Pulm Pharmacol Ther 2010; 23(4): 235-56. [http://dx.doi.org/10.1016/j.pupt.2010.03.011]

[130] Oñatibia-Astibia A, Martínez-Pinilla E, Franco R. The potential of methylxanthine-based therapies in pediatric respiratory tract diseases. Respir Med 2016; 112: 1-9. [http://dx.doi.org/10.1016/j.rmed.2016.01.022]

[131] Diamant Z, Boot JD, Virchow JC. Summing up 100 years of asthma. Respir Med 2007; 101: 378-88. [http://dx.doi.org/10.1016/j.rmed.2006.12.004]

[132] Barnes PJ, Pauwels RA. Theophylline in the management of asthma: time for reappraisal? Eur Respir J 1994; 7: 579-91. [http://dx.doi.org/10.1183/09031936.94.07030579]

[133] Roberts G, Newsom D. Gomez K on behalf of the North West Thames Asthma Study Group. Intravenous salbutamol bolus compared with an aminophylline infusion in children with severe asthma: a randomized controlled trial. Thorax 2003; 58: 306-10. [http://dx.doi.org/10.1136/thorax.58.4.306]

[134] Global strategy for the diagnosis and management of asthma in children 5 years and younger [online]. 2017. [cited 2017-08-24]. Available from: http://www.ginasthma.org

[135] Rehman H, Naveed S, Usmanghani K. Efficacy and safety of Linkus, Aminophylline diphenhydramine and acefyllin piperazine for the treatment of cough in children. Pak J Pharm Sci 2016; 29(3) (Suppl.): 1027-32.

[136] Rogoz R, Dziedzicka-Wasylewska M. Comparison of the effects of antidepressant drugs on the level of cAMP in the rat striatum and nucleus accumbens septi. Pol J Pharmacol 1996; 48(5): 481-7.

[137] Mombereau C, Gur TL, Onksen J, Blendy JA. Differential effects of acute and repeated citalopram in mouse models of anxiety and depression. Int J Neuropsychopharmacol 2010; 13(3): 321-34. [http://dx.doi.org/10.1017/S1461145709990630]

[138] Karish SB, Gagnon JM. The potential role of roflumilast: the new phosphodiesterase-4 inhibitor. Ann Pharmacother 2006; 40(6): 1096-104. [http://dx.doi.org/10.1345/aph.1E651]

[139] Kim SW, Kim JH, Park CK, *et al.* Effect of roflumilast on airway remodelling in a murine model of chronic asthma. Clin Exp Allergy 2016; 46(5): 754-63. [http://dx.doi.org/10.1111/cea.12670]

[140] Bundschuh DS, Eltze M, Barsig J, Wollin L, Hatzelmann A, Beume R. *In vivo* efficacy in airway disease models of roflumilast, a novel orally active PDE4 inhibitor. J Pharmacol Exp Ther 2001; 297(1): 280-90.

[141] Leaker BR, Singh D, Ali FY, Barnes PJ, O'Connor B. The effect of the novel phosphodiesterase-4 inhibitor MEM 1414 on the allergen induced responses in mild asthma. BMC Pulm Med 2014; 14: 166. [http://dx.doi.org/10.1186/1471-2466-14-166]

[142] Clayton RA, Dick CA, Mackenzie A, *et al.* The effect of selective phosphodiesterase inhibitors, alone and in combination, on a murine model of allergic asthma. Respir Res 2004; 5: 4. [http://dx.doi.org/10.1186/1465-9921-5-4]

[143] Al Qadi-Nassar B, Bichon-Laurent F, Portet K, Tramini P, Arnoux B, Michel A. Effects of L-arginine and phosphodiesterase-5 inhibitor, sildenafil, on inflammation and airway responsiveness of sensitized

BP2 mice. Fundam Clin Pharmacol 2007; 21(6): 611-20.
[http://dx.doi.org/10.1111/j.1472-8206.2007.00523.x]

[144] Banner KH, Page CP. Acute *versus* chronic administration of phosphodiesterase inhibitors on allergen-induced pulmonary cell influx in sensitized guinea-pigs. Br J Pharmacol 1995; 114(1): 93-8.
[http://dx.doi.org/10.1111/j.1476-5381.1995.tb14910.x]

[145] Lipworth BJ. Phosphodiesterase-4 inhibitors for asthma and chronic obstructive pulmonary disease. Lancet 2005; 365: 167-75.
[http://dx.doi.org/10.1016/S0140-6736(05)17708-3]

[146] Venkatasamy R, Spina D. Novel relaxant effects of RPL554 on guinea pig tracheal smooth muscle contractility. Br J Pharmacol 2016; 173(15): 2335-51.
[http://dx.doi.org/10.1111/bph.13512]

[147] Sulaiman I, Lim JC, Soo HL, Stanslas J. Molecularly targeted therapies for asthma: Current development, challenges and potential clinical translation. Pulm Pharmacol Ther 2016; 40: 52-68.
[http://dx.doi.org/10.1016/j.pupt.2016.07.005]

[148] Hart PH. Regulation of the inflammatory response in asthma by mast cell products. Immunol Cell Biol 2001; 79(2): 149-53.
[http://dx.doi.org/10.1046/j.1440-1711.2001.00983.x]

[149] Galli SJ, Tsai M, Piliponsky AM. The development of allergic inflammation. Nature 2008; 454(7203): 445-54.
[http://dx.doi.org/10.1038/nature07204]

[150] Djukanović R. Asthma: A disease of inflammation and repair. J Allergy Clin Immunol 2000; 105(2 Pt 2): S522-6.
[http://dx.doi.org/10.1016/S0091-6749(00)90055-1]

[151] Page CP, Spina D. Selective PDE inhibitors as novel treatments for respiratory diseases. Curr Opin Pharmacol 2012; 12(3): 275-86.
[http://dx.doi.org/10.1016/j.coph.2012.02.016]

[152] Martin-Chouly CA, Astier A, Jacob C, Pruniaux MP, Bertrand C, Lagente V. Modulation of matrix metalloproteinase production from human lung fibroblasts by type 4 phosphodiesterase inhibitors. Life Sci 2004; 75: 823-40.
[http://dx.doi.org/10.1016/j.lfs.2004.01.021]

[153] Murad HA, Habib HS, Rafeeq MM, Sulaiman MI, Abdulrahman AS, Khabaz MN. Co-inhalation of roflumilast, rather than formoterol, with fluticasone more effectively improves asthma in asthmatic mice. Exp Biol Med (Maywood) 2017; 242(5): 516-26.
[http://dx.doi.org/10.1177/1535370216685006]

[154] Patel BS, Prabhala P, Oliver BG, Ammit AJ. Inhibitors of Phosphodiesterase 4, but Not Phosphodiesterase 3, Increase β2-Agonist-Induced Expression of Antiinflammatory Mitogen-Activated Protein Kinase Phosphatase 1 in Airway Smooth Muscle Cells. Am J Respir Cell Mol Biol 2015; 52(5): 634-40.
[http://dx.doi.org/10.1165/rcmb.2014-0344OC]

[155] Schuliga M. NF-kappaB Signaling in Chronic Inflammatory Airway Disease. Biomolecules 2015; 5(3): 1266-83.
[http://dx.doi.org/10.3390/biom5031266]

[156] Rajendrasozhan S, Yang SR, Edirisinghe I, Yao H, Adenuga D, Rahman I. Deacetylases and NF-kappaB in redox regulation of cigarette smoke-induced lung inflammation: epigenetics in pathogenesis of COPD. Antioxid Redox Signal 2008; 10(4): 799-811.
[http://dx.doi.org/10.1089/ars.2007.1938]

[157] Fatani SH. Biomarkers of oxidative stress in acute and chronic bronchial asthma. J Asthma 2014; 51(6): 578-84.

[http://dx.doi.org/10.3109/02770903.2014.892965]

[158] Guo CH, Chen PC, Hsia S, Hsu GS, Liu PJ. The relationship of plasma aluminum to oxidant-antioxidant and inflammation status in asthma patients. Environ Toxicol Pharmacol 2013; 35(1): 30-8. [http://dx.doi.org/10.1016/j.etap.2012.10.005]

[159] Al-Afaleg NO, Al-Senaidy A, El-Ansary A. Oxidative stress and antioxidant status in Saudi asthmatic patients. Clin Biochem 2011; 44(8-9): 612-7. [http://dx.doi.org/10.1016/j.clinbiochem.2011.01.016]

[160] Brown DM, Hutchison L, Donaldson K, MacKenzie SJ, Dick CA, Stone V. The effect of oxidative stress on macrophages and lung epithelial cells: the role of phosphodiesterases 1 and 4. Toxicol Lett 2007; 168(1): 1-6. [http://dx.doi.org/10.1016/j.toxlet.2006.10.016]

[161] Lowe APP, Thomas RS, Nials AT, Kidd EJ, Broadley KJ, Ford WR. Route of Administration Affects Corticosteroid Sensitivity of a Combined Ovalbumin and Lipopolysaccharide Model of Asthma Exacerbation in Guinea Pigs. J Pharmacol Exp Ther 2017; 362(2): 327-37. [http://dx.doi.org/10.1124/jpet.117.241927]

[162] Forkuo GS, Kim H, Thanawala VJ, *et al.* Phosphodiesterase 4 Inhibitors Attenuate the Asthma Phenotype Produced by β2-Adrenoceptor Agonists in Phenylethanolamine N-Methyltransferas--Knockout Mice. Am J Respir Cell Mol Biol 2016; 55(2): 234-42. [http://dx.doi.org/10.1165/rcmb.2015-0373OC]

[163] Ciepiela O, Ostafin M, Demkow U. Neutrophils in asthma - a review. Respir Physiol Neurobiol 2015; 209: 13-6. [http://dx.doi.org/10.1016/j.resp.2014.12.004]

[164] Barnes PJ. Therapeutic approaches to asthma-chronic obstructive pulmonary disease overlap syndromes. J Allergy Clin Immunol 2015; 136(3): 531-45. [http://dx.doi.org/10.1016/j.jaci.2015.05.052]

[165] Barnes PJ. Cellular and molecular mechanisms of asthma and COPD. Clin Sci (Lond) 2017; 131(13): 1541-58. [http://dx.doi.org/10.1042/CS20160487]

[166] Thomson NC. Novel approaches to the management of noneosinophilic asthma. Ther Adv Respir Dis 2016; 10(3): 211-34. [http://dx.doi.org/10.1177/1753465816632638]

[167] Bernard GR, Artigas A, Brigham KL, *et al.* The American-European Consensus Conference on ARDS. Definitions, mechanisms, relevant outcomes, and clinical trial The American-European Consensus Conference on ARDS. Definitions, mechanisms, relevant outcomes, and clinical trial coordination. Am J Respir Crit Care Med 1994; 149: 818-24. [http://dx.doi.org/10.1164/ajrccm.149.3.7509706]

[168] ARDS Definition Task Force. Acute respiratory distress syndrome: the Berlin Definition. JAMA 2012; 307: 2526-33.

[169] Fioretto JR, de Carvalho WB. Temporal evolution of acute respiratory distress syndrome definitions. J Pediatr (Rio J) 2013; 89: 523-30. [http://dx.doi.org/10.1016/j.jped.2013.02.023]

[170] Ware LB, Matthay MA. The acute respiratory distress syndrome. N Engl J Med 2000; 342: 1334-49. [http://dx.doi.org/10.1056/NEJM200005043421806]

[171] Finigan JH. The coagulation system and pulmonary endothelial function in acute lung injury. Microvasc Res 2009; 77: 35-8. [http://dx.doi.org/10.1016/j.mvr.2008.09.002]

[172] Matthay MA, Zemans RL. The acute respiratory distress syndrome: pathogenesis and treatment. Annu Rev Pathol 2011; 6: 147-63.

[http://dx.doi.org/10.1146/annurev-pathol-011110-130158]

[173] Castillo RL, Carrasco Loza R, Romero-Dapueto C. Pathophysiological Approaches of Acute Respiratory Distress syndrome: Novel Bases for Study of Lung Injury. Open Respir Med J 2015; 9: 83-91.
[http://dx.doi.org/10.2174/1874306401509010083]

[174] Galani V, Tatsaki E, Bai M, *et al.* The role of apoptosis in the pathophysiology of Acute Respiratory Distress Syndrome (ARDS): an up-to-date cell-specific review. Pathol Res Pract 2010; 206: 145-50.
[http://dx.doi.org/10.1016/j.prp.2009.12.002]

[175] Bhargava M, Wendt CH. Biomarkers in acute lung injury. Transl Res 2012; 159: 205-17.
[http://dx.doi.org/10.1016/j.trsl.2012.01.007]

[176] Mokra D, Kosutova P. Biomarkers in acute lung injury. Respir Physiol Neurobiol 2015; 209: 52-8.
[http://dx.doi.org/10.1016/j.resp.2014.10.006]

[177] Bhatia M, Moochhala S. Role of inflammatory mediators in the pathophysiology of acute respiratory distress syndrome. J Pathol 2004; 202: 145-56.
[http://dx.doi.org/10.1002/path.1491]

[178] Pelosi P, D'Onofrio D, Chiumello D, *et al.* Pulmonary and extrapulmonary acute respiratory distress syndrome are different. Eur Respir J Suppl 2003; 42: 48s-56s.
[http://dx.doi.org/10.1183/09031936.03.00420803]

[179] Rocco PR, Pelosi P. Pulmonary and extrapulmonary acute respiratory distress syndrome: myth or reality? Curr Opin Crit Care 2008; 14: 50-5.
[http://dx.doi.org/10.1097/MCC.0b013e3282f2405b]

[180] Leite-Junior JH, Garcia CS, Souza-Fernandes AB, *et al.* Methylprednisolone improves lung mechanics and reduces the inflammatory response in pulmonary but not in extrapulmonary mild acute lung injury in mice. Crit Care Med 2008; 36: 2621-8.
[http://dx.doi.org/10.1097/CCM.0b013e3181847b43]

[181] Kloot TE, Blanch L, Melynne Youngblood A, *et al.* Recruitment maneuvers in three experimental models of acute lung injury. Effect on lung volume and gas exchange. Am J Respir Crit Care Med 2000; 161: 1485-94.
[http://dx.doi.org/10.1164/ajrccm.161.5.9809014]

[182] González-López A, Albaiceta GM. Repair after acute lung injury: molecular mechanisms and therapeutic opportunities. Crit Care 2012; 16: 209.
[http://dx.doi.org/10.1186/cc11224]

[183] Haberthür C, Seeberger MD. Acute respiratory distress syndrome and mechanical ventilation: ups and downs of an ongoing relationship trap. J Thorac Dis 2016; 8: E1608-9.
[http://dx.doi.org/10.21037/jtd.2016.12.25]

[184] Dushianthan A, Grocott MP, Postle AD, Cusack R. Acute respiratory distress syndrome and acute lung injury. Postgrad Med J 2011; 87: 612-22.
[http://dx.doi.org/10.1136/pgmj.2011.118398]

[185] Boyle AJ, Mac Sweeney R, McAuley DF. Pharmacological treatments in ARDS; a state-of-the-art update. BMC Med 2013; 11: 166.
[http://dx.doi.org/10.1186/1741-7015-11-166]

[186] Standiford TJ, Ward PA. Therapeutic targeting of acute lung injury and acute respiratory distress syndrome. Transl Res 2016; 167: 183-91.
[http://dx.doi.org/10.1016/j.trsl.2015.04.015]

[187] Endres S, Fülle HJ, Sinha B, *et al.* Cyclic nucleotides differentially regulate the synthesis of tumour necrosis factor-alpha and interleukin-1 beta by human mononuclear cells. Immunology 1991; 72: 56-60.

[188] Yoshimura T, Usami E, Kurita C, *et al.* Effect of theophylline on the production of interleukin-1 beta, tumor necrosis factor-alpha, and interleukin-8 by human peripheral blood mononuclear cells. Biol Pharm Bull 1995; 18: 1405-8.
[http://dx.doi.org/10.1248/bpb.18.1405]

[189] Sato K, Stelzner TJ, O'Brien RF, Weil JV, Welsh CH. Pentoxifylline lessens the endotoxin-induced increase in albumin clearance across pulmonary artery endothelial monolayers with and without neutrophils. Am J Respir Cell Mol Biol 1991; 4: 219-27.
[http://dx.doi.org/10.1165/ajrcmb/4.3.219]

[190] Geraets L, Haegens A, Weseler AR, *et al.* Inhibition of acute pulmonary and systemic inflammation by 1,7-dimethylxanthine. Eur J Pharmacol 2010; 629: 132-9.
[http://dx.doi.org/10.1016/j.ejphar.2009.11.064]

[191] Hussein A, Meyrick B, Graber S, Berry L Jr, Brigham KL. Attenuation of endotoxin-induced cytotoxicity and prostacyclin production in cultured bovine pulmonary artery endothelial cells by phosphodiesterase inhibition. Exp Lung Res 1988; 14: 637-54.
[http://dx.doi.org/10.3109/01902148809087834]

[192] Buenestado A, Chaumais MC, Grassin-Delyle S, *et al.* Roflumilast inhibits lipopolysaccharide-induced tumor necrosis factor-α and chemokine production by human lung parenchyma. PLoS One 2013; 8: e74640.
[http://dx.doi.org/10.1371/journal.pone.0074640]

[193] Seldon PM, Barnes PJ, Meja K, Giembycz MA. Suppression of lipopolysaccharide-induced tumor necrosis factor-alpha generation from human peripheral blood monocytes by inhibitors of phosphodiesterase 4: interaction with stimulants of adenylyl cyclase. Mol Pharmacol 1995; 48: 747-57.

[194] Togo S, Liu X, Wang X, *et al.* PDE4 inhibitors roflumilast and rolipram augment PGE2 inhibition of TGF-{beta}1-stimulated fibroblasts. Am J Physiol Lung Cell Mol Physiol 2009; 296: L959-69.
[http://dx.doi.org/10.1152/ajplung.00508.2007]

[195] Dunkern TR, Feurstein D, Rossi GA, Sabatini F, Hatzelmann A. Inhibition of TGF-beta induced lung fibroblast to myofibroblast conversion by phosphodiesterase inhibiting drugs and activators of soluble guanylyl cyclase. Eur J Pharmacol 2007; 572: 12-22.
[http://dx.doi.org/10.1016/j.ejphar.2007.06.036]

[196] Sabatini F, Petecchia L, Boero S, *et al.* A phosphodiesterase 4 inhibitor, roflumilast N-oxide, inhibits human lung fibroblast functions in vitro. Pulm Pharmacol Ther 2010; 23: 283-91.
[http://dx.doi.org/10.1016/j.pupt.2010.02.004]

[197] Konrad FM, Bury A, Schick MA, Ngamsri KC, Reutershan J. The unrecognized effects of phosphodiesterase 4 on epithelial cells in pulmonary inflammation. PLoS One 2015; 10: e0121725.
[http://dx.doi.org/10.1371/journal.pone.0121725]

[198] Ma H, Shi J, Wang C, *et al.* Blockade of PDE4B limits lung vascular permeability and lung inflammation in LPS-induced acute lung injury. Biochem Biophys Res Commun 2014; 450: 1560-7.
[http://dx.doi.org/10.1016/j.bbrc.2014.07.024]

[199] Matute-Bello G, Frevert CW, Martin TR. Animal models of acute lung injury. Am J Physiol Lung Cell Mol Physiol 2008; 295: L379-99.
[http://dx.doi.org/10.1152/ajplung.00010.2008]

[200] Wang HM, Bodenstein M, Markstaller K. Overview of the pathology of three widely used animal models of acute lung injury. Eur Surg Res 2008; 40: 305-16.
[http://dx.doi.org/10.1159/000121471]

[201] Shaver CM, Bastarache JA. Clinical and biological heterogeneity in acute respiratory distress syndrome: direct *versus* indirect lung injury. Clin Chest Med 2014; 35: 639-53.
[http://dx.doi.org/10.1016/j.ccm.2014.08.004]

[202] Escofier N, Boichot E, Germain N, Silva PM, Martins MA, Lagente V. Effects of interleukin-10 and modulators of cyclic AMP formation on endotoxin-induced inflammation in rat lung. Fundam Clin Pharmacol 1999; 13: 96-01.
[http://dx.doi.org/10.1111/j.1472-8206.1999.tb00326.x]

[203] Hsu K, Wang D, Chang ML, Wu CP, Chen HI. Pulmonary edema induced by phorbol myristate acetate is attenuated by compounds that increase intracellular cAMP. Res Exp Med (Berl) 1996; 196: 17-28.
[http://dx.doi.org/10.1007/BF02576825]

[204] Hsu K, Wang D, Shen CY, Chiang CH. Protamine-induced acute lung injury and the protective effect of agents that increase cAMP. Proc Natl Sci Counc Repub China B 1993; 17: 57-61.

[205] Sciuto AM, Strickland PT, Kennedy TP, Gurtner GH. Postexposure treatment with aminophylline protects against phosgene-induced acute lung injury. Exp Lung Res 1997; 23: 317-32.
[http://dx.doi.org/10.3109/01902149709039229]

[206] Ishizaka A, Hatherill JR, Harada H, *et al.* Prevention of interleukin 2-induced acute lung injury in guinea pigs by pentoxifylline. J Appl Physiol 1985; 1989(67): 2432-7.

[207] Guidot DM, Bursten SL, Rice GC, *et al.* Modulating phosphatidic acid metabolism decreases oxidative injury in rat lungs. Am J Physiol 1997; 273: L957-66.

[208] Hybertson BM, Bursten SL, Leff JA, *et al.* Lisofylline prevents leak, but not neutrophil accumulation, in lungs of rats given IL-1 intratracheally. J Appl Physiol 1985; 1997(82): 226-32.

[209] Li J, Li G, Hu JL, *et al.* Chronic or high dose acute caffeine treatment protects mice against oleic acid-induced acute lung injury *via* an adenosine A2A receptor-independent mechanism. Eur J Pharmacol 2011; 654: 295-03.
[http://dx.doi.org/10.1016/j.ejphar.2010.12.040]

[210] Nials AT, Tralau-Stewart CJ, Gascoigne MH, Ball DI, Ranshaw LE, Knowles RG. *In vivo* characterization of GSK256066, a high-affinity inhaled phosphodiesterase 4 inhibitor. J Pharmacol Exp Ther 2011; 337: 137-44.
[http://dx.doi.org/10.1124/jpet.110.173641]

[211] Turner CR, Esser KM, Wheeldon EB. Therapeutic intervention in a rat model of ARDS: IV. Phosphodiesterase IV inhibition. Circ Shock 1993; 39: 237-45.

[212] Miotla JM, Teixeira MM, Hellewell PG. Suppression of acute lung injury in mice by an inhibitor of phosphodiesterase type 4. Am J Respir Cell Mol Biol 1998; 18: 411-20.
[http://dx.doi.org/10.1165/ajrcmb.18.3.2913]

[213] Spond J, Chapman R, Fine J, *et al.* Comparison of PDE 4 inhibitors, rolipram and SB 207499 (ariflo), in a rat model of pulmonary neutrophilia. Pulm Pharmacol Ther 2001; 14: 157-64.
[http://dx.doi.org/10.1006/pupt.2001.0291]

[214] Selige J, Hatzelmann A, Dunkern T. The differential impact of PDE4 subtypes in human lung fibroblasts on cytokine-induced proliferation and myofibroblast conversion. J Cell Physiol 2011; 226: 1970-80.
[http://dx.doi.org/10.1002/jcp.22529]

[215] Udalov S, Dumitrascu R, Pullamsetti SS, *et al.* Effects of phosphodiesterase 4 inhibition on bleomycin-induced pulmonary fibrosis in mice. BMC Pulm Med 2010; 10: 26.
[http://dx.doi.org/10.1186/1471-2466-10-26]

[216] Milara J, Morcillo E, Monleon D, Tenor H, Cortijo J. Roflumilast Prevents the Metabolic Effects of Bleomycin-Induced Fibrosis in a Murine Model. PLoS One 2015; 10: e0133453.
[http://dx.doi.org/10.1371/journal.pone.0133453]

[217] Bursten SL, Federighi DA, Parsons P, *et al.* An increase in serum C18 unsaturated free fatty acids as a predictor of the development of acute respiratory distress syndrome. Crit Care Med 1996; 24: 1129-36.

[http://dx.doi.org/10.1097/00003246-199607000-00011]

[218] The ARDS Network. [No authors listed]. Randomized, placebo-controlled trial of lisofylline for early treatment of acute lung injury and acute respiratory distress syndrome. Crit Care Med 2002; 30: 1-6. [http://dx.doi.org/10.1097/00003246-200201000-00001]

[219] Hashimoto S, Sanui M, Egi M, *et al.* ARDS clinical practice guideline committee from the Japanese Society of Respiratory Care Medicine and the Japanese Society of Intensive Care Medicine. The clinical practice guideline for the management of ARDS in Japan. J Intensive Care 2017; 5: 50. [http://dx.doi.org/10.1186/s40560-017-0222-3]

[220] Harada H, Ishizaka A, Yonemaru M, *et al.* The effects of aminophylline and pentoxifylline on multiple organ damage after Escherichia coli sepsis. Am Rev Respir Dis 1989; 140: 974-80. [http://dx.doi.org/10.1164/ajrccm/140.4.974]

[221] Lechner AJ, Rouben LR, Potthoff LH, Tredway TL, Matuschak GM. Effects of pentoxifylline on tumor necrosis factor production and survival during lethal E. coli sepsis *vs.* disseminated candidiasis with fungal septic shock. Circ Shock 1993; 39: 306-15.

[222] Ridings PC, Windsor AC, Sugerman HJ, *et al.* Beneficial cardiopulmonary effects of pentoxifylline in experimental sepsis are lost once septic shock is established. Arch Surg 1994; 129: 1144-52. [http://dx.doi.org/10.1001/archsurg.1994.01420350042004]

[223] Li H, Tan G, Tong L, *et al.* Pentoxifylline inhibits pulmonary inflammation induced by infrarenal aorticcross-clamping dependent of adenosine receptor A2A. Am J Transl Res 2016; 8: 2210-21. a

[224] Sunil VR, Vayas KN, Cervelli JA, *et al.* Pentoxifylline attenuates nitrogen mustard-induced acute lung injury, oxidative stress and inflammation. Exp Mol Pathol 2014; 97: 89-98. [http://dx.doi.org/10.1016/j.yexmp.2014.05.009]

[225] Li Q, Hu X, Sun R, Tu Y, Gong F, Ni Y. Resolution acute respiratory distress syndrome through reversing the imbalance of Treg/Th17 by targeting the cAMP signaling pathway. Mol Med Rep 2016; 14: 343-8. b [http://dx.doi.org/10.3892/mmr.2016.5222]

[226] Hasegawa N, Oka Y, Nakayama M, *et al.* The effects of post-treatment with lisofylline, a phosphatidic acid generation inhibitor, on sepsis-induced acute lung injury in pigs. Am J Respir Crit Care Med 1997; 155: 928-36. [http://dx.doi.org/10.1164/ajrccm.155.3.9117028]

[227] Oka Y, Hasegawa N, Nakayama M, Murphy GA, Sussman HH, Raffin TA. Selective downregulation of neutrophils by a phosphatidic acid generation inhibitor in a porcine sepsis model. J Surg Res 1999; 81: 147-55. [http://dx.doi.org/10.1006/jsre.1998.5453]

[228] Schick MA, Wunder C, Wollborn J, *et al.* Phosphodiesterase-4 inhibition as a therapeutic approach to treat capillary leakage in systemic inflammation. J Physiol 2012; 590: 2693-08. [http://dx.doi.org/10.1113/jphysiol.2012.232116]

[229] Wollborn J, Schlegel N, Schick MA. Phosphodiesterase 4 inhibition for treatment of endothelial barrier and microcirculation disorders in sepsis. Anaesthesist 2017; 66: 347-52. [http://dx.doi.org/10.1007/s00101-017-0305-5]

[230] Holthoff JH, Wang Z, Patil NK, Gokden N, Mayeux PR. Rolipram improves renal perfusion and function during sepsis in the mouse. J Pharmacol Exp Ther 2013; 347: 357-64. [http://dx.doi.org/10.1124/jpet.113.208520]

[231] Flemming S, Schlegel N, Wunder C, *et al.* Phosphodiesterase 4 inhibition dose dependently stabilizes microvascular barrier functions and microcirculation in a rodent model of polymicrobial sepsis. Shock 2014; 41: 537-45. [http://dx.doi.org/10.1097/SHK.0000000000000152]

[232] Oishi H, Takano K, Tomita K, *et al.* Olprinone and colforsin daropate alleviate septic lung

inflammation and apoptosis through CREB-independent activation of the Akt pathway. Am J Physiol Lung Cell Mol Physiol 2012; 303: L130-40.
[http://dx.doi.org/10.1152/ajplung.00363.2011]

[233] Kuniyoshi T, Kakihana Y, Isowaki S, *et al.* Effects of olprinone on hepatosplanchnic circulation and mitochondrial oxidation in a porcine model of endotoxemia. J Anesth 2005; 19: 295-01.
[http://dx.doi.org/10.1007/s00540-005-0340-2]

[234] Mazzon E, Esposito E, Di Paola R, Impellizzeri D, Bramanti P, Cuzzocrea S. Olprinone, a specific phosphodiesterase (PDE)-III inhibitor, reduces the development of multiple organ dysfunction syndrome in mice. Pharmacol Res 2011; 64: 68-79.
[http://dx.doi.org/10.1016/j.phrs.2010.12.010]

[235] Neviere R, Delguste F, Durand A, Inamo J, Boulanger E, Preau S. Abnormal Mitochondrial cAMP/PKA Signaling Is Involved in Sepsis-Induced Mitochondrial and Myocardial Dysfunction. Int J Mol Sci 2016; 17: E2075.
[http://dx.doi.org/10.3390/ijms17122075]

[236] Bacher A, Mayer N, Klimscha W, Oismüller C, Steltzer H, Hammerle A. Effects of pentoxifylline on hemodynamics and oxygenation in septic and nonseptic patients. Crit Care Med 1997; 25: 795-00.
[http://dx.doi.org/10.1097/00003246-199705000-00014]

[237] Pammi M, Haque KN. Pentoxifylline for treatment of sepsis and necrotizing enterocolitis in neonates. Cochrane Database Syst Rev 2015; 3: CD004205.

[238] Blanch L, Albaiceta GM. Sildenafil for pulmonary hypertension in ARDS: a new pleasant effect? Intensive Care Med 2010; 36: 729-31.
[http://dx.doi.org/10.1007/s00134-010-1771-2]

[239] Vieillard-Baron A, Schmitt JM, Augarde R, *et al.* Acute cor pulmonale in acute respiratory distress syndrome submitted to protective ventilation: incidence, clinical implications, and prognosis. Crit Care Med 2001; 29: 1551-5.
[http://dx.doi.org/10.1097/00003246-200108000-00009]

[240] Moloney ED, Evans TW. Pathophysiology and pharmacological treatment of pulmonary hypertension in acute respiratory distress syndrome. Eur Respir J 2003; 21: 720-7.
[http://dx.doi.org/10.1183/09031936.03.00120102]

[241] Adhikari NK, Burns KE, Friedrich JO, Granton JT, Cook DJ, Meade MO. Effect of nitric oxide on oxygenation and mortality in acute lung injury: systematic review and meta-analysis. BMJ 2007; 334: 779.
[http://dx.doi.org/10.1136/bmj.39139.716794.55]

[242] Gebistorf F, Karam O, Wetterslev J, Afshari A. Inhaled nitric oxide for acute respiratory distress syndrome (ARDS) in children and adults. Cochrane Database Syst Rev 2016; 6: CD002787.

[243] Mizus I, Summer W, Farrukh I, Michael JR, Gurtner GH. Isoproterenol or aminophylline attenuate pulmonary edema after acid lung injury. Am Rev Respir Dis 1985; 131: 256-9.

[244] Hashimoto Y, Hirota K, Yoshioka H, Kudo T, Ishihara H, Matsuki A. A comparison of the spasmolytic effects of olprinone and aminophylline on serotonin-induced pulmonary hypertension and bronchoconstriction with or without beta-blockade in dogs. Anesth Analg 2000; 91: 1345-50.
[http://dx.doi.org/10.1097/00000539-200012000-00007]

[245] Zhao L, Mason NA, Morrell NW, *et al.* Sildenafil inhibits hypoxia-induced pulmonary hypertension. Circulation 2001; 104: 424-8.
[http://dx.doi.org/10.1161/hc2901.093117]

[246] Hashiba E, Hirota K, Yoshioka H, *et al.* Milrinone attenuates serotonin-induced pulmonary hypertension and bronchoconstriction in dogs. Anesth Analg 2000; 90: 790-4.
[http://dx.doi.org/10.1213/00000539-200004000-00004]

[247] Kamamoto H. The effect of PDEIII inhibitors and colforsin daropate on hypoxic pulmonary

vasoconstriction in rabbit lungs. Masui 2008; 57: 1398-07. [Abstract].

[248] Kakura H, Miyahara K, Amitani S, *et al.* Hemodynamic effects of intravenous administration of olprinone hydrochloride on experimental pulmonary hypertension. Arzneimittelforschung 2000; 50: 515-9.

[249] Cornet AD, Hofstra JJ, Swart EL, Girbes AR, Juffermans NP. Sildenafil attenuates pulmonary arterial pressure but does not improve oxygenation during ARDS. Intensive Care Med 2010; 36: 758-64. [http://dx.doi.org/10.1007/s00134-010-1754-3]

[250] Arif SA, Poon H. Tadalafil: a long-acting phosphodiesterase-5 inhibitor for the treatment of pulmonary arterial hypertension. Clin Ther 2011; 33: 993-04. [http://dx.doi.org/10.1016/j.clinthera.2011.06.008]

[251] Hermansen CL, Lorah KN. Respiratory distress in the newborn. Am Fam Physician 2007; 76: 987-94.

[252] Gallacher DJ, Hart K, Kotecha S. Common respiratory conditions of the newborn. Breathe (Sheff) 2016; 12: 30-42. [http://dx.doi.org/10.1183/20734735.000716]

[253] Edwards MO, Kotecha SJ, Kotecha S. Respiratory distress of the term newborn infant. Paediatr Respir Rev 2013; 14: 29-36. [http://dx.doi.org/10.1016/j.prrv.2012.02.002]

[254] Schüller SS, Wisgrill L, Herndl E, *et al.* Pentoxifylline modulates LPS-induced hyperinflammation in monocytes of preterm infants in vitro. Pediatr Res 2017. [Epub ahead of print]. [http://dx.doi.org/10.1038/pr.2017.41]

[255] Häfner D, Germann PG. Additive effects of phosphodiesterase-4 inhibition on effects of rSP-C surfactant. Am J Respir Crit Care Med 2000; 161: 1495-00. [http://dx.doi.org/10.1164/ajrccm.161.5.9905018]

[256] Kosutova P, Mikolka P, Kolomaznik M, Rezakova S, Calkovska A, Mokra D. Effects of roflumilast, a phosphodiesterase-4 inhibitor, on the lung functions in a saline lavage-induced model of acute lung injury. Physiol Res 2017; 66 (Suppl. 2): S237-45.

[257] Fehrholz M, Speer CP, Kunzmann S. Caffeine and rolipram affect Smad signalling and TGF-β1 stimulated CTGF and transgelin expression in lung epithelial cells. PLoS One 2014; 9: e97357. [http://dx.doi.org/10.1371/journal.pone.0097357]

[258] Jing X, Huang YW, Jarzembowski J, Shi Y, Konduri GG, Teng RJ. Caffeine ameliorates hyperoxia-induced lung injury by protecting GCH1 function in neonatal rat pups. Pediatr Res 2017. [Epub ahead of print]. [http://dx.doi.org/10.1038/pr.2017.89]

[259] Teng RJ, Jing X, Michalkiewicz T, Afolayan AJ, Wu TJ, Konduri GG. Attenuation of endoplasmic reticulum stress by caffeine ameliorates hyperoxia-induced lung injury. Am J Physiol Lung Cell Mol Physiol 2017; 312: L586-98. [http://dx.doi.org/10.1152/ajplung.00405.2016]

[260] Weichelt U, Cay R, Schmitz T, *et al.* Prevention of hyperoxia-mediated pulmonary inflammation in neonatal rats by caffeine. Eur Respir J 2013; 41: 966-73. [http://dx.doi.org/10.1183/09031936.00012412]

[261] Schoen K, Yu T, Stockmann C, Spigarelli MG, Sherwin CM. Use of methylxanthine therapies for the treatment and prevention of apnea of prematurity. Paediatr Drugs 2014; 16: 169-77. [http://dx.doi.org/10.1007/s40272-013-0063-z]

[262] Mohammed S, Nour I, Shabaan AE, Shouman B, Abdel-Hady H, Nasef N. High *versus* low-dose caffeine for apnea of prematurity: a randomized controlled trial. Eur J Pediatr 2015; 174: 949-56. [http://dx.doi.org/10.1007/s00431-015-2494-8]

[263] Endesfelder S, Weichelt U, Strauß E, *et al.* Neuroprotection by Caffeine in Hyperoxia-Induced

Neonatal Brain Injury. Int J Mol Sci 2017; 18: E187.
[http://dx.doi.org/10.3390/ijms18010187]

[264] Hütten MC, Wolfs TG, Kramer BW. Can the preterm lung recover from perinatal stress? Mol Cell Pediatr 2016; 3: 15.
[http://dx.doi.org/10.1186/s40348-016-0043-9]

[265] Lista G, Fabbri L, Polackova R, *et al.* Peyona® PASS Group. The Real-World Routine Use of Caffeine Citrate in Preterm Infants: A European Postauthorization Safety Study. Neonatology 2016; 109: 221-7.
[http://dx.doi.org/10.1159/000442813]

[266] Sweet DG, Carnielli V, Greisen G, *et al.* European Consensus Guidelines on the Management of Respiratory Distress Syndrome - 2016 Update. Neonatology 2017; 111: 107-25.
[http://dx.doi.org/10.1159/000448985]

[267] Schulzke SM, Kaempfen S, Patole SK. Pentoxifylline for the prevention of bronchopulmonary dysplasia in preterm infants. Cochrane Database Syst Rev 2014; 11: CD010018.

[268] Harris E, Schulzke SM, Patole SK. Pentoxifylline in preterm neonates: a systematic review. Paediatr Drugs 2010; 12: 301-11.
[http://dx.doi.org/10.2165/11532600-000000000-00000]

[269] de Visser YP, Walther FJ, Laghmani EH, van Wijngaarden S, Nieuwland K, Wagenaar GT. Phosphodiesterase-4 inhibition attenuates pulmonary inflammation in neonatal lung injury. Eur Respir J 2008; 31: 633-44.
[http://dx.doi.org/10.1183/09031936.00071307]

[270] Mokra D, Mokry J. Inflammation in meconium aspiration syndrome: targets for pharmacological modulation. Curr Pediatr Rev 2007; 3: 248-63.
[http://dx.doi.org/10.2174/157339607782411547]

[271] Bhat R, Vidyasagar D. Delivery room management of meconium-stained infant. Clin Perinatol 2012; 39: 817-31.
[http://dx.doi.org/10.1016/j.clp.2012.09.004]

[272] Swarnam K, Soraisham AS, Sivanandan S. Advances in the management of meconium aspiration syndrome. Int J Pediatr 2012; 2012: 359571.
[http://dx.doi.org/10.1155/2012/359571]

[273] Turhan AH, Atıcı A, Muşlu N, Polat A, Helvacı İ. The effects of pentoxifylline on lung inflammation in a rat model of meconium aspiration syndrome. Exp Lung Res 2012; 38: 250-5.
[http://dx.doi.org/10.3109/01902148.2012.676704]

[274] Mokra D, Drgova A, Mokry J, *et al.* Comparison of the effects of low-dose *vs.* high-dose aminophylline on lung function in experimental meconium aspiration syndrome. J Physiol Pharmacol 2008; 59 (Suppl. 6): 449-59. a

[275] Farrow KN, Steinhorn RH. Phosphodiesterases: emerging therapeutic targets for neonatal pulmonary hypertension. Handb Exp Pharmacol 2011; 204: 251-77.
[http://dx.doi.org/10.1007/978-3-642-17969-3_11]

[276] Puthiyachirakkal M, Mhanna MJ. Pathophysiology, management, and outcome of persistent pulmonary hypertension of the newborn: a clinical review. Front Pediatr 2013; 1: 23.
[http://dx.doi.org/10.3389/fped.2013.00023]

[277] Baliga RS, MacAllister RJ, Hobbs AJ. New perspectives for the treatment of pulmonary hypertension. Br J Pharmacol 2011; 163: 125-40.
[http://dx.doi.org/10.1111/j.1476-5381.2010.01164.x]

[278] Tessler RB, Zadinello M, Fiori H, Colvero M, Belik J, Fiori RM. Tadalafil improves oxygenation in a model of newborn pulmonary hypertension. Pediatr Crit Care Med 2008; 9: 330-2.
[http://dx.doi.org/10.1097/PCC.0b013e31816c7035]

[279] Ryhammer PK, Shekerdemian LS, Penny DJ, Ravn HB. Effect of intravenous sildenafil on pulmonary hemodynamics and gas exchange in the presence and absence of acute lung injury in piglets. Pediatr Res 2006; 59: 762-6.
[http://dx.doi.org/10.1203/01.pdr.0000219501.58546.aa]

[280] Kelly LE, Ohlsson A, Shah PS. Sildenafil for pulmonary hypertension in neonates. Cochrane Database Syst Rev 2017; 8: CD005494. [Epub ahead of print].
[http://dx.doi.org/10.1002/14651858.CD005494.pub4]

[281] Ziegler JW, Ivy DD, Wiggins JW, Kinsella JP, Clarke WR, Abman SH. Effects of dipyridamole and inhaled nitric oxide in pediatric patients with pulmonary hypertension. Am J Respir Crit Care Med 1998; 158: 1388-95.
[http://dx.doi.org/10.1164/ajrccm.158.5.9710117]

[282] Bassler D, Kreutzer K, McNamara P, Kirpalani H. Milrinone for persistent pulmonary hypertension of the newborn. Cochrane Database Syst Rev 2010; 11: CD007802.

[283] Lakshminrusimha S, Mathew B, Leach CL. Pharmacologic strategies in neonatal pulmonary hypertension other than nitric oxide. Semin Perinatol 2016; 40: 160-73.
[http://dx.doi.org/10.1053/j.semperi.2015.12.004]

[284] Reuter S, Moser C, Baack M. Respiratory distress in the newborn. Pediatr Rev 2014; 35: 417-28.
[http://dx.doi.org/10.1542/pir.35-10-417]

[285] Shabaan AE, Nasef N, Shouman B, Nour I, Mesbah A, Abdel-Hady H. Pentoxifylline therapy for late-onset sepsis in preterm infants: a randomized controlled trial. Pediatr Infect Dis J 2015; 34: e143-8.
[http://dx.doi.org/10.1097/INF.0000000000000698]

[286] Speer EM, Dowling DJ, Ozog LS, *et al.* Pentoxifylline inhibits TLR- and inflammasome-mediated in vitro inflammatory cytokine production in human blood with greater efficacy and potency in newborns. Pediatr Res 2017; 81: 806-16.
[http://dx.doi.org/10.1038/pr.2017.6]

[287] Sims CR, Singh SP, Mu S, *et al.* Rolipram Improves Outcome in a Rat Model of Infant Sepsis-Induced Cardiorenal Syndrome. Front Pharmacol 2017; 8: 237.
[http://dx.doi.org/10.3389/fphar.2017.00237]

SUBJECT INDEX

A

B

Atta-ur-Rahman & M. Choudhary (Eds.)

C

D

E

F

G

H

P

R

S

www.ingramcontent.com/pod-product-compliance
Lightning Source LLC
LaVergne TN
LVHW070946210926
848482LV00002B/116